Principles of Industrial Chemistry

-Principles of Industrial Chemistry-

CHRIS A. CLAUSEN III

GUY MATTSON

Department of Chemistry
Florida Technological University
Orlando, Florida

A WILEY-INTERSCIENCE PUBLICATION

JOHN WILEY & SONS
New York · Chichester · Brisbane · Toronto

Copyright © 1978 by John Wiley & Sons, Inc.

All rights reserved. Published simultaneously in Canada.

Reproduction or translation of any part of this work
beyond that permitted by Sections 107 or 108 of the
1976 United States Copyright Act without the permission
of the copyright owner is unlawful. Requests for
permission or further information should be addressed to
the Permissions Department, John Wiley & Sons, Inc.

Library of Congress Cataloging in Publication Data:

Clausen, Chris A   1940-
  Principles of industrial chemistry.

  "A Wiley-Interscience publication."
  Includes index.
  1. Chemistry, Technical.   I. Mattson, Guy C., joint
author.   II. Title.

TP145.C67   660.2   78-9450
ISBN 0-471-02774-X

Printed in the United States of America

10 9 8 7 6 5 4 3 2

To Julia and Jean, our wives
for their unfailing patience, devotion, and encouragement

PREFACE

Principles of Industrial Chemistry has been written in response to the frequently expressed concern for the widening gulf between chemistry as it is taught in our colleges and universities, and chemistry as it is practiced commercially. This work is intended as a textbook in a senior or graduate level course in industrial chemistry. It is hoped that it will also be useful, as a guide, to recent graduates who are just starting their industrial careers.

There have been many studies, reports, and discussions regarding the need for providing our students with a better preparation for working in industry. Although the conclusions and suggestions of these studies vary considerably, there seem to be two common general thoughts. The first, and most important, is that most observers feel that the typical, rigorous undergraduate chemistry curriculum is providing a fine foundation in chemical concepts, principles, and theories. Difficulties appear to arise in the application of these principles to specific, practical problems; in a poor understanding of some very basic engineering concepts; and in a lack of awareness of what the chemical industry is, how it operates, and how the chemist fits into the scheme. The second general point of agreement is that most chemists with initiative and experience eventually pick up the tools required to operate successfully in industry.

Our purpose then in writing this book has been to supplement or complement the traditional training of a chemist in order to help him through the change from the academic to the industrial world. To a great extent the content reflects much of the information we wish we had known when we entered industry. Generally we have assumed that the reader has completed the junior year of a typical chemistry curriculum; specifically we have assumed a general knowledge of physical chemistry. We have made no attempt to describe the many industrially important processes currently in use, nor to catalog or describe the many important industrial chemicals. Such descriptive material is available in sources such as the Kirk-Othmer

Encyclopedia of Chemical Technology. Selected topics of this nature should certainly be included in any course in Industrial Chemistry.

The general theme we have used throughout the book is that of process development. This is not to imply that most industrial chemists are involved exclusively with process development. It simply reflects our opinion that other important areas, such as product development and applications work, are very specialized. Process development seems to be a topic that is more general and affords greater opportunities to stress the correlations between classical and industrial chemistries.

We wish to thank Florida Technological University for its support, Etta Jean Smith for assistance in preparing the manuscript, Jan Avis for the illustrations, and friends in industry for their comments and suggestions.

CHRIS A. CLAUSEN III
GUY MATTSON

Orlando, Florida
July 1978

CONTENTS

Principles
of Industrial
Chemistry

1

THE CHEMICAL INDUSTRY

WHAT IT IS AND WHAT IT DOES

Chemistry is an essential part of the manufacture of a large number of products. Only a very small portion of this production is included in what is commonly referred to as "the chemical industry." It is generally regarded that the production of steel, aluminum, paper, and glass, for example, is *not* part of the chemical industry, even though chemical transformations and chemical processing are involved. On the other hand, if the chemical industry is defined so narrowly as to involve only those products considered to be "chemicals" by the general public, then it would exclude the production of such items as polystyrene foam used for insulation and ethylene glycol used in automotive cooling systems. The phrase "the chemical industry" is inherently ambiguous simply because the term "chemical" means different things to different people.

One apparent solution would be to consider that the "chemical industry" consists of the sum total of all of the activities of all of the companies recognized as "chemical companies." Prior to World War II this would have been an acceptable and useful working definition. Today two important developments have changed this picture. First, many of the major chemical companies have diversified their product lines into areas traditionally recognized as nonchemical. Second, many "nonchemical" firms, principally petroleum companies, have become major producers of products that are classified as "chemicals." These two trends are of such magnitude that at least one observer has predicted, by 1992, a reclassification that would phase out the term "chemical industry" as a separate statistical entity.[1] Regardless of the classification or the terminology, the commercial practice of chemistry will continue to grow and become an increasingly important factor in the overall economy.

1.1 THE ECONOMIC IMPORTANCE OF THE CHEMICAL INDUSTRY

In order to gain some insight into the impact of chemistry upon our economy, it is useful to consider some statistics concerning the volume and the value of the products manufactured. One classification, frequently used as a basis for compiling statistical data, is "chemicals and allied products." This is a Standard Industrial Classification devised by the Office of Statistical Standards. It includes, in addition to basic chemicals and chemicals used as intermediates, plastics and resins, synthetic rubber and fibers, pharmaceuticals, soaps and detergents, cosmetics and toiletries, paints and varnishes, fertilizers and pesticides.

Although chemical production represents only a small fraction of the total manufacturing in the United States, it is growing at a rapid rate. Relative growth rates are conveniently compared by production indices that are based on the total production of a particular industry relative to a particular reference year. During the period 1899 to 1970, the production index for all manufacturing rose by a factor of 17 while the chemical production index advanced by a factor of 94.[2] (Fig. 1.1.)

Over a shorter and more recent period (Fig. 1.2), it is noted that the U.S. chemical industry is continuing to grow at a faster rate than manufacturing

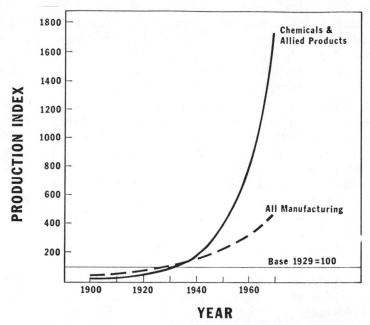

Figure 1.1   NBER production indices for the years 1900–1970. *Source*:   Reference 2.

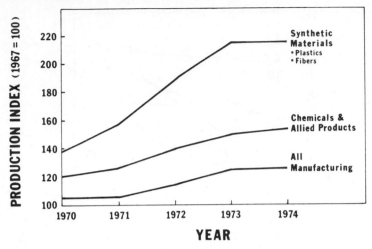

Figure 1.2   Production indices for various industrial classifications. *Source:*   References 3 and 4.

in general. One of the most rapidly expanding sectors in the chemical industry is that of synthetic materials, which includes plastics, synthetic rubber, and man-made fibers. The growth rates of new synthetic materials can be quite dramatic. A comparison of the production of natural and man-made fibers (Fig. 1.3) illustrates the ready acceptance by the consuming public of cellulose derived materials such as acetate rayon and synthetic materials such as nylon and Acrylon. This plot is interesting in that it indicates that this sustained growth has not been at the expense of natural fibers, production of which has remained relatively constant through the years. Total growth in fibers production is apparently due to population growth and increases in living standards.

The use of production indices is a valid and recognized means of comparing relative changes in production output for various industrial classifications. It should be noted, however, that frequently the resultant numbers represent the conglomerate production of some rather diverse products— from locomotives to nylon stockings. Comparisons between industries are awkward because of the lack of a common base. One common denominator of all products is a monetary value. Data relating the annual dollar value of the sales for each industrial classification are readily available.

A comparison of the value of shipments of chemical products with that of all manufactured products in the United States (Fig. 1.4) illustrates the faster growth rate for chemicals noted previously. Also, since it is based on a common unit, dollar value, it gives some measure of the relative size of the

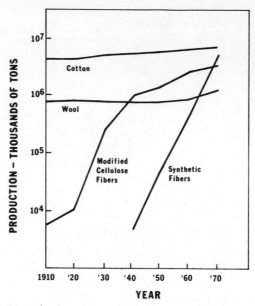

Figure 1.3 World production of natural and man-made fibers. *Source*:   Reference 5.

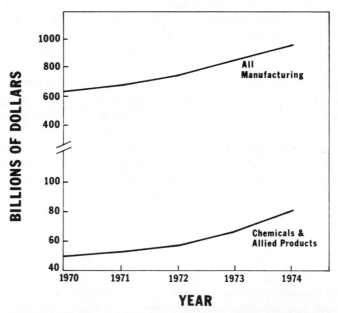

Figure 1.4 Comparison of the value of chemical product shipments with all manufactured products. *Source*:   References 3 and 4.

4

two classifications. Chemicals and allied products account for somewhat less than 10% of the total manufactured goods.

Statistics based on value of shipments, although relatively easy to gather and compile, can also lead to problems. Comparisons of relative growth rates, for example, can be distorted by changes in prices rather than real growth in the industries. A more serious shortcoming is that the simple totaling of sales dollars does not necessarily provide a true index of the health of an industry or the overall economy because no attempt is made to factor out internal sales within an industry. In the chemical industry, where it is not uncommon for a basic raw material to go through five or more intermediates before conversion to the final product purchased by the consumer, this can be a significant factor. To illustrate with a simple example, consider the production of polyethylene film. Company A produces natural gas; it sells $400,000 worth of higher boiling condensate from the gas to Company B. Company B cracks this amount of feedstock producing ethylene which it sells to Company C for $1 million. Company C polymerizes the ethylene, producing polyethylene granules which it sells to Company D for $3 million. Company D produces plastic sandwich bags which it sells to wholesale grocery dealers for $7 million. The total dollar value of these sales transactions to the chemical industry is $11 million; however, the value of these operations to the economy or to society as a whole is probably better represented by $6.6 million. This amount is the value added to the $400,000 worth of raw material by the operations of the chemical industry in producing the $7 million worth of sandwich bags.

This example neglects consideration of the value of by-products formed in the cracking process and the value of the cardboard packaging materials purchased for the final product. But it does illustrate the concept of "value added" which is most useful in the analysis of chemical operations. An appreciation for the complexity of such analyses can be attained by inspection of the product flow charts for some of the basic chemicals such as ethylene or benzene.[6]

Massive studies of these input–output relations have been made by the Department of Commerce.[2] These tables of inter- and intraindustry sales and purchases reveal the patterns of intermediate consumption and production flows within an industry and linking the various industries to one another. They allow calculation of the net value added by each industrial classification. They also give evidence of how pervasive chemistry is in our economy. Seventy-eight of the 82 industries constituting our economy purchase significant amounts of chemicals.

The growth rates of the chemical industry relative to the overall economy discussed previously were based on vast conglomerations of products. Obviously all chemicals did not grow at the same rate. The average growth

Table 1.1 Annual Growth Rates in "Value Added" for Various Sectors of the U.S. Chemical Industry—1958–1964

Chemical Sector	Percent
Organic fibers (non-cellulosic)	15.5
Toilet preparations	12.1
Synthetic elastomers	10.5
Industrial gases	9.9
Basic organic chemicals	9.6
Surface active agents	8.5
Agricultural chemicals	8.0
All "chemical and allied products"	7.5
Plastics and resins	7.5
Pharmaceuticals	6.6
Inorganic chemicals	6.2
Chlorine	6.1
Paints	6.0
Soaps and detergents	5.9
Printing inks	3.9
Cellulosic fibers	3.6
Wood chemicals	3.1
Adjusted, "real" GROSS NATIONAL PRODUCT	3-4

rate for the chemical industry results from a broad spectrum of individual growth rates. The typical pattern is that a newly introduced product enjoys a rapid growth rate; as it matures, the growth rate declines and either falls off as it is displaced by superior products or becomes an established high volume commodity whose growth rate approximates the overall growth rate of the economy. Table 1.1 lists the real growth rates of some product classifications.[7] During the time covered, 1958 to 1964, the gross national product grew at a rate of 3 to 4% per year.

The statistical data considered so far indicates the size of the chemical industry and its importance to the economy. The value added concept can also be used to measure the efficiency of an industry by relating it to the amounts of labor and capital used. The value added per employee can thus be considered a measure of the productivity of the employees; the value added per unit of capital invested can be used as a measure of the skill of management in making investment decisions. Table 1.2 lists these values for the chemical industries for various countries in 1969.[5] It indicates that the United States is competing effectively in the world market.

Another aspect of the chemical industry that is important to our economy is its strong support to a favorable balance of trade with the rest of the world.

Table 1.2 Value Added Indexes for Various Countries in 1969

Country	Sales Million $	Value Added Million $	Invest- ment Million $	Employees Thousand	Value Added per $ Invested	Value Added per Employee in $
U.S.A.	48,760	27,000	3,100	1,049	8.71	25,739
Japan	11,245	4,625	1,892	428	2.44	10,805
W. Germany	10,600	6,095	1,100	503	5.54	12,117
France	7,475	3,130	560	270	5.59	11,593
U. K.	7,470	3,575	616	418	5.80	8,553
Italy	5,680	1,920	480	245	4.00	7,837
Netherlands	2,485	910	359	91	2.53	10,000
Canada	2,274	1,270	206	71	6.17	17,887
Spain	2,175	685	98	107	6.99	6,402
Belgium	1,520	560	226	60	2.48	9,333
Switzerland	1,205	--	--	50	--	--
Sweden	930	485	102	36	4.75	13,472

For a number of years chemical exports have exceeded chemical imports by a wide margin (Fig. 1.5). In 1974 this net positive trade balance for chemicals had grown to $4.8 billion. This bright picture will not continue indefinitely. As foreign markets have grown and as foreign competitive operations have increased in efficiency, the U.S. share of world chemical exports has steadily declined (Fig. 1.6).

The statistics for the U.S. chemical industry as a whole, considering both domestic and worldwide activities, present a picture of massive size and

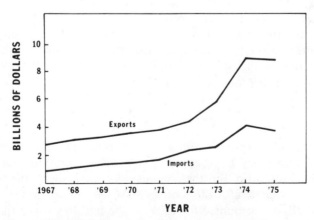

Figure 1.5 Value of United States chemical exports and imports. *Source:* Reference 8.

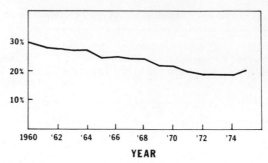

Figure 1.6   United States share of world chemical exports. *Source*:   Reference 8.

strength. It is an important industry. Its wide-based and efficient support to other industries is a critical factor in our economy. The picture is not without some discomforting trends however. Much of the advantage of the U.S. industry has been due to the size of the domestic market, which allows very large and efficient plants, and relatively low cost raw materials and energy. As foreign economies grow and become more sophisticated, their chemical markets will expand, and the effect of these advantages will become less important.

1.2   CHARACTERISTICS OF THE CHEMICAL INDUSTRY

The chemical industry, although marked by great diversity and even defying clear-cut definition, has some special characteristics. Its basic objective—making a profit—is similar to that of other industries. It must buy raw materials, sell products, invest in production facilities, employ workers, and do all the other things that most other industries must do to function. But it is inherently different, for example, from the automobile or the garment industry. Some of the special characteristics, many interrelated, of the chemical industry are considered in this section.

1.2.1   Competition

The chemical industry is very competitive. This fact might be hidden by the existence of some very large corporations such as DuPont, Dow, Union Carbide, and Monsanto. But these giants do not dominate the industry in the manner of a General Motors or an IBM. In 1974 sales for the largest company, DuPont, accounted for 6.2% of the industry total; the 5 largest corporations accounted for 22%; the 10 largest accounted for 31%; the

25 largest accounted for 46.6%. There are hundreds of chemical firms, ranging from very large to very small. The distribution of these sales figures indicates that there is a significant share of the market available to the medium and small sized companies.

This competition is of two types. First is the competition between companies producing and selling the same product. Many of the large volume basic inorganic and organic chemicals are essentially commodities in that they are sold at a price determined by the market. There is little difference between various producers in product quality or technical service available to the customer. There tends to be a small number of producers, large volume production facilities, and intense price competition. A company that lacks the raw material base or the processing technology necessary for competitive production costs simply cannot compete effectively in this sort of market. The other type of competition is that between different products. Many chemicals are purchased to perform a particular function—such as to act as an antioxidant, to kill bacteria, or to absorb ultraviolet light. Frequently there are a number of different products that perform similar functions but with different characteristics, different degrees of effectiveness, and different prices. Since the different products represent alternative ways for the customer to perform the same function, the products are in competition with one another. In this type of market the effectiveness of the product, the price, and the technical service offered to the customer are important elements in the success of the product.

1.2.2 Technology

The chemical industry is very dependent on technical and scientific knowledge for the efficiency of its production capabilities and the profitability of its product line. Few manufacturing industries are based so fundamentally on science. This characteristic contributes to other far-reaching effects. Because of the intense competition and because chemical technology is dynamic and constantly advancing, there is a high rate of obsolescence of processes and of products. This necessitates a heavy and continuing requirement for new capital expenditures and long term commitments to research and development programs.

1.2.3 Research and Development

Compared with other industries, the "chemical and allied products" industry spends large amounts of its money on research and development. In 1970 it spent over $1.6 billion.[9] This represented about one-sixth of the total

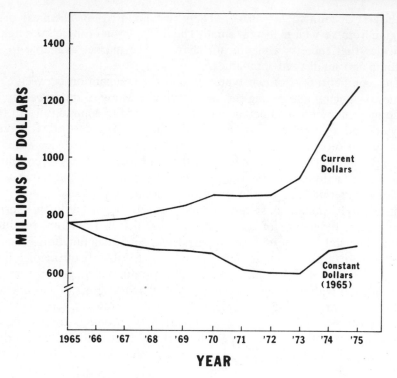

Figure 1.7   Chemical industry outlays for research and development. *Source*:   Reference 3.

industrial expenditure and was exceeded only by the "electrical equipment and communications" industry. It should be noted also that the chemical industry typically funds its own research rather than depending on government support. Of this total amount spent by the "chemical and allied products" industry, the largest expenditures were by the basic chemical industry and the pharmaceutical industry. The recent trends of research and development spending by the basic chemical industry are depicted in Fig. 1.7.

Another statistic that reinforces this picture of heavy emphasis upon chemical research is that, in 1969, 48.3% of the scientists engaged in research funded by the private sector were chemists.[2]

The reasons behind this investment in research and development relate to the rapidly changing technologies and high rates of obsolescence of products and processes. Industrial research is expensive and is becoming more costly each year. However, while this has brought changes in the focus of research and development efforts and in the selection of projects, few companies feel that they can afford to make drastic cuts in their research budgets.

1.2.4   Capital Requirements

The chemical industry requires large amounts of money for the construction of new or expanded production facilities. Table 1.3 depicts the capital spending, in the United States and overseas, by various U.S. industries in recent years. Several trends are reflected in this data: (1) although chemicals represent less than 8 % of total manufacturing, as measured by annual sales, the investment in new plant capacity is high—12 % of capital invested in all manufacturing facilities in the United States; (2) relative to all manufacturing industries the chemical and petroleum industries are investing large amounts of capital overseas.

The sources and distribution of funds by 25 of the leading companies in the chemical and allied products industry in 1974 are summarized in Table 1.4.

The general picture behind these figures is that only a small portion of the capital needs comes from the sale of stock, the bulk of the money comes from operating income and depreciation; only a small portion of the net income is distributed as dividends to the shareholders, the major portion is reinvested in the business.

This characteristic heavy use of capital for production facilities is due to a number of factors. First, it is consistent with the faster rate of growth of the chemical industry relative to all manufacturing. Second, it reflects the high rate of obsolescence of chemical processes and products. Last, it concerns the nature of chemical plants—chemical equipment is normally specialized, complex, and involves a high degree of instrumentation and the use of special materials of construction; in addition, chemical plants require large amounts of utility services such as electrical power, steam, cooling and process water, and waste disposal.

Another factor that has prompted chemical companies to invest in larger and larger production facilities is the "economy of scale." For most processes, the cost of the plant does not increase linearly with an increase in size or production capacity. Some costs such as instrumentation are almost independent of size; others such as the costs of reactors and distillation columns increase approximately with their capacity raised to the two-thirds power. The latter is called the square-cube law and relates to the fact that costs are associated with the surface area of the equipment whereas production capacity is associated with the volume of the equipment. The overall effect therefore is that doubling the capacity of a plant does not double the cost of the plant. So long as there is reason to believe that sales will be sufficient to

Table 1.3 Capital Spending (in $ Billions)

Industry	in the United States					Overseas*				
	1975	1974	1973	1972	1971	1975	1974	1973	1972	1971
Chemicals	6.31	5.69	4.46	3.45	3.44	2.7	2.0	1.3	1.2	1.2
Petroleum	10.19	8.00	5.45	5.25	5.85	10.4	8.6	6.6	5.2	5.0
Rubber	1.41	1.47	1.56	1.08	0.84	0.4	0.4	0.3	0.3	0.2
Paper	2.98	2.58	1.86	1.38	1.25	0.8	0.8	0.6	0.5	0.6
All Mfg.	49.30	46.01	38.01	31.35	29.99	13.0	11.0	8.9	7.3	7.2

*By majority owned affiliates of U. S. companies

Source: U. S. Dept. of Commerce

Table 1.4   1974 Sources and Distribution of Funds of Leading Companies—Chemical and Allied Products Industry

	$ Millions	%
Source of Funds		
Net Income	4,008	45.5
Depreciation, Depletion	2,224	25.2
Deferred Taxes	311	3.5
Other Internal Sources	527	6.0
Long Term Debt	1,622	18.4
Sale of Stock	120	1.4
TOTAL	8,812	100.0
Application of Funds		
Dividends	1,476	16.7
Capital Expenditures	4,953	56.2
Additions to Working Capital	1,293	14.7
Reduction Debt	396	4.5
Other	694	7.9
TOTAL	8,812	100.0

allow operation at near capacity, there will be an economic incentive to build on the largest possible scale.

Problems arise when a number of competitors each attempt to lower their production costs by building very large plants. Such a situation soon results in the total production capacity of a particular product far exceeding the total market demand. The chemical industry has been plagued periodically with such conditions of "overcapacity." The economic return on investment is seriously affected. The only cure for overcapacity appears to be time. Since most markets for chemicals are continuing to grow, the production capacity and the market demand eventually come into balance again.

1.2.5   Labor Requirements

The chemical industry does not employ large numbers of people considering the value of the products produced. In 1974 all manufacturing industries employed 20 million people; the chemical and allied products industry, with annual sales representing about 8% of the value of total manufactured products, employed only 1 million, or 5%, of the total manufacturing employees. The imbalance is more dramatic if only production workers are considered—all manufacturing employed almost 15 million; the chemical industry about 600,000. In many modern chemical plants, the production workers are outnumbered by the maintenance people and the professional and technical employees.

1.2.6   Integration

Growth appears to be an inherent characteristic of a healthy, dynamic company. In the chemical industry, growth through integration is probably more significant than through diversification. This integration can either be vertical or horizontal. Vertical integration may be forward or backward. An example of forward integration would be a basic producer of bulk ethylene glycol entering the consumer market with an antifreeze formulation. The motivation for such a move is the higher profits generally associated with consumer products. Many companies have integrated backward to secure a protected and economical raw material or energy base for their operations. Horizontal integration is most generally related to technology. A company that has built up a technical expertise and customer service function for a particular product area—such as ion exchange resins, antioxidants, or surfactants—will tend to capitalize on this technology base by offering a product line of a number of such functional products.

1.3   COMPANY PHILOSOPHIES

These then are some of the characteristics of the chemical industry. Every successful company has evolved a mode of operation that allows it to survive and prosper in this particular environment. Not surprisingly there are many similarities in the way each company is organized, is managed, and conducts its business. But there are differences, too. There are so many opportunities and so many directions for growth that some choices must be made. The cummulative effects of these choices, through the years, develop into patterns. The growth of a company, from a small enterprise into a large corporation is based upon some philosophy or oriented sense of purpose. This philosophy may not be explicitly stated, and it may change through various phases of the company's growth. From the outside, large corporations may appear to be faceless bureacracies directed by computers and white-collar robots. But each has its particular strengths, its unique spirit, and its distinct personality. The following section includes brief histories of the founding and growth of five companies considered typical of the leaders in the chemical industry. It is hoped that these descriptions suggest the underlying business philosophies of these companies and the spirit of the people who shaped them.

1.3.1   Tennessee Eastman[10]

Tennessee Eastman Company is a manufacturing unit within the Eastman Chemicals Division of Eastman Kodak Company. At its plant in Kingsport, Tennessee, Eastman manufactures over 400 products in three principal categories—fibers, chemicals, and plastics. Tennessee Eastman is one of the leading producers of cellulosic plastics and fibers and holds an important position in the production of other chemicals, dyes, and fibers. It produces materials for Kodak's photographic products as well as for sale to others. It is one of the largest industrial concerns in the Southeastern United States, employing over 13,200 in its Kingsport operation.

The history of the company is linked closely to George Eastman's research which led to the development of photographic films and to Kodak's mass production facilities which required large quantities of methanol and cellulose acetate.

An acute shortage of methanol led Kodak to seek its own source of supply in order to safeguard film production. In 1920 Kodak purchased a partially completed wood alcohol plant in Kingsport from the U.S. Government. Tennessee Eastman was founded and two years later began the production of methanol by a wood distillation process. The raw material was waste from the lumbering operations in the surrounding forests. Charcoal, acetic acid, and wood creosote were sold as by-products. In 1930 Kodak transferred its production of cellulose acetate to a large, modern plant in Tennessee This markedly reduced the cost of cellulose acetate to Kodak. Excess production capacity was available for external sales. By 1931 Tennessee Eastman had developed a process and a market for cellulose acetate yarn.

From this beginning many Eastman products were to emerge. Cellulose acetate could be molded as well as spun into fibers. In 1932 Tennessee Eastman introduced Tenite acetate, a plastic molding composition. This was followed by Tenite butyrate in 1938 and Tenite propionate in 1957. Tennessee Eastman's developing cellulose acetate technology gave rise to related products. For example, the traditional dyes did not work well on the new man-made fibers, so a search for new dyes was initiated. In 1938, after much research, the company began the manufacture of a new line of dyes for acetate yarns. Today Tennessee Eastman is the leader in the field of disperse dyes for acetate and polyester fibers. In 1952 two important additions to Eastman's fiber probuction became realities. Chromspun yarn, a solution-dyed acetate textile fiber, went into large-scale production. Estron acetate filter tow was introduced and its use in cigarette filters has become an important part of Eastman's business.

From the knowledge and experience gained in manufacturing and marketing acetate fibers, Eastman went into the manufacture of other man-made fibers—Verel modacrylic fiber and Kodel polyester fiber. These are used in apparel, carpets, and home furnishings. Kodel polyester filament yarn was added later.

In addition, Eastman has added to its product line by marketing certain intermediates. Thus, in the case of Kodel polyester fiber, terephthalic acid, dimethyl terephthalate, and polyester polymer are sold as industrial chemicals to other fiber producers.

Other Eastman Chemicals Division manufacturing units have grown out of Tennessee Eastman's operation. In 1950 Texas Eastman Company was established at Longview to produce alcohols and aldehydes required in large quantities at Tennessee Eastman. Expansions in Texas have added other petrochemicals and polyolefin plastics. Other manufacturing units have been established in South Carolina and Arkansas.

The success and growth of Tennessee Eastman are based upon establishing a solid, thorough technological base; growing and diversifying into areas directly related to this technology; increasing sales and reducing costs by finding new applications and new markets for Eastman products.

1.3.2   Syntex[11]

Syntex Corporation is a diversified, multinational organization with sales approaching $300 million and 6000 employees. Many companies are larger, but few have grown faster than Syntex, which began just 32 years ago with a major breakthrough in the field of steroid chemistry.

Increasing clinical use of steroid sex hormones set off an intensive search for a low cost raw material for the synthesis of these compounds. In the 1930s the starting materials for steroids were animal glands, horse urine, and cattle bile. Many believed that these limited animal sources would not be sufficient to meet the expected demand. The key to the development of a major, new raw material was work done by Russell E. Marker at Pennsylvania State College. Marker developed an efficient method of degrading the side chain of certain plant materials called sapogenins to convert them into steroids from which progesterone could be synthesized. Thus began an incredible botanical hunt which involved over 80,000 lb of plant materials from 400 different species. After a two-year search Marker left Penn State and established a laboratory in Mexico City. In the jungles of Veracruz in southern Mexico he had found a wild yam, cabeza de negro which met his requirements.

Marker established a commercial venture with two naturalized Mexicans of European origin, Dr. Emeric Somlo from Hungary and Dr. Federico Lehmann, from Germany. This was the start of Syntex Corporation, in Mexico City in January of 1944. Less than two years later, Marker had a falling out with his two partners, and sold his interests, taking the secrets of his work with him.

Following Marker's departure Syntex hired Dr. George Rosenkranz, a chemist trained in Switzerland who is now Chairman of the Board of Syntex. Within two months Syntex was back in production. Rosenkranz gathered together a research group that, by 1948, had solved most of the technical problems, enabling Syntex to sell on the world markets all of the natural hormones that at the time were known to be medically valuable.

From the original use of sex hormones in replacement therapy, the clinical applications of steroids have increased manyfold. Steroid drugs are now used as oral contraceptives, anti-inflammatory agents, sex hormones, adrenocortical hormones, and anabolic agents. These developments strengthened Syntex's position and led to a rapid growth. Three of the first four oral contraceptives available in the United States were developed by Syntex or were manufactured from materials produced by Syntex. Throughout the 1950s the research team assembled by Rosenkranz, which included such outstanding scientists as Djerassi, Ringold, Zaffaroni, Bowers, Halpein, and Alvarez, worked out the details for the commercial production of the steroid hormones needed for the increasing medicinal applications.

The early growth of Syntex was closely related to the developing field of steroids and the supply of raw materials located in Mexico. More recently the company has diversified its activities. To bolster its manufacturing capabilities Syntex acquired Arapahoe Chemicals in Boulder, Colorado, and later purchased a manufacturing plant in Tennessee. Other manufacturing facilities are located in Mexico, the Bahamas, and Ireland. By acquisitions Syntex has diversified into other product areas such as veterinary products, feed additives, dental instruments. Subsidiaries making sophisticated instruments used in medical X-ray technology and X-ray crystallography were started within the company.

There are perhaps three keys to the success of Syntex—astute management, an international outlook, and research. Many of the older domestic firms have, through determined efforts, evolved into international corporations. In contrast, Syntex was in a sense a multinational organization at its inception. In the field of research Syntex is truly outstanding. This might be expected in a company whose top management has been drawn in large measure from research scientists. The strong commitment to research is indicated by the current research expenditures which approach $25 million per year. About 70% of this amount is devoted to human pharmaceutical research; approximately 20% supports the veterinary and agricultural research; the remaining 10% is invested in the development of medical and scientific instrumentation. The research budget, which has increased every year of Syntex's history, has enabled the company to expand from steroids into a dozen different therapeutic areas.

1.3.3   Du Pont[12]

E. I. du Pont de Nemours and Company is one of the largest industrial firms in the world. It employs over 130,000 people and has sales of over $8 billion per year; it operates more than 120 manufacturing plants and over 100 research and development, sales, service, and plant laboratories in the United States; it has 44 foreign subsidiaries or affiliated companies; it manufactures about 1700 product lines.

The company was started in 1802 by E. I. du Pont, a young French chemist who built a small powder mill on the banks of Brandywine Creek in Delaware. For the first 100 years Du Pont remained primarily a manufacturer of black powder, with dynamite and smokeless powder being introduced toward the end of the century. In the early 1900s the company began to diversify its product line into paints, lacquers, plastics, and some heavy chemicals such as sulfuric acid. At first this technology was purchased; but in 1903, Du Pont organized the Experimental Station with a budget of $300,000. This was one of the first industrial laboratories in the United States. It was broadly commissioned to search the field of chemistry for commercial opportunities. By 1909 programs were underway in artificial leather, plastics, photographic film, and synthetic fibers.

The first major product of these efforts came in 1923 with the introduction of nitrocellulose lacquers for automobile finishes. Research on polymers, led by Wallace Carothers, resulted in neoprene in 1931 and nylon in 1938. Continued polymer research has yielded "Dacron" polyester fiber, "Delrin" acetal resin, "Surlyn" ionomer resin, and "Teflon" FEP fluorocarbon resin. In other areas product development programs have led to "Freon" refrigerants, antifreeze formulations, and many other high volume, well respected product lines.

There are many facets to Du Pont's success. One obvious aspect is that Du Pont was able to evolve an organizational structure to efficiently manage such a massive and diverse corporation. The company is managed through an Executive Committee, a Finance Committee, and 12 industrial departments set up along logical product lines or areas. To a surprising degree these industrial departments function as independent entities. They buy and sell among themselves and, on occasion, their products compete with each other in the same market. The Executive Committee is composed of the chief executive officer, the president, and six senior vice-presidents. This group is the policymaking body as well as the supervisory body for all operations. If an industrial department wishes to expand its activities, the Executive Committee must consider and approve the proposal before it can be put into effect. The organization and detailed operation of the individual industrial departments is delegated to the general managers of these departments. The Executive Committee keeps the whole organization coordinated and in proper balance; it achieves this through setting broad company policy and by controlling, with the Finance Committee, the allocation of capital expenditures. Du Pont's shift to decentralized organizational structure was a pioneering step in business and has provided a model that has since been emulated by many companies.

The steady flow of new products that originate in Du Pont's research and development groups is a vital component in the continued success of the company. Its exploratory research has created broad areas of technical expertise that have resulted in many profitable products. Its applications groups have been particularly successful at finding new uses for existing products, which maximizes their profitability. Its customer service work, which backs up the products, is perhaps unequaled in the industry.

1.3.4   Dow[13,14]

Dow Chemical is a major international corporation with annual sales of over $4 billion, with current assets of over $2 billion, and with over 50,000 employees. The history of Dow's growth is fascinating because it is tied so closely to the philosophies and personalities of the people who have managed this growth.

Herbert Henry Dow was 24 years old when he came to Midland, Michigan, in 1890. He believed he could extract bromine from brine using an electric current and a "blowing out" process that he discovered while teaching chemistry in a small hospital school in Ohio. Dow had graduated from Case School of Applied Science in Cleveland several years before, and he received strong support in his early efforts from faculty members at Case. His choice of Midland, a small Michigan lumber town, was because of the area's vast underground brine deposits, which were particularly rich in bromine, and the availability of cheap scrap wood fuel to generate electricity.

It took several years and a number of technical and business failures before his electrolysis process became commercially practical. The Dow Chemical Company was founded in 1897. The simple but unique electric cell that he developed was the backbone of the new company. The electrolysis of brine is still a vital technological base of the company's business.

Dow's first generation products were bromine based. His second generation products were based on chlorine and caustic. He then developed products based on the magnesium and calcium chloride contained in the Midland brine.

The first 50 years were marked by steady economic growth. The management style under Herbert H. Dow, and later his son, Willard, was a very informal relationship of one strong leader backed up by strong, technically competent managers. Many of the policies laid down during this period are still evident in Dow's operations today. One is the policy of basic self-sufficiency; of relying as little as possible on others for raw materials and other basic needs. Herbert Dow built his own electrical power sources. Dow still generates much of its own power. It established its Brazos Oil and Gas Division in 1947 to assure a supply of petroleum and gas for raw materials and energy. Another policy is that of developing the technology to utilize all of the potential of the available raw materials. Dow did this originally with brine and is currently doing it with its petrochemical resources. A third policy is that of emphasis on the high volume, basic commodity chemicals. Although Dow has diversified its products and services and has entered the consumer market, the strength of the company is still its capability of producing basic chemicals on a very large scale.

The history of Dow is conveniently divided into two phases. The first phase covers 50 years and brought the company through World War II and the start of the postwar expansion. The gains made during this period were impressive. The company had more than $200 million in assets; sales of $130 million; 12,500 employees; and the fastest growth rate of any major U.S. chemical company. Dow technology had supplied the magnesium and synthetic rubber needed for the war. From the standpoint of the future of the company, probably the most significant event was establishing a production plant on the Gulf Coast in Freeport, Texas. The immediate purpose was to extract critically needed magnesium from the sea. The result was to firmly establish Dow in a location rich in raw materials—natural gas, oil, brine, lime, and sea water. Dow, which had grown very rapidly, was about to enter a new phase of even more rapid expansion. But it had one leader. Willard Dow was the general manager, president, and chairman of the board as well as the director of research, public relations, and almost every other function. In 1949 Willard Dow died in an airplane accident.

The new president, Lee Doan, handled the expansion. In the period 1949 to 1959, assets rose from $294 million to $859 million; annual sales rose from $200 million to $860 million. But he did more than that, he started a reorganization of the company. Dow Chemical at that time was dominated by very strong, very competent production managers. Organization was quite informal; responsibilities overlapped. Each manager had his sphere of influence—based on geography or area of technology. Each had his production plants, engineering support, and research groups. Lee Doan broadened this power base and elevated sales to a position of equal importance. He recognized that chemical marketing was changing and required strong technical support. He developed an overseas sales organization. Throughout this period of change and expansion, Lee Doan gradually shifted control of the operations to the younger managers.

In 1962 Ted Doan became the president of Dow. His primary concern was to complete the organizational plan he had developed—the company was divided into product departments—chemicals, plastics, metal products, bioproducts, packaging, and consumer products. Each department manager had a staff of managers who headed business teams responsible for a particular group of products. Marketing, production, and research and development were all represented on these business teams. The effect was to create relatively small, manageable businesses. They afford the flexibility and maneuverability of a small business with the strength and depth of a large corporation. It is a rather loose, freedom-oriented organization which gives people the opportunity to develop their initiative, their ideas, and their abilities.

With Ted Doan handling the reorganization, Carl Gerstacker looking after the finances, and Ben Branch in charge of operations, Dow Chemical became a truly global operation. The company has a history of change, innovation, growth, and diversification. But its roots to the past are strong. The company was founded upon Herbert Dow's process for the electrolysis of brine. Of the 1100 or more products manufactured today, over 80% involve brine-based chemicals in some way.

1.3.5 Ethyl Corporation[15]

Ethyl Corporation (Delaware) was formed in 1924 as a 50-50 joint company of General Motors and Standard Oil of New Jersey to capitalize on the combination of two technological breakthroughs—the discovery of the antiknock effect of tetraethyllead (TEL) and a process for its manufacture. Thomas Midgley, Jr., and T. A. Boyd, under the direction of Charles F. Kettering, discovered in 1921 (at what later became the General Motors Research Laboratory) that the addition of TEL to gasoline allowed the use of high-compression engines without the power-robbing knock that occurred in the absence of TEL. Shortly after General Motor's discovery, Standard Oil of New Jersey developed the most economical synthesis of TEL—the high pressure reaction of ethyl chloride with a sodium-lead alloy. Since 1924 Ethyl has grown from a one-product company into a diversified manufacturer with 1976 sales of more than $1.1 billion. Most of this diversification has occurred in the last decade. Ethyl now manufactures several hundred products, operates more than 50 U.S. and 9 foreign plants, and employs approximately 16,000 persons.

In initial use it was found that TEL produced lead oxide during combustion, which deposited in the engine and on spark plugs. Early TEL fluids included organic halogen compounds to produce volatile lead halides which were carried out with the exhaust. This problem was not adequately solved until Ethyl defined the optimum ratios of 1,2-dibromoethane(EDB) and 1,2-dichloroethane(EDC) needed with TEL for "scavenging" the lead.

As the market for TEL grew, the demand for bromine in the antiknock fluid soon surpassed the available bromine supply. Ethyl searched the world for bromine sufficient for its needs. Finally, Ethyl turned to the sea where bromine is present only in about 65 parts per million. From 1934 through 1969 Ethyl obtained EDB scavenger from Ethyl-Dow Chemical Company, a joint venture established to extract bromine from sea water.

From the time of the formation of Ethyl Corporation, chemists and engineers have searched for a better antiknock compound. In 1952 Ethyl chemists conceived of the preparation of a new class of organometallic compounds by using the pi bonding properties of various unsaturated organic molecules and radicals such as benzene, cyclopentadienyl, and even ethylene. By combining the pi bonded organic groups and the traditional coordinating groups such as carbonyls, nitrosyls, and amines on a common metal atom, Ethyl chemists were able to synthesize gasoline-soluble stable compounds of all of the transition metals and many nontransition metals. A major effort was made to synthesize and test as antiknock agents a wide variety of organometallics of this class. Several effective antiknock compounds were found and patented including an outstanding product, methylcyclopentadienyl manganese tricarbonyl (MMT).

In the mid 1950s Ethyl research developed a laboratory route to TEL based on the reaction of triethyl aluminum (TEA) with a lead salt. A better route to TEA was needed, and Ethyl personnel developed a direct synthesis of TEA from aluminum metal, ethylene, and hydrogen. At about the same time, European scientists discovered that polypropylene and high-density polyethylene could be produced using TEA or aluminum alkyl halides as part of a catalyst system. The timing of these two discoveries made it possible for Ethyl to become today's leader in providing commercial quantities of these catalysts which are used to produce a variety of new large volume polymers.

For some years Albemarie Paper Manufacturing Company of Richmond, Virginia, had recognized plastics as the prime competitor to paper in the packaging field and had sought to acquire a chemicals manufacturer. In 1962 Albemarie acquired Ethyl from General Motors and Standard Oil of New Jersey. The new company, renamed Ethyl Corporation (Virginia), immediately began a greatly expanded diversification effort.

During this period a major raw material for some of the large volume detergents was the alcohol mixture derived from hydrogenation of coconut oil. In the early 1960s Ethyl discovered a process for reacting ethylene with TEA to produce a mixture of long-chain fatty alcohols that essentially duplicated the carbon number distribution of the coconut. The mechanism of this process involves using ethylene to grow alkyl chains on the aluminum metal, separating the desired aluminum alkyls and recycling the undesired, oxidizing the desired aluminum alkyls to alkoxides, and thereafter hydrolyzing the alkoxides to alcohols. In 1965 Ethyl started up a 100 million lb/year fatty alcohol plant in Houston, Texas, using this technique.

Ethyl's initial entry into plastic products came with the purchase of the Visqueen polyethylene film operations of Union Carbide in 1963. Then Ethyl's oxychlorination technology, which had been developed for its ethyl chloride, ethylene dichloride, and vinyl chloride production, proved to be very valuable in providing a means for further diversification into other plastics. The technology was provided to the two largest European manufacturers of polyvinyl chloride (PVC) in exchange for their advanced technology for the production of PVC. Ethyl modified and adapted the acquired technology for the U.S. market and in 1965 began production of PVC resins and compounds. The same year a small PVC bottle operation was formed, and Ethyl also began production of PVC pipe. In 1968 Ethyl expanded their plastic bottle manufacture by purchasing Imco Container Company, which now operates 11 U.S. and two Canadian plastic bottle plants.

In 1969 Ethyl built a new bromine and EDB plant at Magnolia, Arkansas, and withdrew from Ethyl-Dow. Ethyl's Magnolia plant is based on brines having unusually high bromine content. In 1971 carpet manufacturers needed flame resistant fibers. Ethyl developed a new process for manufacture of vinyl bromide from EDB and currently is the world's major producer of vinyl bromide, used for flame retarding acrylic fibers.

In 1971 Ethyl again used the earlier-developed aluminum chain growth technology, this time to produce alpha olefins (butene-1 through octadecene-1) which are sold as plasticizer and detergent intermediates. Ethyl built the world's largest alpha olefin plant at Houston, Texas, using this technology.

The use of catalytic converters on many new automobiles starting with the 1975 model year has forced refiners to eliminate the use of lead from some of their gasoline. This has greatly increased the importance of the manganese-based antiknock (MMT) which provides the gasoline refiners with an important alternative method to meet the octane requirements for the new cars that require unleaded gasoline. Ethyl has expanded its MMT plant in Orangeburg and in January 1977 began work on a MMT plant at another location.

Ethyl Corporation's growth in the chemicals business is guided by a well-defined set of principles. Since the 1962 acquisition by Albemarle, Ethyl's sales have more than quadrupled. Internally generated projects have contributed significantly to this growth. Ethyl was born of research and plans to continue its growth via research. Ethyl has deliberately sought to be reasonably diverse in the end markets it serves in order to minimize the effects of unavoidable cycles in any one business or product area. Ethyl concentrates on projects that are technologically difficult so that successful developments will make Ethyl a unique technology leader—a position that others can then not easily duplicate.

1.4 SUMMARY

The commercial practice of chemistry, in the chemical industry itself and in the many associated industries, is a very important factor in our culture and in our economy. The chemical industry has some special characteristics. It is a dynamic, growing, competitive industry. It is capital intensive rather than labor intensive. The production plants are generally highly specialized, complex, and expensive. The underlying technologies are subject to rapid change. Working in this environment can be challenging and stimulating. Each company has developed its particular style of doing business. As the new employee gains an awareness of the patterns and directions of a company's growth, he is better able to view his job in perspective and to make more significant contributions to the growth.

1.5 REFERENCES

1. American Chemical Society, *Chem. Eng. News*, July 10, 1972, p. 6.
2. *Chemistry in the Economy*, American Chemical Society, Washington, D.C., 1973.
3. *Chem. Eng. News*, June 2, 1975, pp. 59, 65.
4. *Chem. Eng. News*, June 3, 1974, p. 45.
5. B. G. Reuben and M. L. Burnstall, *The Chemical Economy*, Longman, London, 1973.
6. *Chemical Origins and Markets*, 4th ed., Chemical Information Services, Stanford Research Institute, California, 1971.
7. *Riegel's Handbook of Industrial Chemistry*, 7th ed., James A. Kent, Van Nostrand Co., New York.
8. *Chem. Eng. News*, December 22, 1975.
9. *National Science Foundation, Industrial R & D Spending*, Science Resources Studies Highlights (NSF 71-39), Washington, D.C., 1970.
10. Calvin M. Anderson, Tennessee Eastman Company, Kingsport, Tennessee, private communication, 1977.
11. William L. Spencer, Syntex Corporation, Palo Alto, California, private communication, 1977.
12. Carl B. Kaufmann, E. I. du Pont de Nemours & Company, Wilmington, Delaware, private communication, 1977.
13. Terry Marks, Dow Chemical Company, Midland, Michigan, private communication, 1977.
14. D. Whitehead, *The Dow Story: The History of the Dow Chemical Company*, McGraw-Hill, New York, 1968.
15. John T. Balhoff and J. E. Brown, Ethyl Corporation, Baton Rouge, Louisiana, private communication, 1977.

2

BASIC CONSIDERATIONS

ENVISIONING THE PROCESS

There is frequently considerable difference between "classical," laboratory scale chemistry and the practice of chemistry on an industrial scale. Even in those cases where the equations describing the reactions appear similar, the actual operations and reaction conditions are often quite different.

Consider the formation of ethyl alcohol by the hydration of ethylene:

$$CH_2{=}CH_2 + H_2O \xrightarrow[H^+]{} CH_3{-}CH_2{-}OH$$

In the laboratory this reaction might be run most conveniently by bubbling ethylene into 98% sulfuric acid and then diluting and warming the reaction mixture to hydrolyze the resultant sulfate ester. In the industrial process a stream of ethylene is mixed with steam at 325°C and 1000 psi and passed over a solid catalyst consisting of phosphoric acid absorbed on diatomaceous earth; the process is run continuously, and unreacted ethylene is separated from the reaction product and recycled to the feed stream.[1]

The differences in approach reflect basic differences in objectives. In the commercial process the objective is to produce product at minimum total cost on a scale that will generate the maximum economic return. In the laboratory the objective is frequently to synthesize a substance in the most convenient manner in terms of the chemist's time and the laboratory equipment available. Since this usually involves running the reaction in glassware, the temperature and pressure limits of this equipment act as constraints upon the choice of reaction conditions.

In addition to a greater latitude in the choice of conditions of temperature and pressure, industrial processes may be based upon either batch or continuous operation and may involve reactants in either vapor phase or liquid phase. Laboratory procedures are most commonly, but not always, liquid phase reactions conducted in batchwise manner.

The goal in the development of an industrial process is to devise a procedure that has been optimized from the standpoint of total cost. Since the total cost includes capital costs and indirect costs as well as the costs of utilities, maintenance, labor, and raw materials, the optimum process from the standpoint of economics might not be identical with the process with the highest percentage yield.

Generally, however, industrial chemistry is concerned with minimizing raw material costs, either by improvements in yield or by utilizing less costly reactants. The use of solvents is commonly minimized, not only to decrease raw material costs and the costs of separation and purification of the solvent for reuse, but also to maximize the effective reactor volume and thus decrease the equipment costs. The industrial chemist must also be concerned with the compositions of vent gases and waste streams from the process to minimize air and water pollution and to assess accurately the costs of waste disposal and pollution control.

2.1   CONVERSION, EFFICIENCY, AND YIELD

Many reactions do not go to completion within a reasonable period of time. Frequently one reactant is charged to the reactor in greater than the stoichiometric amount in order to improve the reaction rate, or to shift the equilibrium in a favorable direction, or to limit by-product formation. For these reasons unchanged reactants frequently appear in the reaction mixture or in the product stream. These unconverted reactants may be disposed of as waste or, as is most often the case, they may be separated from the product and recycled to the process.

These alternatives have a profound effect upon the economics of the process. The terms "conversion," "efficiency," and "yield" are used to describe the amount of reactant consumed and the amount of product formed in a process. Conversion is expressed as a percentage and relates the amount of reactant that is chemically converted to another substance or substances to the amount of reactant initially charged to the process. Efficiency is also expressed as a percentage and relates the amount of reactant that has been converted to the desired product to the amount that has been chemically converted. The term "selectivity" may be used in place of "efficiency," particularly when discussing catalytic reactions. Unfortunately, the term "yield" can cause some confusion. Yield may refer to the amount of product formed with relation to the amount of reactant consumed; or it may refer to the amount of product formed with relation to the amount of reactant charged to the reaction. In the latter usage the yield is equal to the product of conversion times efficiency. In order to avoid this ambiguity, the basis should be noted.

If, in the acid catalyzed hydration of ethylene, 6.000 moles of ethylene and 10.000 moles of water were fed to a reactor, and the reactor effluent contained 0.244 mole of ethanol, 9.750 moles of water, and 5.748 moles of ethylene; then:

1. The *conversion* of ethylene is

$$\frac{6.000 - 5.748}{6.000} \times 100 = 4.20\%$$

2. The *efficiency* of the process, or the *selectivity* of the catalyst, is

$$\frac{0.244}{6.000 - 5.748} \times 100 = 97.0\%$$

3. The *yield* of ethanol *based upon ethylene converted* is

$$\frac{0.244}{6.000 - 5.748} \times 100 = 97.0\%$$

4. The *yield* of ethanol *based upon ethylene charged* is

$$\frac{0.244}{6.000} \times 100 = 4.07\%$$

2.2 A COMPARISON OF ACADEMIC AND INDUSTRIAL CHEMISTRIES

There are a large number of commercially important processes that are based upon the same familiar reactions described in textbooks. Examples include:

1. The reduction of nitrobenzene to aniline.

2. The manufacture of aspirin.

3. The formation of magnesium metal.

$$MgCl_2 \xrightarrow{\text{electrolysis}} Mg + Cl_2$$

4. The production of calcium hydroxide from limestone or oyster shells.

$$CaCO_3 \longrightarrow CaO + CO_2$$

$$CaO + H_2O \longrightarrow Ca(OH)_2$$

Other industrial processes involve reactions which, although based upon classical reactions, might at first glance appear to be different. For example:

Pentaerythritol is manufactured by the base catalyzed reaction of formaldehyde and acetaldehyde:

$$4H-\overset{\overset{\displaystyle O}{\|}}{C}-H + CH_3-\overset{\overset{\displaystyle O}{\|}}{C}-H + NaOH$$

$$\longrightarrow HO-CH_2-\underset{\underset{\displaystyle CH_2}{|}}{\overset{\overset{\displaystyle OH}{|}}{\underset{\displaystyle |}{C}}}-CH_2OH + H\overset{\overset{\displaystyle O}{\|}}{C}-ONa$$

$$ OH$$

Closer inspection reveals that this process consists of three mixed aldol condensations followed by a crossed Cannizzaro reaction, which are familiar to all first-year students of organic chemistry.

The chelating agent ethylenediamine tetraacetic acid is produced by the reaction of ethylenediamine, formaldehyde and sodium cyanide:

$$NH_2-CH_2CH_2-NH_2 + 4H-\overset{\overset{\displaystyle O}{\|}}{C}-H + 4NaCN$$

$$\xdownarrow{H^+, H_2O}$$

$$\begin{array}{ccc}
HO-\overset{O}{\overset{\|}{C}}-CH_2 & & CH_2-\overset{O}{\overset{\|}{C}}-OH \\
& N-CH_2-CH_2-N & \\
HO-\overset{O}{\overset{\|}{C}}-CH_2 & & CH_2-\overset{O}{\overset{\|}{C}}-OH
\end{array}$$

This process is based upon a modification of the Strecker synthesis of α-amino acids which is described in most organic texts.

In many other industrial processes the reactions appear strange. They have no readily apparent relation to the familiar "classical" chemistry described in most undergraduate textbooks. Consider the following oxidations of alkenes, for example:

$$CH_2{=}CH_2 + \tfrac{1}{2}O_2 \xrightarrow[Ag]{250°} CH_2\!\!\diagdown\!\!\underset{O}{}\!\!\diagup\!\!CH_2$$

$$CH_2{=}CH_2 + \tfrac{1}{2}O_2 \xrightarrow[PdCl_2,\,CuCl_2]{150°} CH_3{-}\overset{\overset{\textstyle O}{\|}}{C}{-}H$$

$$CH_2{=}CH_2 + CH_3\overset{\overset{\textstyle O}{\|}}{C}{-}OH + \tfrac{1}{2}O_2$$

$$\xrightarrow[PdCl_2,\,CuCl_2]{150°} CH_2{=}CH{-}O{-}\overset{\overset{\textstyle O}{\|}}{C}{-}CH_3 + H_2O$$

$$CH_2{=}CH{-}CH_3 + O_2 \xrightarrow[Bi\text{-}Mo]{250°} CH_2{=}CH{-}\overset{\overset{\textstyle O}{\|}}{C}{-}H + H_2O$$

$$CH_2{=}CH{-}CH_3 + \tfrac{3}{2}O_2 + NH_3 \xrightarrow[Bi\text{-}Mo]{450°} CH_2{=}CH{-}C{\equiv}N + 3H_2O$$

Several points should be noted concerning these industrially important reactions. The first is the use of gaseous oxygen or air as the oxidant. Potassium permanganate and sodium dichromate, which are so familiar in the laboratory, are simply not used on an industrial scale because of the high cost and disposal problems. The second point is the use of selective catalysts, both heterogeneous and homogeneous, and the use of high temperatures and pressures.

The principal objective of this discussion is simply to illustrate that the chemistry studied in college is useful in the industrial world; but the industrial chemist cannot limit his thinking to the reactions described in the typical textbook. The purpose of a textbook is to teach. In order to teach effectively, textbooks must emphasize general reactions and broad unifying concepts. Reactions that do not fit conveniently into these general patterns are only occasionally mentioned and very infrequently discussed. However, many of the most successful and most economically attractive processes have involved unusual or unique reactions.

2.3 EVALUATION OF A REACTION

It is useful for an industrial chemist to develop a strategy for considering various reaction paths to a desired product. Evaluation for economic and technical feasibility is a continuous task during the course of any industrial research or development project. Frequently this evaluation is formalized into a series of reviews occurring at various stages of the project. The evaluation discussed here is at the conceptual stage. This is not a formal evaluation procedure but rather is a method of viewing an idea for a process before any serious literature search or laboratory work is started.

This strategy involves considering all of the various options available; hopefully including not only the obvious, well-known approaches, but also the unfamiliar, weird and far-out possibilities. These candidate processes can then be assessed on the basis of (1) a rough, "ball park" economic analysis; (2) a consideration of the technical feasibility based upon knowledge of the reaction or similar reactions. These assessments are not meant to stifle creativity but simply to prevent wasting effort on a path that is clearly uneconomical or has practically no chance of success. Industrial research and development is somewhat like a poker game. No company, no matter how large, has unlimited resources. Part of the challenge of industrial research is in deciding which projects to bet on. These decisions should be based not only upon the size of the pot or the economic return if a project is successful, but also upon the odds or the chances of success.

As an example let us consider the development of a process for the manufacture of ethylamine. A natural first step would be to list all of the possible reactions that could conceivably yield the product. This listing would include the known preparative reactions for primary amines such as would be listed and described in various surveys of synthetic methods.[2-5] In addition, and most importantly, it would include any other possibilities that might occur to the chemist. This part of the process is highly dependent upon the chemist's broad knowledge of chemistry and prior experience, his intuition, and his creativity. This is the part of research that makes chemistry an intellectually exciting and rewarding discipline; it relates to one of the goals of industrial research, which is to conceive and develop a truly unique concept into an economically sound reality. The patent system was devised to reward such a venture—the reduction to practice of an idea that is not readily apparent to one skilled in the state of the art. To be of significant interest in industrial research, in addition to being unique and workable, the idea must also possess some inherent economic advantage over the known technology.

In the case of the ethylamine process a partial listing of candidate reactions might include:

1. $CH_3CH_2Cl + 2NH_3 \longrightarrow CH_3CH_2NH_2 + NH_4Cl.$

2. $CH_3-C\equiv N + 2H_2 \longrightarrow CH_3CH_2NH_2.$

3. $CH_3CH_2NO_2 + 3H_2 \longrightarrow CH_3CH_2NH_2 + 2H_2O.$

4. $CH_3-\overset{\displaystyle O}{\overset{\displaystyle \|}{C}}-H + NH_2OH + 2H_2 \longrightarrow CH_3CH_2NH_2 + 2H_2O.$

5. $CH_3-\overset{\displaystyle O}{\overset{\displaystyle \|}{C}}-H + NH_3 + H_2 \longrightarrow CH_3CH_2NH_2 + H_2O.$

6. $CH_3CH_2OH + NH_3 \longrightarrow CH_3CH_2NH_2 + H_2O.$

7. $CH_2{=}CH_2 + NH_3 \longrightarrow CH_3CH_2NH_2.$

8. $CH_3-CH_3 + \frac{1}{2}N_2 + \frac{1}{2}H_2 \longrightarrow CH_3CH_2NH_2.$

2.4 ECONOMIC FEASIBILITY

The relative economic potential of these reactions can then be considered by estimating the difference between the market value of the products and the reactants. One source of the required price data for many common chemicals is the Chemical Marketing Reporter, which is published weekly.[6] Frequently, a purchasing department or a sales group can furnish pricing information that is more realistic for a particular situation. As a first approximation, these estimates might be made on the common bases of (1) an assumed 100% yield, (2) no costs for solvents or catalysts, and (3) no value for co-products.

Table 2.1 lists the market prices for raw materials that would be of interest in an ethylamine process.

In the ethylamine process evaluation, one could proceed as follows for Reaction 1:

The published market values for the product and the raw materials are:

ethylamine	$0.430/lb
ethyl chloride	0.090/lb
ammonia	0.075/lb

$$CH_3CH_2Cl + 2NH_3 \longrightarrow CH_3CH_2NH_2 + NH_4Cl$$

64.5 amu	17.0 amu	45.1 amu	53.5 amu
64.5 lb	34.0 lb	45.1 lb	53.5 lb

Table 2.1   Representative Market Values

Chemical	Value ($/lb)
1. Acetaldehyde	0.120
2. Acetonitrile	0.280
3. Ammonia	0.075
4. Ethane	0.030
5. Ethyl alcohol	0.120
6. Ethyl chloride	0.090
7. Ethylene	0.070
8. Hydrogen	0.020
9. Hydroxylamine sulfate	0.76
10. Nitroethane	0.41
11. Nitrogen	0.010

To produce 1 lb of $CH_3CH_2NH_2$ requires

$$\frac{64.5}{45.1} = 1.43 \text{ lb of } CH_3CH_2Cl$$

which has a cost of 1.43 lb $\times$ \$0.09/lb = \$0.129.

$$\frac{34.0}{45.1} = 0.76 \text{ lb of } NH_3$$

which has a cost of 0.76 lb $\times$ \$0.075/lb = \$0.057.

The market value of the product is \$0.430/lb; the total calculated raw material cost is \$0.186/lb (\$0.129 + \$0.057). The difference represents the money that would be available for all of the other direct and indirect costs associated with production and marketing, plus the potential profit:

Value of product	\$0.430/lb
Calculated raw material costs	0.186
Difference	\$ + 0.244/lb

Proceeding in a similar manner, the market value differences would be calculated for each of the candidate reactions. Using the representative prices listed in Table 2.1, the following economic comparison results:

Reaction	Raw Materials	Market Value Difference
1	Ethyl chloride, ammonia	$ + 0.244/lb
2	Acetonitrile, hydrogen	+ 0.173
3	Nitroethane, hydrogen	− 0.258
4	Acetaldehyde, hydroxylamine, hydrogen	− 1.069
5	Acetaldehyde, ammonia, hydrogen	+ 0.276
6	Ethyl alcohol, ammonia	+ 0.279
7	Ethylene, ammonia	+ 0.369
8	Ethane, nitrogen, hydrogen	+ 0.407

Because of the inherent error in the use of published market prices as the value of a chemical and because of the simplifying assumptions made in this approach, small differences in the values have no significance. However, large differences can be considered and some broad generalizations can frequently be made. In the case of ethylamine, we might conclude:

1. Processes based upon nitroethane and hydroxylamine are not economically feasible.
2. A process based upon ethylene has a distinct raw material cost advantage over processes using acetonitrile, acetaldehyde, ethyl chloride, or ethanol.
3. A process utilizing ethane would have the lowest raw material costs.

2.5 TECHNICAL FEASIBILITY

The odds or chances for the technical success of a proposed project can range from the well-known "almost sure thing" to the proverbial snowball. Reasoning by analogy is perhaps the most useful approach. The closer the relationship between the proposed process and known technology the more accurate this assessment can be. Although this judgment might be based upon prior knowledge and past experience, frequently a preliminary literature search can be very helpful.

One approach that can sometimes be used in predicting the feasibility of a postulated reaction is to consider the change in Gibbs free energy (ΔG).

For an isothermal reaction,

$$\Delta G_R = \Delta H_R - T \Delta S_R$$

The change in Gibbs' energy for a reaction (ΔG_R) is an index of the driving force of a reaction or process. In general, a spontaneous reaction must be accompanied by a decrease in Gibbs energy of the system. The values used in calculating the Gibbs energy of reaction are standard Gibbs energies of formation ΔG_f°. This is the energy increment accompanying the formation of a compound in its standard reference state from the elements in their standard reference states. If the ΔG_f° for each reactant and each product is available, then

$$\Delta G_{reaction} = \sum \Delta G_{f\,products}^\circ - \sum \Delta G_{f\,reactants}^\circ$$

As an example of this type of calculation and a comparison of the change of Gibbs' energy with the changes in enthalpy and entropy, consider the dissociation of ethyl chloride at two temperatures.

$$CH_3CH_2Cl \longrightarrow CH_2{=}CH_2 + HCl$$

At 298°K

	CH_3CH_2Cl	$CH_2{=}CH_2$	HCl
ΔH_f°	−26.70	12.50	−22.06
ΔG_f°	−14.34	16.28	−22.77
S°	65.93	52·45	44.64

At 1000°K

ΔH_f°	−30.43	9.21	−22.56
ΔG_f°	18.60	28.85	−24.08
S°	93.80	72.07	53.25

The changes in enthalpy, entropy, and Gibbs' energy can then be calculated at each temperature:

$$\Delta H_{reaction} = \Delta H_{f\,CH_2=CH_2}^\circ + \Delta H_{f\,HCl}^\circ - \Delta H_{f\,CH_3CH_2Cl}^\circ$$

$$\Delta G_{reaction} = \Delta G_{f\,CH_2=CH_2}^\circ + \Delta G_{f\,HCl}^\circ - \Delta G_{f\,CH_3CH_2Cl}^\circ$$

$$\Delta S_{reaction} = S_{CH_2CH_2}^\circ + S_{HCl}^\circ - S_{CH_3CH_2Cl}^\circ$$

At 298°K:

$$\Delta H_{reaction} = +17.14 \text{ kcal/mole}$$

$$\Delta G_{reaction} = +7.86 \text{ kcal/mole}$$

$$\Delta S_{reaction} = +31.15 \text{ cal/mole}$$

At 1000°K:

$$\Delta H_{reaction} = +17.08 \text{ kcal/mole}$$

$$\Delta G_{reaction} = -14.43 \text{ kcal/mole}$$

$$\Delta S_{reaction} = 31.52 \text{ cal/mole}$$

Now, the experimental facts are that ethyl chloride is stable at lower temperatures and spontaneously dissociates at higher temperatures. The validity of the use of $\Delta G_{reaction}$ as a measure of the driving force of the reaction is confirmed by the change from a positive value to a negative value as the temperature is increased from 298° to 1000°. Note that over this temperature range $\Delta H_{reaction}$ and $\Delta S_{reaction}$ have changed very little and obviously do not represent good criteria for predicting the feasibility of the reaction. Note also that the large decrease in ΔG at the higher temperature results from the increased importance of the $T \Delta S$ term.

The $\Delta G_{reaction}$ criteria are subject to oversimplification. In order to keep some sense of perspective, it should be kept in mind that the kinetics of the proposed reactions have not been considered and that the proposed reaction is assumed to be the *only* possible reaction. In the absence of other information, however, the following qualitative guidelines are useful:

Change in Free Energy	Indication
$-\Delta G$	Promising
Small $+ \Delta G$	Worthy of further investigation
Large $+ \Delta G$	Possible only under unusual conditions

Using the thermochemical data listed in Table 2.2, the energy change for several of the candidate reactions for the ethylamine process can be calculated.

$$CH_3{-}CH_3 + \tfrac{1}{2}N_2 + \tfrac{1}{2}H_2 \longrightarrow CH_3CH_2NH_2$$

$$\Delta G_{reaction} = \sum \Delta G_{products} - \sum \Delta G_{reactants}$$

$$\Delta G_{298°} = +16.78 \text{ kcal/mole}$$

$$\Delta G_{1000°} = +34.83 \text{ kcal/mole}$$

These values indicate that the chances for utilizing this reaction are slight, and laboratory work would not appear to be justified.

$$CH_3CH_2OH + NH_3 \longrightarrow CH_3CH_2NH_2 + H_2O$$

$$\Delta G_{298°} = -1.65 \text{ kcal/mole}$$

$$\Delta G_{1000°} = -1.91 \text{ kcal/mole}$$

These values indicate the reaction appears promising and warrants further consideration.

$$CH_2{=}CH_2 + NH_3 \longrightarrow CH_3CH_2{-}NH_2$$

$$\Delta G_{298°} = -3.51 \text{ kcal/mole}$$

$$\Delta G_{1000°} = +17.86 \text{ kcal/mole}$$

Table 2.2   Selected Thermochemical Data

Substance	ΔG_f° Kcal/mole	
	298°	1000°
$CH_2=CH_2$	16.28	28.25
CH_3-CH_3	-7.87	26.13
$CH_3CH_2-NH_2$	8.91	60.96
CH_3CH_2OH	-40.22	1.98
NH_3	-3.86	14.85
H_2O	-54.64	-46.04
H_2	0	0
N_2	0	0

This reaction is worth looking into further. Consideration of the change in ΔG with temperature, the decrease in the moles of gas as the reaction proceeds, and the stoichiometry of possible side reactions would suggest that the most promising reaction conditions would be low temperatures, high pressures, and a large excess of ammonia.

2.6   OTHER CONSIDERATIONS

This use of thermochemical data in considering the feasibility of a reaction can be instructive and interesting. The picture presented, however, should be viewed with a certain amount of scepticism. The evaluation cannot be reduced to a procedure for mechanically inserting values into an equation. This is not only due to the basic simplifying assumptions mentioned earlier but also because the approach omits any consideration of the characteristics of the chemistry of the reactions involved. The following two examples illustrate how knowledge of the chemistry of a reaction might dictate the direction of a development effort.

It is desired to develop a process for the manufacture of 2-chloro-2,4,4 trimethylpentane. The synthetic scheme can be divided into two parts (1) formation of the rather unusual carbon skeleton from low cost raw materials, and (2) introduction of the chlorine. Isobutylene and isobutane would make ideal starting points for the process, either through an alkylation reaction or the dimerization of an alkene. The chlorine could then be introduced into the alkane by direct halogenation with chlorine or, in the case of the alkene, by the addition of hydrogen chloride. There are thus two distinct routes to the desired compound:

Route 1   Route 2

$$CH_3-\overset{\overset{\displaystyle CH_3}{|}}{C}=CH_2 + CH_3-\overset{\overset{\displaystyle CH_3}{|}}{CH}-CH_3 \qquad\qquad 2\,CH_3-\overset{\overset{\displaystyle CH_3}{|}}{C}=CH_2$$

alkylation   dimerization

$$CH_3-\underset{\underset{\displaystyle CH_3}{|}}{\overset{\overset{\displaystyle CH_3}{|}}{C}}-CH_2-\overset{\overset{\displaystyle CH_3}{|}}{CH}-CH_3 \qquad\qquad CH_3-\underset{\underset{\displaystyle CH_3}{|}}{\overset{\overset{\displaystyle CH_3}{|}}{C}}-CH=\overset{\overset{\displaystyle CH_3}{|}}{C}-CH_3$$

and

$$CH_3-\underset{\underset{\displaystyle CH_3}{|}}{\overset{\overset{\displaystyle CH_3}{|}}{C}}-CH_2-\overset{\overset{\displaystyle CH_3}{|}}{C}=CH_2$$

substitution   HCl   addition

Cl_2

$$CH_3-\underset{\underset{\displaystyle CH_3}{|}}{\overset{\overset{\displaystyle CH_3}{|}}{C}}-CH_2-\underset{\underset{\displaystyle Cl}{|}}{\overset{\overset{\displaystyle CH_3}{|}}{C}}-CH_3$$

In comparing these two possible routes, the characteristics of the reactions should be considered. The second step in Route 1 is a free radical substitution reaction. Such reactions are notoriously nondiscriminating. It is true that substitution of a tertiary H does occur more readily than a primary or secondary H. However, in this case, substitution in the desired 2 position is competing with the substitution of 17 other H atoms. It could be anticipated then, that even if only the monosubstituted possibilities are considered, a mixture of chlorinated products would be obtained. The yield of desired product would be decreased and a challenging separation would be involved. In Route 2, the dimerization step yields a mixture of alkenes. Both of these intermediates, however, would react with hydrogen chloride to yield the desired 2-chloro isomer as the main product. From the standpoint of the chemistry, Route 2 is preferred. Another very important consideration is that Route 1 generates hydrogen chloride as an undesirable by-product.

In many places hydrogen chloride has no value because the supply exceeds the demand. Route 2, on the other hand, uses hydrogen chloride and would be the first choice in a development effort.

In the second example, it is desired to develop a process for the manufacture of *p*-aminophenol for use as a pharmaceutical intermediate. Among the many schemes possible, two that are especially interesting are the amination of *p*-chlorophenol and the reduction of *p*-nitrophenol. Both possibilities involve a one-step process starting with a commercially available, reasonably priced material.

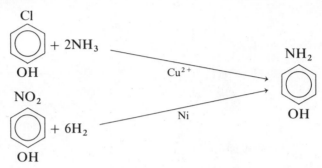

A preliminary evaluation of the raw material costs differential does not clearly indicate a distinct advantage to either process. A consideration of the properties of the product, however, does give a clue as to the more desirable scheme. *p*-Aminophenol is similar to hydroquinone in that it is very susceptible to oxidation to quinone and other products that are intensely colored. These can be air oxidations, and they occur particularly readily in the presence of metal catalysts such as copper(II) ion. The reaction mixture from the amination reaction would be particularly prone to such degradation. The reaction mixture from the reduction process would not suffer from this disadvantage and would very likely give less troublesome isolation and purification steps.

These two examples are simply to illustrate that there are many factors that enter into the consideration of a process in addition to chemical costs and Gibbs' energy change. A thorough knowledge of the chemistry of the reactions, a healthy respect for the economics of the situation, a lot of common sense, and a little imagination are all required.

2.7 SUMMARY

In this chapter we have attempted to illustrate that "classical" chemistry, and "industrial" chemistry are remarkably similar. Both are firmly based upon the same chemical principles. The differences arise naturally from the differences in objectives.

The academic synthetic chemist's objective is to advance the science of chemistry and his own professional career by discovering and reporting new routes to new or difficult-to-synthesize compounds. The judgment of his efforts is by his peers and the referees to respected journals. The industrial chemist's objective typically might be to advance the science of chemistry and his personal career by discovering lower cost synthetic routes to compounds having a value or a potential value to society. The judgment of his efforts is by economic reality.

The basic difference in these two approaches is perhaps best illustrated by a consideration of the synthetic work respected by the two groups. In the world of "classical" chemistry one of the highest accolades is to have a synthetic procedure described as "elegant." This usually implies a reaction or series of reactions marked by high yields; short, convenient work-up of product; and frequently utilizing a novel or recently introduced, highly selective reagent. The successful industrial process is usually marked by simplicity—stark simplicity. The simplest conceivable synthetic route is envisioned. Every practical condition of temperature, pressure, and catalytic promotion is studied to reduce this to practice. Any steps that can be avoided are eliminated. Any unnecessary solvents are eliminated. Any raw materials that do not show up in the final product are eliminated. It is frequently necessary to operate at very low conversions and, as a consequence, to recycle large quantities of reactants. The plants constructed to use these synthetic methods are typically quite large and are characterized by high capital costs, but very low costs per pound of product.

There are certainly differences in the two endeavors but a perceptive person can appreciate the intellectual challenges and satisfaction inherent in both.

2.8   REFERENCES

1. A. M. Brownstein, *U.S. Petrochemicals*, The Petroleum Publishing Company, Tulsa, Oklahoma, 1972.

2. R. B. Wagner and H. D. Zook, *Synthetic Organic Chemistry*, Wiley, New York, 1953.

3. H. D. Weiss, *Guide to Organic Reactions*, Burgess Publishing Company, Minneapolis, Minnesota, 1969.

4. C. A. Buehler and D. E. Pearson, *Survey of Organic Synthesis*, Wiley-Interscience, New York.

5. I. T. Harrison and S. Harrison, *Compendium of Organic Synthetic Methods*, Wiley-Interscience, New York.

6. *Chemical Marketing Reporter*, Schnell Publishing Co., New York.

7. Stull, Westrum, and Sinke, *The Chemical Thermodynamics of Organic Compounds*, Wiley, New York, 1969.

3
MATERIAL ACCOUNTING

THE LAW OF CONSERVATION OF MASS REALLY WORKS

In attempting to carry out a reaction in the laboratory, every chemistry student has learned that a vital step in the procedure deals with the calculation of the amount of starting materials required and the amount of product that is expected to be produced. Such calculations are also required when chemical reactions are carried out on an industrial scale. However, the format of the calculations differs from those that the chemist normally utilizes in the laboratory. The calculations differ because chemists in the laboratory normally work with closed systems, whereas most industrial processes are invariably open systems. Open systems are those in which there is a flow of materials into and out of the process equipment. In such systems heat or work energy may be added to certain process units and may be removed from others not only to satisfy operational requirements, but also to operate most economically. The complete accounting of all mass and energy in such chemical processes is referred to as a material and energy balance.

The material balance is frequently referred to as a mass balance or weight balance in industrial chemistry practice. Therefore, these terms are used interchangeably throughout this book. A material balance can be effected without an energy balance, but an energy balance requires a knowledge of the mass and composition of all streams. This combination, the material and energy balance, is one of the most powerful tools used in the sequence of steps necessary to bring a chemical reaction from the idea stage to a viable large-scale commercial process. It is also an essential tool in the effective evaluation of the day-to-day operation of an existing chemical process. The concept of the material balance is so simple, however, that the student is prone to assume erroneously that he can apply these balances skillfully without much training in their application. Thus, it is the purpose of this chapter to discuss the basic principles of material balances and how they can be applied to industrial problems. Energy balance concepts are considered in Chapter 4.

3.1   THE MATERIAL BALANCE EQUATION

In general, there are two types of chemical processes with which the industrial chemist deals, the "batch process" and the "continuous process." In the batch process the chemicals are added to the processing vessel in one operation and then the process is carried out. In some cases the products are removed during this period, but in others they are removed after the processing is completed. In the continuous process the chemical charge and the products enter and leave continuously. The material balance can be applied to either a batch or a continuous process. The biggest difference in applying the material balance to these two processes involves the element of time. A material balance applied to a batch process usually does not include a time variable. The balance in such a case is generally made over a complete cycle, which involves merely the processing of a single charge. In the case of a continuous process, however, the time variable must enter the material balance. The balance must be made over a specified period of time.

   The material balance is based upon the concept of the law of conservation of matter, which in effect says that, except for situations involving nuclear reactions, atoms are neither created or destroyed. Atoms that enter a system must either accumulate in the system or must leave. This observation leads to the balance expressed in Eq. 3.1, which is valid for all atomic species in the system.

$$\begin{Bmatrix} \text{Accumulation of} \\ \text{atomic species } j \\ \text{within the system} \end{Bmatrix} = \begin{Bmatrix} \text{total atomic} \\ \text{species } j \text{ entering} \\ \text{system} \end{Bmatrix} - \begin{Bmatrix} \text{total atomic} \\ \text{species } j \\ \text{leaving system} \end{Bmatrix} \quad (3.1)$$

By summing over all the atomic species entering and leaving a system, the total material balance is obtained:

$$\begin{Bmatrix} \text{Total accumu-} \\ \text{lation within} \\ \text{the system} \end{Bmatrix} = \begin{Bmatrix} \text{total mass} \\ \text{entering system} \end{Bmatrix} - \begin{Bmatrix} \text{total mass} \\ \text{leaving system} \end{Bmatrix} \quad (3.2)$$

When there is no accumulation within the system, Eq. 3.2 reduces to the following:

$$\begin{Bmatrix} \text{Total mass} \\ \text{entering system} \end{Bmatrix} = \begin{Bmatrix} \text{total mass} \\ \text{leaving system} \end{Bmatrix} \quad (3.3)$$

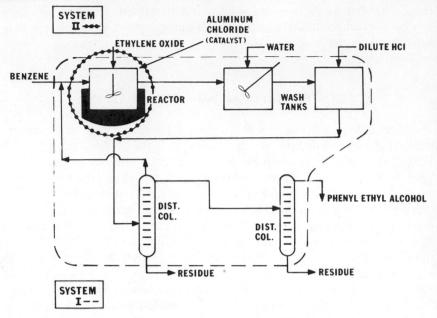

Figure 3.1   Flow sheet for the production of phenylethyl alcohol by Friedel-Crafts reaction.

Inherent in the formulation of each of the above balances is the concept of a system for which the balance is made. By system we mean any arbitrary portion or whole of a process as set out specifically by the chemist for analysis.

Figure 3.1 shows a process in which flow and reaction take place. We may be interested in performing a material balance on the entire process, in which case our system would be that enclosed by the boundary line for system I. On the other hand, our interest may only involve a material balance on the reactor, in which case our system would be that enclosed by the boundary line for system II.

In most continuous processes that the chemist encounters, the mass accumulation term will be zero; that is, the process is in a steady-state. In this book, we consider only steady-state processes. This eliminates the necessity for measuring the accumulation of material in the system so that input may be equated to output, and consequently we can say, "what goes in must come out."

One of the many useful applications of the material balance is during the pilot plant stage when the performance of a given process is being investigated. If each item of the input and output streams is measured, the material balance will serve to check the accuracy of the experimental measurements. If a balance is not obtained within the desired accuracy, it is apparent that one or more of the measurements is in error. In many of the processes with which the chemist will deal, it will be impossible or uneconomical to measure directly all the items of input and output. However, if sufficient other data are available, by making a material balance on the process, it is possible to get the needed information about the quantities and compositions at the inaccessible locations.

Another use of the material balance is to make calculations for prospective processes. For example, if we choose to prepare ethylamine by the ethyl alcohol-ammonia reaction discussed in the previous chapter, we must devise a plan for routing the various materials throughout the emerging process. Information about the quantities of the various materials (reactants, products, unconverted reactants, waste, etc.) routed through the process can be used to determine process equipment sizes, and the specific paths through which the materials are routed will determine the types of separation problems that must be solved. Therefore, the key to the successful design and development of a chemical process begins with an understanding of the material flow in the process.

3.2 REPRESENTATION OF MATERIAL FLOW IN A CHEMICAL PROCESS

A chemical process consists of an integrated series of reactions and associated operating steps, such as mixing, separation by distillation, and the like, whereby available raw materials are converted into a desired product. The process is carried out in an assembly of equipment through which material and energy travel batchwise or continuously. This equipment may consist of heat exchangers; material handling devices like pumps, compressors and conveyors; in addition to equipment used for materials mixing, reaction, or separation. Even the simpler processes involve several pieces of equipment. The movement of material through the various pieces of equipment is most often represented by a flow sheet.

A flow sheet is usually presented in the form of a line and block diagram, as illustrated in Fig. 3.1. Sometimes quantities of material and energy in the various streams are shown. Within the chemical industry there has been little success in developing a standard set of symbols for general use on flow sheets representing different types of equipment, and usual practice still involves indicating each equipment stage by a square, a rectangle, or a circle suitably labeled.

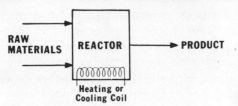

Figure 3.2   A flow sheet for a single reactor producing two product streams.

Flow sheets of processes involving a single chemical reactor can normally be classified into one of the five categories described below. For more complex processes involving two or more reactors the flow sheet may be constructed by using combinations of these five basic types.

Type 1.   Single Reactor Product Stream without Separation.   Some chemical manufacturing processes, for which a typical flow sheet is shown in Fig. 3.2, involve no equipment other than a reactor to which reactants are fed directly without purification, and from which the product stream exits as a single phase, ready for use. Energy may also be added or removed from the reactor as needed.

An example of a process that can be represented by a Type 1 flow sheet is the production of polyethylene glycol by the reaction shown in Eq. 3.4.

$$HOCH_2CH_2OH + nCH_2\!\!-\!\!CH_2 \longrightarrow HO(CH_2CH_2O)nCH_2CH_2O$$
$$\diagdown\!\!O\!\!\diagup$$

$$(3.4)$$

where $n = 6 - 9$. The reaction is run in the liquid phase in the presence of a sodium hydroxide catalyst at temperatures in the region of 120–140°C. With some formulations, the cooled product is ready for use in the manufacture of stearate emulsifiers for the food, drug, and cosmetic industries. For more information on this specific process consult Reference 1.

Type 2.   Multiple Reactor Product Streams without Separation.   This process is obtained when the reactants and products remain in different phases as shown in Fig. 3.3. Unreacted feed materials may leave in either one or both phases. An example of a process that can be represented by this type of flow sheet is as follows:

Aniline can be produced from chlorobenzene according to Eq. 3.5.

$$C_6H_5Cl + 2NH_3 \longrightarrow C_6H_5NH_2 + NH_4Cl \qquad (3.5)$$

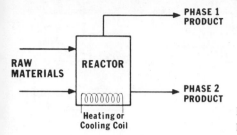

Figure 3.3 A flow sheet for a single reactor producing two product streams.

Ammonia is added in the form of a 28 % solution where it reacts with the chlorobenzene in the presence of a cuprous oxide catalyst in a mild carbon steel reactor at temperatures between 220 and 250°C and pressures in the region of 2000 psi. The aniline and any unreacted chlorobenzene leave the reactor as an organic layer, and the coproduct ammonium chloride and any unreacted ammonia leave in the aqueous layer. The aniline produced is used as an intermediate in the manufacture of rubber chemicals and dyes. For more information on this process consult Reference 2.

Type 3. Single Reactor Product Stream with Separation. In many chemical processes the desired product must be separated and recovered from the mixture leaving the reactor. Many different separation schemes are used. Some of the more common separation techniques used are crystallization, distillation, solvent extraction, absorption, adsorption, filtration, and ion exchange. In the flow sheet of Fig. 3.4, recovery is by absorption in a suitable solvent followed by stripping in a distillation column. The unconverted reactants may leave in the tail gas, one of the product streams, or a combination of both product streams plus tail gas. An example of a process that can be represented by this flow sheet is as follows:

Methylamine is made by reacting methanol and ammonia over a silica-alumina catalyst. The stoichiometry of the reaction is given in Eq. 3.6.

$$CH_3OH + NH_3 \longrightarrow CH_3NH_2 + H_2O \qquad (3.6)$$

The reaction takes place in the vapor phase at 350 to 450°C and 200 psi pressure. The methylamine vapor leaving the reactor is recovered as shown in Fig. 3.4 by water scrubbing and purification by stripping and distillation in the still. The methylamine produced is used in surfactants, photographic developers, rubber accelerators, drugs, and agricultural chemicals. For more information on the manufacture of methylamine consult Reference 3.

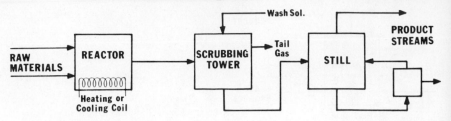

Figure 3.4   A flow sheet for a single reactor product stream with separation.

Type 4.   Multiple Separations Involving Reactor Feed and Product Streams.
This flow sheet represents an extension of Type 3, in that separations are
also performed on the raw materials. In the illustrative flow sheet for this
case, shown in Fig. 3.5, raw material purification is by adsorption and
product purification by absorption and stripping. Any other means of
separation would involve the same type of block diagram. An example of
a process that is represented by this flow sheet is the manufacture of H_2 by
steam reforming of methane. The reaction is represented in Eq. 3.7.

$$CH_4 + 2H_2O \xrightarrow[\text{catalyst}]{Ni} CO_2 + 4H_2 \qquad (3.7)$$

As indicated, the reaction is carried out over a nickel catalyst. In order to
prevent the deactivation of the catalyst, the methane obtained from
natural gas is desulfurized by contact with activated carbon. The CO_2 is
removed from the product H_2 by scrubbing with a monoethanolamine
solution. For more details on the manufacture of H_2 by the steam-hydro-
carbon process consult Reference 4.

Type 5.   Reactor with Recycle.   This flow representation is very common,
for a large number of chemical processes contain some form of recycle

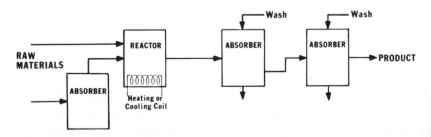

Figure 3.5   A flow sheet with multiple separations of reactor feed and product streams.

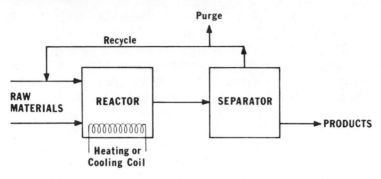

Figure 3.6 A flow sheet with recycle.

stream. Recycle as shown in Fig. 3.6 involves the return of part or all of a process stream from a later to an earlier stage in the process. The general object of recycle in a chemical process is to minimize materials consumption per unit of product produced. Thus, when a reaction does not go to completion because of unfavorable kinetics or equilibrium, the unconsumed material is separated from the product stream. When material is recycled in this fashion, a purge must be provided, as shown, to prevent accumulation of impurities. The Haber process for producing ammonia is an example of a recycle system that can be represented by the flow sheet shown in Fig. 3.6.

The stoichiometry of the reaction is shown in Eq. 3.8.

$$N_2 + 3H_2 \longrightarrow 2NH_3 \tag{3.8}$$

Ammonia is made by passing a hydrogen–nitrogen mixture over a promoted iron catalyst at elevated pressure. The ammonia is recovered from the reactor off-gases by cooling and condensation or by water scrubbing. However, because of equilibrium limitations encountered at the operating conditions usually used, only a small amount of the hydrogen and nitrogen reacts to form ammonia. The economics of the process require that the gases be recycled. A purge is required to eliminate the buildup of any undesirable gases such as argon or methane in the reactor. For more information on the manufacture of ammonia, consult Reference 5.

Complex Flow Sheet. As indicated earlier, most complex flow sheets may be considered as combinations of the five types just discussed. There are a number of reasons why such combinations occur. One of the most common is due to the need for multiple product separations. For example, suppose that a product stream containing a number of components is produced in some process, and recovery of the individual components in relatively pure form requires many steps. The separation of crude oil into its useful products is a very complex process that results from the need to separate a mixture containing many components into mixtures containing fewer components. However, the entire refining process of crude oil can be represented by a flow sheet made up of a composite of the five types discussed above.

Another common reason for the occurrence of complex flow sheets is that frequently a process requires a number of reactions to be carried out in series. Such reactions often cannot be made to occur in a single reaction vessel, since the conditions favoring one reaction may not favor the others. A flow sheet of such a process might look like that shown in Fig. 3.7. The flow sheet as depicted is a combination of Types 1 and 4. The heat exchangers between the reactors allow for the reactors to be operated at different temperatures. This would be necessary if the temperature dependency of the stagewise reactions are different.

In all of the flow sheets that we have presented thus far, no information as to quantities of material flow has been recorded on the flow sheet. A complete representation of the process, of course, requires that this information be presented. It is at this point that we must rely on the material balance to give a complete description of the process.

In reviewing detailed process flow sheets where the material balance information has been recorded, the chemist soon learns that the units representing material flow and process conditions are not the same as those commonly used in the laboratory. The next section deals with the units problem that the chemist must confront in handling material flow and balance problems.

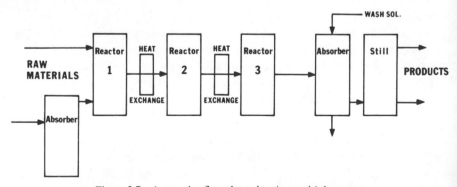

Figure 3.7   A complex flow sheet showing multiple reactors.

3.3 UNITS OF MATERIAL FLOW AND BALANCING

The material balance principle may be applied either on the basis of units of weight, such as pounds or kilograms, or on the basis of gram-moles or pound-moles. Since the mole is merely a unit of mass that is numerically equal to the molecular weight, it is a very convenient unit to use in a material balance. If no chemical change takes place in the system on which a material balance is being made, the weight unit can ordinarily be used most conveniently. There are many applications, however, in which the mole unit is the most convenient even though no chemical change takes place. Of course, if a chemical change takes place then the mole unit is almost always the most convenient.

Within the chemical industry in the United States two systems of units are used, the metric system and the American engineering system. Table 3.1 compares the units of measurement in the two systems. The metric system is used primarily by chemists, while practicing engineers use the American engineering system. A major effort is currently underway to convert both scientific and engineering communities to the International System of Units (SI). The SI units may soon be predominant in our journals and eventually may be officially adopted for professional and lay usage. However, that day has not yet arrived. Practicality, therefore, dictates that an industrial chemist possess the ability to work and communicate effectively in both the metric and engineering systems. Accordingly, we use both systems in this book. For your convenience, a short set of essential conversion factors is given in Appendix A. At this point, we are now ready to take up the task of developing a method for solving material balance problems.

Table 3.1   Units of Measurement in Chemical Industry

System	Length	Mass	Force	Energy
Metric	cm	gram	dyne	erg, joule or calorie
Amer. Engr.	ft	pound (mass)	pound (force)	ft·lb$_f$, Btu or hp·hr

3.4 TECHNIQUES IN MATERIAL BALANCING

In this section we demonstrate the technique of setting up and solving problems that call for material balances. Most material balance problems contain the same basic elements, although the details of the application of the method of solution may be slightly different. Therefore first consider a generalized method for analyzing such problems that can be applied to the solution of almost any type of material balance problem. For some problems the method of approach is relatively simple, for others it will be more complicated; but the important point is to regard problems in chemical synthesis, distillation, crystallization, evaporation, mixing, gas absorption, or combustion not as being different from each other but as being related from the viewpoint of how to proceed to solve them.

The material balance may be applied to a process as an overall balance of the incoming mass with the outgoing mass. Or, it may be applied to intermediate parts of a continuous process. In general, for either of these systems, data dealing with two fundamental pieces of information is required. One of these is the mass in all streams of the material entering and leaving the system and present in the system. Information about the composition of all the streams entering and leaving the system and the composition of the material in the system must also be available. If a chemical reaction takes place inside the system, information about the reaction rate will be required.

In making a material balance there are a number of preliminary steps that are very helpful and desirable. The steps presented are flexible and can be modified as the chemist develops his own style.

1. Draw a simplified flow sheet of the process as described in Section 3.2.
2. Place all the available data on the flow sheet using consistent units as discussed in Section 3.3.
3. Set down all chemical equations for the chemical reactions that occur in the process.
4. Select a convenient basis for the calculations.

Even in relatively simple problems it is worthwhile to draw a diagram, a schematic, or a flow sheet that depicts the problem at hand. A simplified flow sheet of the system that includes all the features pertinent to the material balance will help to keep careless errors to a minimum. It is well to develop the habit of always sketching a simplified flow sheet even for the simple problem. In complex problems, a flow sheet is essential.

All the data that will be used in the material balance can be located on the flow sheet. The streams connecting the units may be one or more of several different types, such as reactor feed, recycle, overhead vapors, and so on. The species known to be present in a stream are stated near the stream label, along with known flow rates and compositions. Gases, pressures, and temperatures are also indicated since they can be used to calculate the mass flow rate if the volumetric flow rate is known. However, other stream data not directly connected with the material balancing tend to clutter the flow sheet and should not be mentioned at this point in the analysis. In this way, connections between the data and the various items to be taken into account in the balance can be seen at a glance. Furthermore, this procedure will be helpful in determining what data must be obtained in the case of preparations for a material balance on an operating process. It is also desirable to locate on the flow sheet, the equations of the chemical reactions that occur in the process if these equations will be used in the material balance.

Choosing a convenient basis in the fourth step is vitally important both to your understanding of how to solve the problem and also to your solving it in the most expeditious manner. The basis is the reference chosen by you for the calculations you plan to make. The basis may be a period of time, a unit of weight, or a flow rate of one of the items in the process. It simplifies our thinking to specify one hour of operation, a ton of product, 100 lb/hr of fuel, or whatever is suitable for a basis. As your study of the material balance progresses, it will become quite apparent why the selection of a common basis of calculation is essential.

After choosing a basis, you are ready to make the necessary number of material balances. In general, for any process you can write a total material balance and component material balance for each component present. However, all the balances will not be independent as explained in the following.

In writing a component balance for a problem involving a chemical reaction, the component to choose is an atomic specie or multiple thereof (i.e., chlorine expressed as Cl_2). For problems without chemical reactions, any compound or atomic species may be selected for the component balance. If, for example, there are four components present, it is possible to write four component material balances. If we include the total material balance, it is possible to write five equations. But, these equations are not independent inasmuch as the sum of the four component material balances will add up to the total material balance (i.e., equations are independent if no one of them can be derived by some algebraic combination of the remaining equations in the set). Consequently, the number of degrees of freedom or the number of independent equations will be equal to the number of components. However, the total material balance can be substituted for any one of the component material balances if two or more equations are solved simultaneously.

Many of the problems you will encounter involving the use of material balances may require the solution of more than one unknown. Therefore, you should remember that for each unknown, you need to have at least one independent material balance or other independent piece of information. Otherwise, the solution to the problem is indeterminate. If the number of independent equations exceeds the number of unknowns to be computed, it becomes a matter of judgment to determine which independent set of equations should be selected to solve the problem. If all the data in each equation were perfect, it would not matter which independent set of equations were selected. However, since such data are usually uncertain because of errors in sampling and measurement, it is generally best to use equations based on the species present in the greatest concentrations.

Simple addition or subtraction can be used to solve those problems in which only one weight and one composition are unknown. Problems in which all the compositions are known and two or more of the weights are unknown require more detailed calculations. However, certain patterns that appear commonly in these problems can be used to simplify the calculations. One pattern involves tie components. A tie component is an element, molecule, species, or even a part of a molecule that goes through the balance region intact, in one place and out another place. Under steady-state conditions, if the flow rate of a tie component is known where it crosses the balance boundary in one place, it is known at the other place. When there is no direct or indirect tie element available, algebraic methods must be used to relate the unknown weights to the known weights. The manner in which these methods are applied is illustrated in the following sections.

3.5 MATERIAL BALANCING BY USING A DIRECT APPROACH

Many material balance problems are encountered in which only one weight and one composition are unknown. Such problems can be solved by direct addition or subtraction as shown in the examples below.

Example 3.1 The Manufacture of Starch

One of the products obtained from the refining of corn is commercial starch. In the final stages of the refining process the starch contains 44% water. After drying, it is found that 73% of the original water has been removed. The resulting material is sold as pearl starch. Calculate:

(a) the weight of water removed per pound of wet starch.
(b) the composition of the pearl starch.

Solution

Step 1. First draw a simplified flow diagram of the drying process and place all of the available data on it such as we have done below.

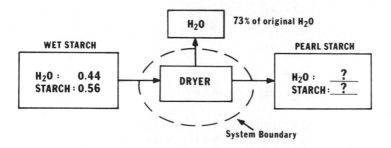

Step 2. Since one of the questions calls for an answer based upon a pound of wet starch, let us choose this as a basis of calculation.

Basis: 1 lb of wet starch.

Step 3. The system that we are working with includes the dryer with an incoming stream of wet starch and outgoing streams of water and pearl starch.

At this point we know only the composition and weight (the chosen basis) of the wet starch. However, the water stream is 100% H_2O, and we can calculate its weight since it is 73% of the original water present.

Step 4. Water removed = (0.73)(0.44 lb) = 0.32 lb. This is the answer to question (*a*).

Step 5. It is now possible to perform a water balance for the purpose of calculating the amount of water remaining in the pearl starch. Water balance.

$$\text{in} = \text{out}$$

$$(H_2O \text{ in wet starch}) = (H_2O \text{ stream}) + (H_2O \text{ in pearl starch})$$

$$(0.44 \text{ lb}) = (0.32 \text{ lb}) + (x)$$

$$x = (0.44 \text{ lb}) - (0.32 \text{ lb}) = 0.12 \text{ lb}$$

Step 6. The answer to question (*b*) can now be calculated. The calculated composition of the pearl starch is:

	lb	%
Starch	0.56	82.4
H_2O	0.12	17.6
Total	0.68	100.0

Example 3.2   *Crystallization of KCl*

In a pilot plant experiment, 1000 kg of a saturated KCl solution at 80°C is produced. We want to crystallize 100 kg of KCl from this solution. To what temperature must the solution be cooled?

Solution

Step 1. A flow diagram of the process is drawn.

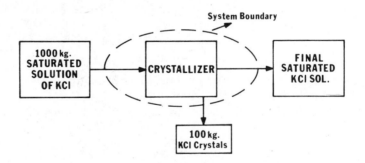

Step 2. In addition to the data given in the problem, data on the solubility of KCl as a function of temperature are also required. Data obtained from Reference 6 are given in the tabulation at the top of page 55.

Step 3. Basis: 1000 kg saturated KCl solution at 80°C.

We know both the weight of the initial solution and the weight of the crystals. We also know the composition of the crystals (100% KCl), and from the solubility data we can calculate the initial composition of the saturated solution.

| | Solubility |
Temperature (°C)	(g KCl/100 g H$_2$O)
80	51.1
70	48.3
60	45.5
50	42.6
40	40.0
30	37.0
20	34.0
10	31.0
0	27.6

Step 4.   Initial composition

$$\frac{(51.1 \text{ g KCl})}{(51.1 \text{ g KCl} + 100 \text{ g H}_2\text{O})}(100) = 33.8\% \text{ KCl}$$

$$100 - 33.8 = 66.2\% \text{ H}_2\text{O}$$

Step 5.   A component balance on the KCl gives the amount of KCl in the final solution.

$$\text{Initial KCl} - \text{crystals KCl} = \text{KCl in final solution}$$

$$(0.338)(1000) - 100 \text{ kg} = 238 \text{ kg}$$

Step 6.   The final composition and weight of resulting solution is

$$\text{H}_2\text{O} = (0.662)(1000 \text{ kg}) = 662 \text{ kg}$$

$$\text{KCl} = \qquad\qquad\qquad 238$$

$$\overline{}$$

$$\text{Total} = 900 \text{ kg}$$

Step 7.   This solution is still saturated and has a KCl solubility of

$$\frac{238 \text{ kg KCl}}{662 \text{ kg H}_2\text{O}} = \frac{36 \text{ g KCl}}{100 \text{ g H}_2\text{O}}$$

Step 8.   The temperature to which the solution must be cooled in order to produce this level of solubility can be calculated by linear interpolation of the KCl solubility data.

$$\begin{bmatrix} 30°\text{C} \!-\!\!-\!\!-\! 37.0 \\ \quad x \!-\!\!-\!\!-\! 36.0 \\ 20°\text{C} \!-\!\!-\!\!-\! 34.0 \end{bmatrix} \equiv \frac{10°\text{C}}{3} = \frac{x - 20°\text{C}}{2} \qquad x = 26.6°\text{C}$$

Therefore the temperature to which the initial solution must be cooled in order to crystallize 100 kg of KCl is 26.6°C.

Example 3.3   Combustion of a Fuel Gas

The combustion of fuel is the chief source of heat energy in the chemical process industries, and consequently material balancing of combustion processes is a common problem. In dealing with problems involving combustion, it is necessary that you become acquainted with a few special terms, such as:

(a)   Stack gas—consists of all gases resulting from a combustion process including the water vapor, sometimes known as combustion on a "wet basis."

(b)   Flue gas—consists of all the gases resulting from the combustion process not including the water vapor. Sometimes known as combustion on a "dry basis."

(c)   Theoretical air (or theoretical oxygen)—the amount of air (or oxygen) required for complete combustion.

(d)   Excess air (or excess oxygen)—the amount of air (or oxygen) in excess of that required for complete combustion as computed in (c).

The Problem

A fuel gas containing 97 vol % methane (CH_4) and 3 vol % N_2 is burned in a boiler furnace with 200% excess air. Eighty-five percent of the methane goes to CO_2, 10% to CO, and 5% remains unburned. Calculate the composition of the stack gas.

Solution

We know the composition of the air (79% N_2; 21% O_2) and fuel gas in terms of volume percent. Assuming ideal gas behavior allows for the direct conversion of volume percent to mole percent. With the composition represented in these units a mole basis is the most convenient to use in the solution of this problem. The arithmetic can be simplified by choosing a basis of 100 moles of entering fuel gas. With this basis and balanced equations for the reactions, we can outline a flow diagram for the combustion process as shown in the diagram that follows.

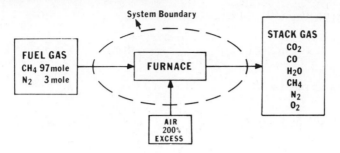

Step 1.   Basis: 100 moles of fuel gas

$$CH_4 + 2O_2 \longrightarrow CO_2 + 2H_2O$$

$$CH_4 + \tfrac{3}{2}O_2 \longrightarrow CO + 2H_2O$$

Step 2.   The total oxygen entering is 3.0 times the required oxygen (100% required plus 200% excess). From the information given above, we calculate the required amount of oxygen.

O_2 required for complete combustion:

$$(97 \text{ moles } CH_4)\frac{(2 \text{ moles } O_2)}{(1 \text{ mole } CH_4)} = 194 \text{ moles } O_2$$

This is the theoretical amount of O_2 required. When supplied in 200% excess, the total number of moles of O_2 required is

O_2 entering:

$$(3)(194 \text{ moles } O_2) = 582 \text{ moles } O_2$$

N_2 entering:

$$(582 \text{ moles } O_2)\frac{(0.79 \text{ mole } N_2)}{(0.21 \text{ mole } O_2)} = 2189 \text{ moles } N_2$$

Step 3.   The balanced combustion equations along with the degree of completeness of the combustion reaction can now be used to find the amounts of components generated within the system:

CO_2 generated (85% of CH_4 goes to CO_2):

$$(97 \text{ moles } CH_4)\frac{(1 \text{ mole } CO_2)}{(1 \text{ mole } CH_4)}(0.85) = 82.5 \text{ moles } CO_2$$

CO generated (10% of CH_4 goes to CO):

$$(97 \text{ moles } CH_4)\frac{(1 \text{ mole } CO)}{(1 \text{ mole } CH_4)}(0.10) = 9.7 \text{ moles } CO$$

H_2O generated with CO_2:

$$(97 \text{ moles } CH_4)\frac{(2 \text{ moles } H_2O)}{(1 \text{ mole } CH_4)}(0.85) = 164.9 \text{ moles } H_2O$$

H_2O generated with CO:

$$(97 \text{ moles } CH_4)\frac{(2 \text{ moles } H_2O)}{(1 \text{ mole } CH_4)}(0.10) = 19.4 \text{ moles } H_2O$$

O_2 consumed in CO_2 generation:

$$(97 \text{ moles } CH_4)\frac{(2 \text{ moles } O_2)}{(1 \text{ mole } CH_4)}(0.85) = 164.9 \text{ moles } O_2$$

O_2 consumed in CO generation:

$$(97 \text{ moles } CH_4)\frac{(1.5 \text{ moles } O_2)}{(1 \text{ mole } CH_4)}(0.10) = 14.6 \text{ moles } O_2$$

Step 4. To determine the composition of the stack gas, we need to perform a balance on each component:
O_2 balance:

$$O_2 \text{ out} = O_2 \text{ in} - O_2 \text{ consumed} + O_2 \text{ generated}$$
$$= 582 - (164.9 + 14.6) + 0 = 402.5 \text{ moles } O_2$$

H_2O balance:

$$H_2O \text{ out} = H_2O \text{ in} - H_2O \text{ consumed} + H_2O \text{ generated}$$
$$= 0 - 0 + (164.9 + 19.4) = 184.3 \text{ moles } H_2O$$

CO_2 balance:

$$CO_2 \text{ out} = CO_2 \text{ in} - CO_2 \text{ consumed} + CO_2 \text{ generated}$$
$$= 0 + 0 + 82.5 \text{ moles } CO_2$$

CO balance:

$$CO \text{ out} = CO \text{ in} - CO \text{ consumed} + CO \text{ generated}$$
$$= 0 + 0 + 9.7 = 9.7 \text{ moles } CO$$

N_2 balance:

$$N_2 \text{ out} = N_2 \text{ in} - N_2 \text{ consumed} + N_2 \text{ generated}$$
$$= 3 + 2189 - 0 + 0 = 2192 \text{ moles } N_2$$

CH_4 balance:

$$CH_4 \text{ out} = CH_4 \text{ in} - CH_4 \text{ consumed} + CH_4 \text{ generated}$$
$$= 97 - (82.5 + 9.7) + 0 = 4.8 \text{ moles } CH_4$$

Step 5. Summarizing these calculations we have:

	Moles in			% in
Component	Fuel	Air	Stack Gas	Stack Gas
CH_4	97	0	4.8	0.17
O_2	0	582	402.5	14.00
N_2	3	2189	2192	76.22
CO_2	0	0	82.5	2.87
CO	0	0	9.7	0.33
H_2O	0	0	184.3	6.41
Total	100	2771	2875.8	100.00

3.6 MATERIAL BALANCING BY USING AN ALGEBRAIC TECHNIQUE

The problems in the previous section were rather easy to solve because the missing information pertained to a single stream. Therefore, only simple addition or subtraction was required to find the unknown quantities. When the missing information pertains to more than one stream, the method of solution becomes somewhat more complicated. Such problems are generally solved by assigning a letter to replace the unknown value in the total material balance or the component material balance, as the case may be. Then an independent material balance is written for each unknown so that the set of resulting equations will have a unique solution.

In solving problems with multiple streams, the calculations are often made easier by splitting a big problem into smaller parts. This splitting is often accomplished by making a material balance around a "mixing point." As illustrated in Fig. 3.8, a mixing point is a junction of three or more streams and can be designated as a system in exactly the same manner as any other piece of process equipment. You should remember that in splitting the larger problems into smaller parts, material balances can be

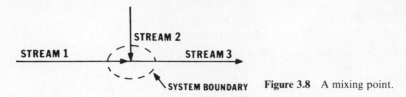

STREAM 2

STREAM 1

STREAM 3

SYSTEM BOUNDARY   **Figure 3.8**   A mixing point.

written for each piece of equipment and each mixing point as well as a balance around the whole process. However, since the overall balance is nothing more than the sum of the balances about each piece of equipment, not all of the balances will be independent.

Illustrations of the use of the algebraic technique in solving material balance problems are given in the following problem set.

Example 3.4   Distribution of a Benzene-Toluene Mixture

A blend of 40 liquid vol % toluene and 60 liquid vol % benzene is separated by distillation in a solvent recovery plant. The overhead distillate product contains 10.0 mole % toluene and the bottoms product contains 0.5 mole % benzene. The blend is fed to the distillation column at 20°C and at a rate of 1000 gal/day. How many gallons of distillate and bottoms are produced?

Solution

Step 1.   A flow sheet for this process is diagrammed below:

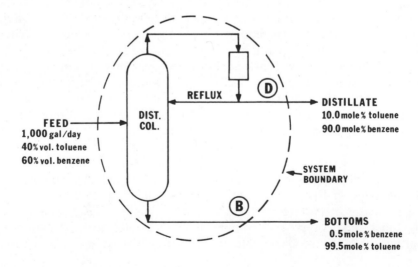

Step 2.   Basis: 1000 gal of feed

Inspection of the diagram shows that the compositions of the feed and the products are represented by two different conventions, since volume-percent and mole-percent for liquids are not the same. Consequently, we either must convert the mole-percent to volume-percent or vice versa. Since only one stream must be converted if we choose to use mole-percent, let's make the feed composition compatible with mole-percent. To do this, it is necessary to know the densities of benzene and toluene. These values are obtained from Table B.1 in Appendix B and are 0.879 g/ml for benzene and 0.866 g/ml for toluene. In converting from a volume composition to a weight composition, we will assume that volume is conserved. In this example the assumption is valid since benzene and toluene form near ideal solutions. For non-ideal mixtures, the failure to account for loss in conservation of volume can lead to large errors in converting from a volume basis to a weight basis. We can now calculate the number of pound-moles of benzene and toluene in 1000 gal of feed as follows:

Step 3

	gal	g/ml	lb/gal	lb	mol wt	lb·mole
Benzene	600	0.879	7.33	4398	78	56.4
Toluene	400	0.866	7.23	2892	92	31.4
Total	1000			7290		87.8

Step 4.   Let $D = $ lb · mole of distillate
$B = $ lb · mole of bottoms

By using a component material balance around our chosen system, we can write two independent simultaneous equations involving the two unknowns D and B.

$$\text{lb} \cdot \text{mole benzene in feed} = 56.4 = (0.900)D + (0.005)B$$

$$\text{lb} \cdot \text{mole toluene in feed} = 31.4 = (0.995)B + (0.10)D$$

Solving the simultaneous equations for B and D gives:

$$B = 25.4 \text{ lb} \cdot \text{mole}$$

$$D = 62.4 \text{ lb} \cdot \text{mole}$$

Step 5. We can now convert the pound-moles of distillate to gallons by use of the density as shown below:

Distillate: 62.4 lb · mole

	mole %	moles	mol wt	lb	lb/gal	gal
Benzene	90%	56.2	78	4384	7.33	598.1
Toluene	10%	6.2	92	570	7.23	78.8
Total		62.4		4954		676.9

Step 6. By difference, the volume of the bottoms can be calculated:

$$\text{bottoms} = 1000 - 676.9 = 323.1 \text{ gal}$$

Example 3.5   Absorption Process

Carbon dioxide can be removed from gas streams by absorption in a mono-ethanolamine (MEA) solution. Absorption is carried out by countercurrent (flow in opposite directions) contact of the gas and liquid in a packed column. The MEA solution is regenerated by heating in a stripping column. The absorption and regeneration process for removing CO_2 from a H_2 stream in a pilot plant experiment is shown in the diagram below.

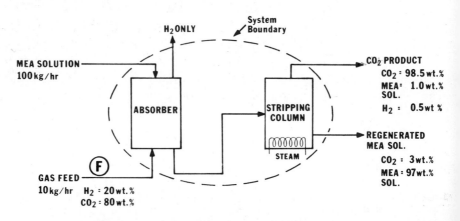

From the information provided, calculate the flow rates (in kg/hr) of the CO_2 product stream and the regenerated MEA stream.

Solution

Step 1. Since all of the compositions in the process are given in weight-percent, a weight basis would be the most convenient to use.

Basis: 10 kg of feed.

Step 2. There are five streams entering and leaving the entire system. The flow rates of two of these are known, leaving as unknown three flow rates. To solve for these three unknowns, we will have to set up at least three independent equations. Let:

F = feed to absorber (kg)
H = hydrogen leaving absorber (kg)
M = MEA feed to absorber (kg)
C = CO_2 product stream from stripper (kg)
R = Regeneration MEA from stripper (kg)

The unknowns are H, C, and R. Setting up balances involving these unknowns leads to the following equations:

Total balance $F + M = H + C + R$

Substituting values for F and M reduces this equation to

$$10 \text{ kg} + 100 \text{ kg} = H + C + R$$

Hydrogen balance $(0.2)(F) = H + (0.005)(C)$

this reduces to the following when the value for F is substituted into the expression:

$$2 \text{ kg} = H + (0.005)(C)$$

CO_2 balance $(0.8)(F) = (0.985)(C) + (0.03)(R)$

Substituting 10 kg for F leads to

$$8 \text{ kg} = (0.985)(C) + (0.03)(R)$$

MEA balance $M = (0.01)(C) + (0.97)(R)$

Substituting 100 kg for M leads to

$$100 \text{ kg} = (0.01)(C) + (0.97)(R)$$

We now have four equations and three unknowns; however, only three of the equations are independent. Therefore, any combination of three of these equations can be used to solve for the unknowns H, C, and R.

Step 3. When the equations above were solved, the following values were obtained for the flow rates of the streams leaving the pilot plant.

$H = 1.98 \text{ kg/hr}$
$C = 4.98 \text{ kg/hr}$
$R = 103.04 \text{ kg/hr}$

3.7 MATERIAL BALANCING BY USING TIE ELEMENTS

As stated in Section 3.4, a tie element is a material that goes from one stream into another without changing in any respect or without having like material added to it or lost from it. To identify a tie element, ask yourself the question, "What species passed through the process unchanged with constant mass?" The answer is the tie element. Often problems are encountered in which several components pass through a process with continuity so that there are several choices of tie elements. In such a case you should choose the component that has the smallest percentage error in its analysis. Or sometimes several of the components can be added together to give an overall tie element.

It should be understood that tie-element problems can, of course, be worked by the algebraic method, but in some of the more complicated industrial chemical calculations so many unknowns and equations are involved that the use of a tie element greatly simplifies the calculations. Therefore, in some of the following examples the algebraic solution will be given along with the tie-element solution.

Example 3.6   The Production of Absolute Ethanol

Absolute ethanol can be produced on a continuous basis by the distillation of a ternary mixture of ethanol, water, and benzene. On a trial pilot plant run the calculated data shown in the diagram below were obtained. From the other data obtained during the run, calculate the benzene flow in the feed in units of pounds per hour.

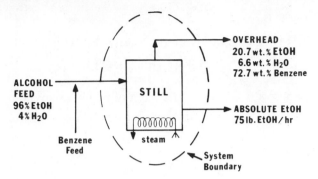

"Tie-Element Solution"

Step 1. An inspection of the flow diagram shows that there are two tie elements from the feed to the overhead stream: water and benzene. Either one of these could normally be used, but in this case we will use the water tie.

A convenient basis for this problem is 100 lb of the overhead stream. Since the benzene composition is unknown 100 lb of total feed would involve more work. However, 100 lb of alcohol feed or absolute EtOH product would also be suitable bases.

Basis: 100 lb of overhead

Step 2. Let x = lb of benzene in the total feed per 100 lb of total feed.

Component	Total Feed (lb)	Overhead (lb) = %
EtOH	$(100 - x)(0.96)$	20.7
H$_2$O	$(100 - x)(0.04)$	6.6
Benzene (BZ)	x	72.7
		100.0 lb

Step 3. Next, let us calculate the quantity of feed per 100 lb of overhead. We know the water in the feed (6.6 lb) and the benzene in the feed (72.7 lb); since these two materials only appear in the overhead and not in the product alcohol they can act as the tie element. Thus, we need only calculate the EtOH in the feed. As stated before, we will use water as the tie element to calculate this quantity as shown below.

$$\frac{\textbf{(6.6 lb H}_2\textbf{O)}}{\text{(100 lb overhead)}} \xleftarrow[tie]{\times} \frac{\textbf{(0.96 lb EtOH)}}{\textbf{(0.04 lb H}_2\textbf{O)}} = \frac{158.4 \text{ lb EtOH}}{100 \text{ lb overhead}}$$

The product now is

$$\text{EtOH in feed} - \text{EtOH in overhead} = \text{EtOH in product}$$

$$(158.4) - (20.7) = 137.7 \text{ lb}$$

Step 4. Choose a new basis to get the production on an hourly basis.

Basis: 1 hr ≡ 75 lb EtOH product

$$\frac{(72.7 \text{ lb benzene in feed})}{(137.7 \text{ lb EtOH in product})} \frac{(75 \text{ lb EtOH product})}{(1 \text{ hr})} = 39.6 \text{ lb BZ/hr}$$

This is the rate at which benzene enters the feed stream.

"Algebraic Solution"

Step 1. To employ an algebraic solution to this problem, we need an independent material balance for each unknown. The unknowns are:

Y = overhead
F = total feed
x = lb benzene/100 lb feed

Step 2. Basis: 1 hr ≡ 75 lb EtOH/hr

The equations are

Total balance $\qquad F = Y + 75$

H_2O balance $\qquad (F) \left[\dfrac{(0.04)(100 - x)}{100} \right] = (Y)(0.066)$

EtOH balance $\qquad (F) \left[\dfrac{(0.96)(100 - x)}{100} \right] = (Y)(0.207) + 75$

Benzene balance $\quad (F) \left(\dfrac{x}{100} \right) = (Y)(0.727)$

The solution of any combination of three of the above equations will lead to a value of the benzene feed rate x. Of course, the value of x will be the same as determined by the earlier tie-element solution. However, you can see that the simultaneous solution of three equations involves more work than the tie-element method.

Example 3.7   Production of $NaHCO_3$

Data are presented in the diagram below of a process for obtaining crystalline $NaHCO_3$ from a 15% $NaHCO_3$ solution. Calculate the rate of H_2O removal and crystalline $NaHCO_3$ production in kilograms per hour.

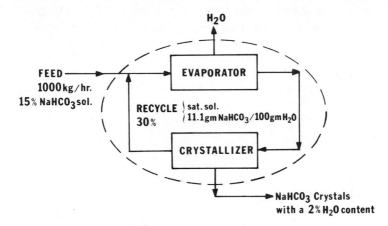

"*Tie Solution*"

Step 1.   In this problem a search for a tie element shows that $NaHCO_3$ exists on an overall basis (e.g., all the $NaHCO_3$ enters in the feed and leaves in the product). Ignoring the interior streams, we can draw a simplified picture of the process as shown below:

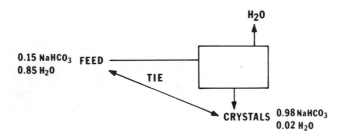

Step 2.   Now we can calculate the weight of crystals produced and the amount of water evaporated.

Basis: 1000 kg feed ≡ 1 hr of operation

$$(1000 \text{ kg } F)\frac{(0.15 \text{ kg } NaHCO_3)}{(1 \text{ kg } F)}\frac{(1 \text{ kg crystals})}{(1.98 \text{ kg } NaHCO_3)} = 153.1 \text{ kg/hr crystals}$$

Step 3. The H_2O evaporated is equal to the difference between the feed rate and the rate of crystal production

$$1000 \text{ kg/hr} - 153.1 \text{ kg/hr} = 846.9 \text{ kg } H_2O/\text{hr}$$

"Algebraic Solution"
To employ an algebraic method in the solution of this problem requires that we set up an independent material balance for each unknown.
Let

F = feed rate = 1000 kg/hr
W = H_2O evaporated
C = Crystals

Total balance	$F = W + C$
$NaHCO_3$ *balance*	$(0.15)(F) = (0.98)(C)$
H_2O *balance*	$(0.85)(F) = (0.02)(C) + W$

Any two of the above equations can be used to solve for the rate of crystal production C and the weight of water evaporated W. However, again it should be obvious that the tie-element method involves less work than the solution of a set of algebraic equations.

Example 3.8   Sodium Phosphate Production

As head chemist of a small chemical company, the plant superintendent brings you the following problem: He must produce 10 metric tons per day

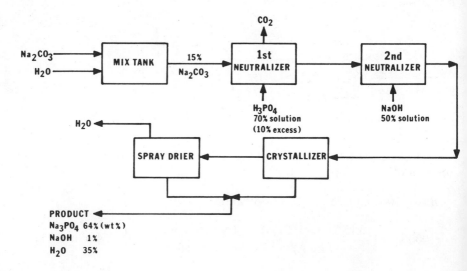

of the solid sodium phosphate product indicated in the process diagrammed below. He wants you to calculate (*a*) the weight of Na_2CO_3 and H_2O per day that must be added to the mix tank and (*b*) the weight of 70% phosphoric acid and 50% NaOH required per day.

Solution

Step 1. As the flow sheet indicates, a solution of sodium carbonate in water is pumped to the first neutralization tank where it is reacted with 10% excess phosphoric acid according to the following reaction:

$$Na_2CO_3 + H_3PO_4 \longrightarrow Na_2HPO_4 + CO_2 + H_2O$$

The carbon dioxide produced leaves as a gas. The disodium phosphate solution produced is pumped to a second neutralizer where the excess phosphoric acid and the disodium phosphate are converted to trisodium phosphate by the addition of a 50% NaOH solution. The solution leaving the second neutralizer contains 20% by weight trisodium phosphate, Na_3PO_4. The trisodium phosphate solution is cooled, and some solid product crystallizes out. The solution remaining is spray dried, and the solid products of the crystallizer and the spray drier are combined.

A search for a tie element indicates that all of the PO_4 that enters the first neutralizer as H_3PO_4 leaves as Na_3PO_4 in the product. Therefore, the PO_4 group can be used as a tie element. If we use the PO_4 group as a tie element, then a convenient basis to choose is 10 tons of product/day.

Basis: 10 metric tons of product $\equiv$ 1 day of operation

Step 2. Calculate the PO_4 content in the product.

$$(0.64)(10 \text{ tons})\left(\frac{95\,PO_4}{164\,Na_3PO_4}\right) = 3.707 \text{ tons of } PO_4$$

All of this PO_4 enters in the H_3PO_4 stream, therefore, we can calculate the rate of addition of the H_3PO_4 solution.

$$(3.707 \text{ tons } PO_4)\frac{(98\,H_3PO_4)}{(95\,PO_4)}\frac{(\text{ton of } H_3PO_4 \text{ sol})}{(0.70 \text{ tons } H_3PO_4)}$$

$$= 5.463 \text{ tons of } 70\% \; H_3PO_4 \text{ solution/day}$$

Step 3. Since the H_3PO_4 is added at a rate that is in 10% excess of the stoichiometric amount, we can calculate the Na_2CO_3 flow:

$$(3.707 \text{ ton } PO_4)\left(\frac{100\%}{110\%}\right)\left(\frac{106\,Na_2CO_3}{95PO_4}\right) = 3.76 \text{ tons of } Na_2CO_3/\text{day}$$

Step 4. Since the stream leaving the mix tank is 15% Na_2CO_3, the water added to the mix tank must be

$$(3.76 \text{ tons } Na_2CO_3/\text{day})\left(\frac{0.85 \text{ H}_2\text{O}}{0.15 \text{ Na}_2\text{CO}_3}\right) = 21.31 \text{ tons } H_2O/\text{day}$$

Step 5. We can now calculate the amount of 50% NaOH solution added by noting that

$$\text{ton} \cdot \text{moles NaOH} = \text{moles } Na_2CO_3 + 3 \text{ (moles excess } H_3PO_4)$$

$$= \left(\frac{3.76}{106}\right) + (3)(0.1)\left(\frac{3.76}{106}\right)$$

$$= 0.046$$

$$\text{wt of } 50\% \text{ NaOH sol} = (0.046 \text{ ton} \cdot \text{moles})\left(\frac{40 \text{ tons}}{\text{ton} \cdot \text{mole}}\right)\left(\frac{1}{0.5}\right)$$

$$= 3.68 \text{ tons of } 50\% \text{ NaOH sol}$$

The algebraic solution of this problem is left as an exercise for you to do.

3.8 COMPLEX MATERIAL BALANCES

The material balance problems that we have discussed up to now have been relatively simple. The problems have been kept simple by focusing our attention primarily on processes that involved a small number of reactant and product streams and only one or two reactions. As the number of streams and reactions in the process increase, so does the complexity of the total material balance. Even when a moderate-sized plant such as shown in Fig. 3.9 is being balanced, the number of streams and pieces of equipment involved is very large. This small benzene chlorination plant has 17 basic equipment elements, 42 process streams, 6 raw materials, and 10 products. It is obvious that a stepwise or simultaneous solution of material balances for each phase of this plant would truly be a staggering task. However, an overall plant balance can be composed of a number of individual, interlocking material balances that, however tedious they are to set up and solve, can be set down according to the principles and techniques discussed in this chapter. In addition, the actual work that the chemist must do in carrying out the large number of routine calculations required for problems of this complexity can be greatly reduced by the use of high-speed computers.

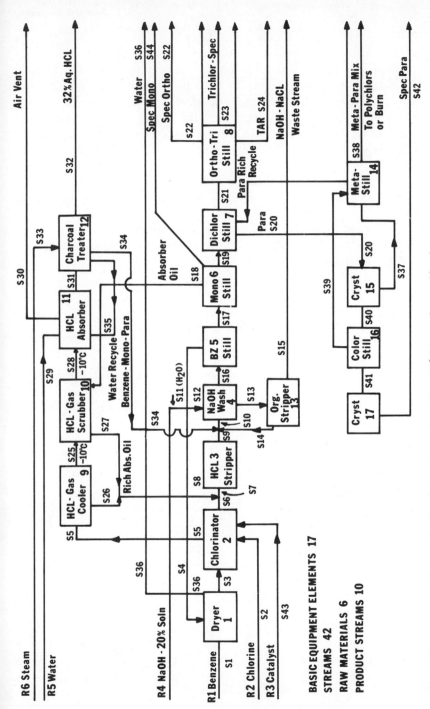

Figure 3.9 A block flow diagram of a plant designed to produce 100×10^6 lb/year of monochlorobenzene and 12×10^6 lb/year of ortho plus paradichlorobenzene.

71

3.9 SUMMARY

In this chapter we have learned that a quantitative description of a chemical process includes knowledge about the mass flow of materials in all process streams. This quantitative description is referred to as a material balance, which can be obtained by a systematic method of analysis. The method of analysis that we developed in this chapter involves the application of the following steps:

1. Draw a simplified flow sheet of the process.
2. Put down all the known compositions of the entering and leaving streams.
3. Decide which weights are known and unknown.
4. Find a tie element and/or set up mass balances involving the unknown.
5. Solve for the required values.

3.10 REFERENCES

1. E. J. Mills et al. (to Union Carbide Corp.), U.S. Patent 2,988,572, June 13, 1961.
2. W. H. Williams et al. (to Dow Chemical Co.), U.S. Patent 2,432,551, December 16, 1947.
3. J. F. Olin (to Sharples Chemicals, Inc.), U.S. Patent 2,377,511, June 5, 1945.
4. R. E. Kirk and D. F. Othmer, *Encyclopedia of Chemical Technology*, Vol. 2, 2nd ed., Wiley, New York, 1965, p. 275.
5. *Ibid.*, p. 266 ff.

Supplementary References on Material Balancing

Anderson, H. V., *Chemical Calculations*, McGraw-Hill, New York, 1955.

Benson, S. W., *Chemical Calculations*, 2nd ed., Wiley, New York, 1963.

Himmelblau, D. M., *Basic Principles and Calculations in Chemical Engineering*, Prentice-Hall, Englewood Cliffs, New Jersey, 1967.

Hougen, O. A., K. M. Watson, and R. A. Ragaty, *Chemical Process Principles*, Part I, 2nd ed., Wiley, New York, 1956.

Rudd, D. F., G. J. Powers, and J. J. Siirola, *Process Synthesis*, Prentice-Hall, Englewood Cliffs, New Jersey, 1973.

Schmidt, A. X., and H. L. List, *Material and Energy Balances*, Prentice-Hall, Englewood Cliffs, New Jersey, 1962.

Whitwell, J. C., and R. K. Toner, *Material and Energy Balances*, McGraw-Hill, New York, 1973.

3.11 PROBLEMS

1. Hydrogen sulfide gas is to be removed from a natural gas stream by an absorber-stripper process. The hydrogen sulfide is absorbed in a solvent in the absorber, and the loaded solvent is sent to the stripper, where the hydrogen sulfide is removed. The stripped solvent is recycled to the absorber to pick up more hydrogen sulfide. How many pounds per hour of hydrogen sulfide leaves the stripping column?

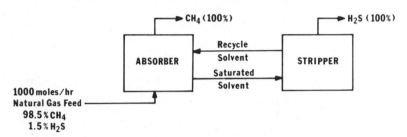

2. A furnace burns a fuel oil, the composition of which can be represented as $(CH_2)_n$. It is planned to burn this fuel with 10% excess air. Assuming complete combustion, calculate the composition of the flue gas on a dry basis.

3. A natural gas that contains $85\% \ CH_4$, $5\% \ C_2H_6$, and $10\% \ N_2$ is burned in a boiler for steam generation. Most of the CO_2 is scrubbed out of the flue gas and is used in the manufacture of dry ice. The exit gas from the scrubber analyzes CO_2, 1.1%; O_2, 4.6%; and N_2, 94.3%. Calculate the:
   (a) Percentage of the CO_2 absorbed.
   (b) Percentage of excess air used.

4. A contact sulfuric acid plant burns pure sulfur with air. The products of combustion pass to a converter where some of the SO_2 is converted to SO_3 at a fast rate owing to the high temperature of the gas. However, because of an unfavorable equilibrium at high temperatures, the gas is cooled and sent to a final converter where the oxidation can be carried further. The lower temperature favors a greater conversion of SO_2 to SO_3. The gas from the first converter was found to contain 4.5 vol $\%$ SO_2. The gas from the final converter contained $0.1\% \ SO_2$ and $8\% \ O_2$.
   (a) What percentage of the total SO_2 was converted to SO_3 in the first and second converters?
   (b) What was the excess air used to burn the sulfur?

5. How much water must be evaporated from 1000 gal of Na_2SO_4 solution, containing 30 g/liter at 49°C, so that 80% of the Na_2SO_4 will crystallize out when the solution is cooled to 10°C?

6. An evaporator is used to produce a 70 wt % sugar solution from a 30 wt % sugar cane solution. The sugar cane solution is fed at the rate of 100 metric tons per day. How many kilograms of water do these units evaporate per day?

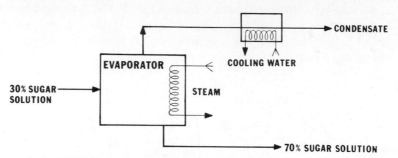

7. The following feed to a distillation column is separated into an overhead product that contains nothing boiling higher than isobutane and bottoms that contain nothing boiling lower than propane.

Feed Component	Mole %
Ethylene	1.9
Ethane	2.1
Propylene	5.2
Propane	14.8
Isobutane	25.0
n-Butane	34.0
n-Pentane	16.0
Total	100.0

The concentration of isobutane in the overhead product is 4.8 mole %, and the concentration of propane in the bottoms is 0.7 mole %. Calculate the moles of the various components in the overhead and bottoms per 100 moles of feed.

8. The feed to a distillation column contains 36% benzene by weight, the remainder is toluene. The overhead distillate is to contain 52% benzene by weight, while the bottoms are to contain 5% benzene by weight. Calculate the following:

(a) The percentage of the benzene feed contained in the distillate.

(b) The percentage of the total feed that leaves as distillate.

9. Low-grade phosphate rock can be upgraded by flotation processing. In this process the low-grade rock is ground into a powder and suspended as a water slurry. The addition of small amounts of oil and other surface-active agents causes the phosphate to adhere to air bubbles. By frothing

the slurry with air, a phosphate-rich foam floats off the top of a froth flotation unit, while the heavier waste rock leaves at the bottom. A two-stage flotation unit designed to produce a 20% P_2O_5 concentrate from a 5% low-grade rock is illustrated below. (Note: That the phosphate content of the samples is reported as P_2O_5.) Laboratory data for the process gave the following stream analysis:

Stream	Wt % P_2O_5 on a Dry Basis
Feed F	5
Waste W	1
Recycle R	10
Stream Q	12
Product P	20

All the streams consist of a slurry of approximately 30% solids. The weight of the frothing agents is negligible compared to the weight of the ore processed.

(a) Calculate the percentage of the phosphate in the feed material that is recovered as the product.

(b) Calculate the flow rates of all the streams if 100 metric tons/hr of feed enters the process.

(c) How much water leaves with the waste-rich stream per hour?

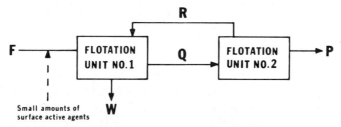

10. Pure carbon dioxide may be prepared by treating limestone with aqueous sulfuric acid. The flow scheme for the process is given below:

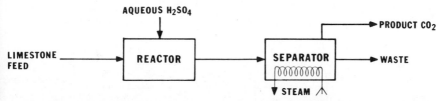

The limestone used in such a process contained calcium carbonate and magnesium carbonate; the remainder was inert, insoluble materials. The acid used contained 10% H_2SO_4 by weight. The residue from the process had the following composition: $CaSO_4$, 7.41%; $MgSO_4$, 4.81%; H_2SO_4, 0.96%; inerts, 0.65%; CO_2, 0.10% water, 86.07%. During the process the mass was warmed and carbon dioxide and water vapor removed in the separator.

(*a*) Calculate the analysis of the limestone used.

(*b*) What percentage of excess acid was used?

(*c*) Calculate the weight and analysis of the CO_2 product distilled from the separator per 1000 lb of limestone feed.

11. It is desired to measure the rate at which a waste gas stream is passing through a discharge line in a pilot plant experiment. There is no convenient way to attach a flow-measuring device to the line. However, it is possible to introduce and remove samples from the line. Analysis shows that gases entering the discharge line contain 1.5% carbon dioxide by weight. Pure carbon dioxide is introduced at the entrance of the discharge line at a measured rate of 50 g/min. It is found that the gases leaving the discharge line contain 2.5% carbon dioxide by weight. Calculate the rate of flow, in grams per minute, of the waste gases leaving the discharge line.

12. A solid containing 17% water by weight is dried with air to a moisture content of 5%. The fresh-air feed contains 0.005 lb of H_2O/lb of dry air, the recycle stream 0.09 lb of dry air, and the air to the drier at point A 0.02 lb H_2O/lb of dry air. Calculate:

(*a*) How many pounds of air are recycled per 100 lb of solid feed.

(*b*) How much fresh air is required?

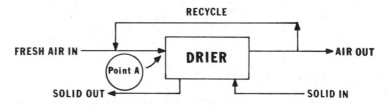

13. An NH_3 absorption column is designed to reduce the NH_3 content of a nitrogen stream. If the entering water contains 1% NH_3 and the exit water contains 5% NH_3 by weight, how much 1% solution is needed to make 100 kg of 5% NH_3 solution?

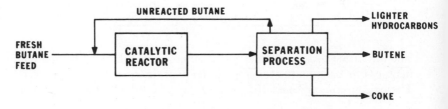

14. Butane is to be dehydrogenated to butene in a catalytic reactor. A flow diagram of the process is shown below. Five percent of the butane entering the reactor is converted to butene and the other products shown. The selectivity of the process is 50%. If the fresh butane feed is 100 moles/hr, calculate the following:

   (a) The total feed to the reactor per hour.
   (b) The moles of butene produced per hour.

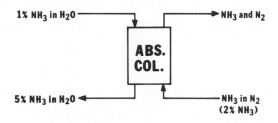

15. Xylene and other aromatic compounds are recovered from hydrocarbon feedstocks in petroleum refineries by solvent extraction. Several different solvents have been successfully employed as extractants, such as the use of sulfur dioxide in one of the earlier commercial processes. The diagram below shows a three-stage countercurrent SO_2 extraction of a catalytic reformer product.

   One thousand pounds of sulfur dioxide are fed to 250 lb of reformate in the system per hour. The xylene product stream contains 0.10 lb of sulfur dioxide per pound of xylene. The raffinate stream contains all the initially charged nonxylene material as well as 0.06 lb of xylene per pound of nonxylene material. The remaining component in the raffinate stream is the sulfur dioxide.

   (a) Calculate how many pounds of xylene are extracted per hour.
   (b) Calculate the pounds of extract and pounds of raffinate produced per hour.

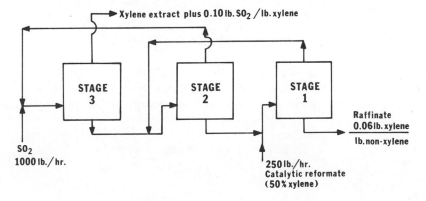

4

ENERGY ACCOUNTING IN CHEMICAL PROCESSES

THE LAW OF CONSERVATION OF ENERGY WORKS TOO

One important problem that chemists often encounter in the industrial world is the need to know what energy changes are occurring in a chemical process. As stated at the beginning of Chapter 3, energy changes and energy flows in a chemical process can be quantified by means of an energy balance. The energy balance is similar to the material balance inasmuch as it deals with a written account of energy in a process compared with a written account of mass.

In the discussion that follows, the energy balance is presented along with the methods that have been developed to measure and evaluate various common types of energy. Following this the energy balance, along with the material balance, is applied to practical, as well as some hypothetical, problems of interest to chemists. Again, as was done in Chapter 3, only steady-state problems are considered.

4.1 THE ENERGY BALANCE

The energy balance is based on the first law of thermodynamics (i.e., a statement of the conservation of energy). In applying this law to the energy balance, we can equate the energy input to energy output plus energy accumulation in the system expressed as follows:

$$\frac{\text{accumulation of}}{\text{energy in system}} = \frac{\text{energy}}{\text{input}} - \frac{\text{energy}}{\text{output}} \tag{4.1}$$

The best procedure for applying the energy balance equation is to use the format prescribed for the material balance equation, namely:

1. Draw a simplified flow sheet of the process.
2. Insert on the flow sheet all pertinent data.
3. Write all chemical equations of reactions involved.
4. Fix a convenient basis for calculation.
5. Solve for the required unknowns.

In order to apply the energy balance equation, all forms of energy must be expressed in the same units. Appendix A presents conversion factors for the most common expressions of energy. To facilitate the use of the energy balance equation, it is desirable to subdivide it into the various forms of energy that will be commonly encountered in industrial chemical processes. These divisions are discussed below.

Potential energy is a result of the position of a mass as a whole relative to some arbitrarily selected reference plane. The product of the distance above the reference plane and the force of gravity on the mass of each item entering, leaving, or accumulating in a process is the external potential energy of each item. The force of gravity is the product $(m \cdot g)$, where g is the acceleration due to gravity and m the mass. If x is the vertical distance of the mass above the reference plane, then the external potential energy related to this plane is the product $m \cdot g \cdot x$ because the force of gravity acts through the distance x. In the United States, however, the force of gravity is usually expressed in terms of pounds. Since the force of one pound is the force of gravity acting on a mass of one pound at sea level, this force is equivalent to 32.2 pdl. Thus, the expression of external potential energy in foot-pounds is numerically equivalent to $m \cdot x$, but it must be understood that m is numerically equal to the force involved only when the force is expressed in units of the force of gravity. If m is expressed in units of kilograms and x in units of meters, then the potential energy $m \cdot x$ will be in units of kilogram-meters.

The kinetic energy of each item of mass entering and leaving a process is a result of the motion of the mass. Accumulation of material within a process entails no significant motion, and thus only the streams entering and leaving contain kinetic energy. If v is the velocity of an item of mass entering or leaving, its kinetic energy is $\frac{1}{2}mv^2$. Where m is the mass, in the foot-pound or kilogram-meter unit system, the formula would be written as $\frac{1}{2}(mv^2)/g$ where g is the gravitational constant. Thus the kinetic energy is expressed in foot-pounds if one-half the product of the mass in pounds and velocity squared $(ft/s)^2$ is divided by the acceleration of gravity, 32.2 ft/s^2. If the mass is expressed in kilograms, the velocity in meters per second, and the acceleration of gravity as 9.81 m/s^2, then the kinetic energy will be in units of kilogram-meters.

Thermal energy is the heat Q supplied to the process and must also be included in the energy balance. Heat is commonly given in British thermal units or kilocalories. British thermal units can be converted to foot-pounds by multiplying by 778, and kilocalories can be converted to kilogram-meters by multiplying by 427.

Work energy appears in several ways in a process. An entering stream must be forced into a continuous process against a constant pressure P. The work done on the process in this instance is $m \cdot P \cdot V$, where V is the specific volume of the mass. If pressure is expressed in pounds (of force) per square foot, V in cubic feet per pound, and m in pounds, then mPV will be in foot-pounds. If P is in kilograms (of force) per square meter, V in cubic meters per kilogram, and m in kilograms, then mPV will be in kilogram-meters. It is likewise true that work is done by the process in forcing an effluent stream out of the process. Thus the net work done by the process due to a stream entering and leaving is $m \cdot (P_2 V_2 - P_1 V_1)$, where the subscript 2 indicates the leaving stream and subscript 1 the entering stream. In the case of a process where pressure is not constant the work energies are expressed as $\int P_1 \, dv_1$ and $\int P_2 \, dv_2$. The relationship between volume and pressure must therefore be known in order to calculate the net work done by the process in charging and discharging of a given item of mass. Work may also be performed by the process itself in the form of mechanical or electrical energies. For example, expanding gases in a process may be used to run a turbine, or a liquid stream may be pumped around the process by an external pump. All of these forms of work energy that the process expends or consumes must be considered. Such forms of work are denoted as W. A positive W indicates work done *by* the system.

Internal energy is contained by each item of mass that enters and leaves a process. The internal energy E is the macroscopic observation of the molecular, atomic, and subatomic energies. The change in internal energy of each item of mass is $m(E_2 - E_1)$, where E is given in units of energy per unit mass. This expression represents the summation of the changes in the molecular atomic and subatomic energies of the species entering and leaving the process.

We have now considered all of the energy terms that are normally found in chemical processes. Thus, as far as this book is intended to extend, the total energy balance may be written as Eq. 4.2 for any given item of mass that enters and leaves a system. This equation says that the total energy of the entering stream plus the heat added to the system is equal to the total energy of the leaving stream plus the work done by the system.

$$mx_1 + \frac{mv_1^2}{2g} + mP_1 V_1 + mE_1 + Q = mx_2 + \frac{mv_2^2}{2g} + mP_2 V_2 + mE_2 + W$$

$$(4.2)$$

It is possible to combine some of the terms in Eq. 4.2. For example, the variable enthalpy, represented by the symbol H, is defined as the combination of two variables that appear in the energy balance.

$$H = E + PV \tag{4.3}$$

As with internal energy, enthalpy has no absolute value; only changes in enthalpy can be calculated. Consequently, a reference set of conditions will most often be used in computing enthalpy changes. In calculating changes in enthalpy for the system, the reference conditions cancel out. Thus, the terms $m(E_2 + P_2 V_2)$ and $m(E_1 + P_1 V_1)$ may be represented by enthalpy values above a common reference plane or as $m \, \Delta H_2$ and $m \, \Delta H_1$, respectively. Equation 4.2 may therefore be expressed as

$$mx_1 + \frac{mv_1^2}{2g} + m \, \Delta H_1 + Q = mx_2 + \frac{mv_2^2}{2g} + m \, \Delta H_2 + W \tag{4.4}$$

Equations 4.2 and 4.4 were written for a single item of mass, but for a total energy balance on any process, Eq. 4.2 or 4.4 must be applied to every item of mass that enters and leaves the process. If $\sum$ represents the summation of each energy item such as mx_1, $mv_1^2/2g$, and so on, then when the summation is applied to all items of mass that enter and leave the process Eq. 4.4 becomes

$$\sum mx_1 + \frac{\sum mv_1^2}{2g} + \sum m \, \Delta H_1 + Q = \sum mx_2 + \frac{\sum mv_2^2}{2g} + \sum m \, \Delta H_2 + W$$

$$\tag{4.5}$$

4.2 APPLICATION OF THE ENERGY BALANCE

We have found that in the application of the total energy balance equation (i.e., Eq. 4.5) to many of the different types of processes that chemists encounter, it can be greatly simplified. For example, in a majority of the continuous processes the potential and kinetic energies of the entering and leaving streams are so small that they can be disregarded. Thus, in such cases, Eq. 4.5 reduces to

$$Q - W = \sum M \, \Delta H_2 - \sum M \, \Delta H_1 \tag{4.6}$$

The symbol M represents moles if H is in molar units and mass if H is in mass units. A sketch of a generalized flow process to which Eq. 4.6 might be applied is shown in Fig. 4.1. The overall energy balance for this process (ignoring kinetic and potential energy changes) is written as shown in Eq. 4.7.

$$Q - W = (M_C \, \Delta H_C + M_D \, \Delta H_D) - (M_A \, \Delta H_A + M_B \, \Delta H_B) \tag{4.7}$$

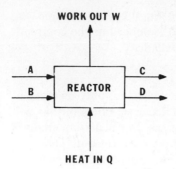

WORK OUT W

HEAT IN Q

Figure 4.1 Sketch of a generalized flow process with chemical reaction.

In many processes, the only work items involved are a result of forcing the various streams into and out of the process. In such cases Eq. 4.6 reduces to Eq. 4.8. Thus, the heat transferred

$$Q = \sum M \, \Delta H_2 - \sum M \, \Delta H_1 \qquad (4.8)$$

to or from a process for all practical purposes is equal to the difference in enthalpy between the streams that enter and leave the process. The sign of Q determines whether heat is given to or absorbed from the surroundings. When Q is negative, the system (i.e., the process or part of the process under study) gives off heat to the surroundings. When Q is positive, the system absorbs heat from the surroundings. In the application of Eq. 4.8 we have reduced the energy balance to a "heat balance." This is very convenient inasmuch as enthalpy values are integral quantities. Therefore tables or charts may be available or can be prepared that give the enthalpy values for each specified type of mass as a function of pressure, temperature, and phase condition. In addition, changes in enthalpy can be calculated from heat capacity data. Refer to Appendix C for a review of such calculations.

In applying Eq. 4.8 it must be remembered that the heat balance expression as written is valid for a continuous process only when potential, kinetic, and all forms of work energy except those included in enthalpy are negligible. It is also valid for any batch process at constant pressure. However, it is not valid for a batch process at constant volume if the pressure is changing. Example 4.1 illustrates the accuracy of the heat balance as compared with a total energy balance on a typical continuous process.

Example 4.1   Energy Balance in a Steam Heat Exchanger

Steam used to heat a reaction vessel enters the exchanger, which is segregated from the reactants, at a linear velocity of 100 ft/s. The steam is saturated and enters at an absolute pressure of 169 lb/in.^{2} and a temperature of 204°C. The steam is completely condensed and leaves the steam chest at 164°C with a linear velocity of 1 ft/s. The difference in vertical height between the steam inlet at the top and condensate exit at the bottom is 10 ft. How much energy is given up per pound of condensate?

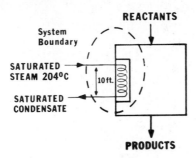

REACTANTS

System
Boundary

SATURATED
STEAM 204°C

10 ft.

SATURATED
CONDENSATE

PRODUCTS

Solution

Step 1. The steam tables in Reference 1 give the enthalpy of steam at 169 lb/in.² (abs.) and 204°C (400°F) as 1201.0 Btu/lb above water at 32°F. The enthalpy of the condensate at 164°C (328°F) is given as 298.7 Btu/lb above water at 32°F. If we ignore, for the moment, the kinetic and potential energies of the streams, then a heat balance can be used to calculate the amount of energy given up per pound of steam condensate.

Step 2. Basis $\equiv$ 1 lb of condensate.

Step 3. $Q = \Delta H_{\text{cond out}} - \Delta H_{\text{steam in}}$
         $Q = 298.7 \text{ Btu/lb} - 1201.0 \text{ Btu/lb}$
         $Q = -902.3 \text{ Btu/lb}$

The negative sign means that our chosen system, the exchanger, gives up heat to the surroundings at the rate of 902.3 Btu per pound of condensate.

Let us now calculate the energy given up per pound of condensate by using the total energy expression, where both the kinetic and potential energies are included. For this system, Eq. 4.5 can be written as follows:

$$mx_{\text{steam in}} + \frac{mv^2}{2g_{\text{steam in}}} + \Delta H_{\text{steam in}} + Q = mx_{\text{cond out}} + \frac{mv^2}{2g_{\text{cond out}}} + \Delta H_{\text{cond out}}$$

(Note: since the system does no work, W is zero.)

Solution

Step 1. Again, choose as a basis 1 lb of condensate.

Step 2. Calculate the potential energy terms by choosing as a reference plane a line running along the condensate stream.

$$mx_{\text{steam in}} = (1 \text{ lb})(10 \text{ ft}) = (10 \text{ ft} \cdot \text{lb})\frac{(1 \text{ Btu})}{(778 \text{ ft} \cdot \text{lb})} = 0.0129 \text{ Btu}$$

$$mx_{\text{cond out}} = (1 \text{ lb})(0 \text{ ft}) = 0$$

Step 3. Calculate the kinetic energy terms.

$$\frac{mv^2}{2g_{\text{steam in}}} = \frac{(1 \text{ lb})(100 \text{ ft/s})^2}{(2)(32.2 \text{ ft/s}^2)} = (155.3 \text{ ft} \cdot \text{lb})\frac{(1 \text{ Btu})}{(778 \text{ ft} \cdot \text{lb})} = 0.20 \text{ Btu}$$

$$\frac{mv^2}{2g_{\text{cond out}}} = \frac{(1 \text{ lb})(1 \text{ ft/s})^2}{(2)(32.2 \text{ ft/s}^2)} = (0.016 \text{ ft} \cdot \text{lb})\frac{1 \text{ Btu}}{(778 \text{ ft} \cdot \text{lb})} = 0.0002 \text{ Btu}$$

Step 4. We can now place the above values along with the enthalpies into the total energy expression and solve for Q.

$$Q = 0 + 0.00002 + 298.7 - 0.0129 - 0.20 - 1201.0 = -902.51288 \text{ Btu/lb}$$

When all figures beyond the first digit to the right of the decimal point are ignored, Q becomes equal to -902.5 Btu/lb. This value differs by only 0.02 % from the value of Q calculated from the heat balance expression. Thus for all practical purposes the heat balance was as good as the total energy balance in this case, because the correction for the differences in external potential and external kinetic energies is far outside the limits of accuracy of measuring the quantity of steam entering the system or the condensate leaving.

Example 4.2   Energy Balance around a Sulfur Dioxide Pilot Plant

Sulfur dioxide is being manufactured in a small pilot plant experiment by the direct oxidation of sulfur. The sulfur is fed in an air stream to the burner as a powder in the rhombic form at a temperature of 25°C. Air is supplied to

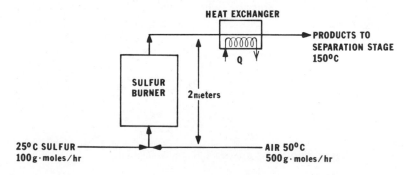

the burner at a rate of 500 g · moles/hr at 50°C. Before the products are fed to the separation stage where the SO_2 is recovered, they are cooled to 150°C in a heat exchanger. The reactants enter the burner at a linear velocity of 5 m/s and leave the heat exchanger at a linear velocity of 30 m/s. The product exit is 2 m higher than the reactant entrance. Calculate the heat removed per hour by the heat exchanger.

Solution

Step 1. Basis: 100 moles of sulfur ≡ to 1 hr of operation.

Step 2. Material input:

Component	Moles	Grams
Sulfur	100	(32)(100) = 3,200
N_2	(0.79)(500) = 395	(28)(395) = 11,060
O_2	(0.21)(500) = 105	(32)(105) = 3,360
		Total 17,620

Material output (complete conversion of S to SO_2):

Component	Moles	Grams
SO_2	100	(64)(100) = 6,400
N_2	395	(28)(395) = 11,060
O_2	5	(32)(5) = 160
		Total 17,620

Step 3. Enthalpy input (enthalpy and heat capacity data are obtained from Appendix C. Let the reference temperature be 25°C):

$S_{rhombic}$: $\quad \Delta H^\circ_{25°C} = 0$

O_2: $\qquad \Delta H^\circ_{50°C} = \Delta H^\circ_{25°C} + \int_{25°C}^{50°C} C_p \, dT$

$\qquad C_p$ in units cal/(g · mole)(°C) $= 7.129 + 0.1407 \times 10^{-2}T$
$$-0.6438 \times 10^{-5}T^2$$

$$\Delta H^\circ_{50°} = 0 + \int_{25}^{50} (7.129 + 0.1407 \times 10^{-2}T - 0.6438 \times 10^{-5}T^2) \, dT$$

$$\Delta H^\circ_{50°} \text{ for } O_2 = 178.64 \text{ cal/g} \cdot \text{mole}$$

$$N_2: \quad \Delta H^\circ_{50^\circ C} = \Delta H^\circ_{25^\circ C} + \int_{25}^{50} C_p \, dT$$

$$\Delta H^\circ_{25^\circ C} = 0$$

$$C_p = 6.919 + 0.1365 \times 10^{-2}T - 0.0227 \times 10^{-5}T^2$$

$$\Delta H^\circ_{50^\circ} = 0 + \int_{25}^{50} (6.919 + 0.1365 \times 10^{-2}T - 0.0227 \times 10^{-5}T^2) \, dT$$

$$\Delta H^\circ_{50^\circ} \text{ for } N_2 = 173.41 \text{ cal/g} \cdot \text{mole}$$

Component	ΔH at Input Temperature (kcal/g $\cdot$ mole)	Gram Moles	ΔH Input (kcal)
S	0	100	(100)(0) $= 0$
O_2	0.1786	105	(105)(0.1786) $= 18.8$
N_2	0.1734	395	(395)(0.1734) $= 68.5$

$$\Sigma \, \Delta H_{input} = \text{total} = 87.3 \text{ kcal}$$

Step 4. Enthalpy output:

$$SO_2: \quad \Delta H_{150^\circ C} = \Delta H^\circ_{25^\circ C} + \int_{25^\circ C}^{150^\circ C} C_p \, dT$$

$$\Delta H^\circ_{25^\circ C} = -70.96 \text{ kcal/g} \cdot \text{mole}$$

$$C_p = 9.299 + 0.933 \times 10^{-2}T - 0.7418 \times 10^{-5}T^2$$

$$\Delta H_{150^\circ C} = -70,960 \text{ cal/g} \cdot \text{mole} + \int_{25^\circ}^{150^\circ} (9.299 + 0.933 \times 10^{-2}T$$

$$-0.7418 \times 10^{-5}T^2) \, dT$$

$$\Delta H_{150^\circ C} \text{ for } SO_2 = -69,730 \text{ cal/g} \cdot \text{mole}$$

$$O_2: \quad \Delta H_{150^\circ C} = \Delta H^\circ_{25^\circ C} + \int_{25^\circ C}^{150^\circ C} C_p \, dT$$

$$\Delta H_{150^\circ C} = 0 + \int_{25^\circ}^{150^\circ} (7.129 + 0.1407 \times 10^{-2}T$$

$$-0.6438 \times 10^{-5}T^2) \, dT$$

$$\Delta H_{150^\circ C} = 897.9 \text{ cal/g} \cdot \text{mole}$$

$$N_2: \quad \Delta H_{150°C} = \Delta H_{25°C}^{\circ} + \int_{25°C}^{150°C} C_p \, dT$$

$$\Delta H_{150°C} = 0 + \int_{25}^{150} (6.919 + 0.1365 \times 10^{-2}T$$

$$- 0.0227 \times 10^{-5}T^2) \, dT$$

$$\Delta H_{150°C} = 875.5 \text{ cal/g} \cdot \text{mole}$$

Component	ΔH at Output Temperature (kcal/g·mole)	Gram Moles	ΔH Output(kcal)
SO_2	−69.73	100	$(100)(-69.73) = -6973$
O_2	0.898	5	$(5)(0.898) \quad = 4.5$
N_2	0.876	395	$(395)(0.876) \quad = 346$

$$\Sigma \Delta H_{output} = \text{total} = -6622 \text{ kcal}$$

Step 5. For the moment, let us ignore the kinetic and potential energies of the inlet and outlet streams. Also in this process, the system does no work so that W is zero. Thus Eq. 4.8 is applicable.

$$Q = \sum M \, \Delta H_2 - \sum M \, \Delta H_1 = \sum \Delta H_{output} - \sum \Delta H_{input}$$

Substituting our calculated input and output enthalpies into the above equation allows us to solve for Q.

$$Q = -6622 \text{ kcal} - 87.3 \text{ kcal} = -6710 \text{ kcal}$$

This is the amount of heat that must be removed by the heat exchanger per hour.

Step 6. Let us now calculate the contributions that the potential and and kinetic energies make to the total energy balance:

Potential energy input. If we choose the reference plane to lie along the entering streams, then the potential energy of the entering streams is zero.

Potential energy output. The output stream lies 6 m above the reference plane, and its total weight is 17.620 kg.

$$\sum mx_2 = (17.62 \text{ kg})(6\text{m})\left(\frac{1 \text{ kcal}}{426.7 \text{ kg}}\right) = 0.24 \text{ kcal}$$

Kinetic energy input. The velocity input stream = 5 m/s, mass input stream = 17.620 kg.

$$\frac{\sum mv_1^2}{2g} = \frac{(17.62 \text{ kg})(5 \text{ m/s})^2}{(2)(9.81 \text{ m/s}^2)} \times \left(\frac{1 \text{ kcal}}{426.7 \text{ kg} \cdot \text{m}}\right) = 0.05 \text{ kcal}$$

Kinetic energy output. The velocity output stream = 30 m/sec, mass output stream = 17.62 kg.

$$\frac{\sum mv_2^2}{2g} = \frac{(17.62 \text{ kg})(30 \text{ m/s})^2}{(2)(9.81 \text{ m/s}^2)} \times \left(\frac{1 \text{ kcal}}{426.7 \text{ kg} \cdot \text{m}}\right) = 2.48 \text{ kcal}$$

Step 7.   We can now solve the total energy balance equation for Q.

$$Q = \sum mx_{\text{output}} + \frac{\sum mv_2^2}{2g_{\text{output}}} + \sum \Delta H_{\text{output}} - \sum mx_{\text{input}} - \frac{\sum mv_1^2}{2g_{\text{input}}} - \sum \Delta H_{\text{input}}$$

$$Q = 0.24 + 2.48 - 6622 - 0 - 0.05 - 87.3$$

$$Q = -6707 \text{ kcal}$$

Comparison of this value of Q to that calculated by the heat balance equation shows that only a 0.04 % error was introduced by ignoring the potential energy and kinetic energy terms. Thus again the heat balance was as good as the total energy balance in this case, because the error introduced by ignoring the potential and kinetic energy terms is outside the limits of accuracy of measuring the heat removed by the heat exchanger.

The heat balance is by far the most common application of the energy balance. However, there are situations where the total energy balance equation must be used. One classical example of such a case is in problems dealing with the unit operation known as fluid flow. We consider this application in Chapter 5.

In Example 4.2 our initial material balance provided the groundwork for writing the energy balance. In fact, in all energy balance problems, however simple they may seem, you must know the amount of material entering and leaving the process if you are to apply successfully the appropriate energy balance equations. However, there are situations that require the simultaneous solution of the material and energy balance. Example 4.3 illustrates such a problem.

Example 4.3   Simultaneous Solution of Material and Energy Balance

The distillation column shown in the figure below separates 100 kg/hr of a 40 wt % methylcyclohexane (MCH), 60 wt % toluene (Tol) liquid solution which is at 25°C. The liquid product from the top of the column is 99.5 wt %

MCH, while the bottom stream from the reboiler contains 1% MCH. The condenser uses water that enters at 15°C and leaves at 35°C, while the reboiler uses saturated steam at 140°C. The reflux ratio (i.e., the ratio of the liquid overhead returned to the column to the liquid overhead product removed) is 10 to 1. Both the condenser and reboiler operate at 1 atm pressure. The temperature of the product stream leaving the condenser is 50°C, while the temperature of the streams leaving the reboiler is 108°C. The fraction of methylcyclohexane in the vapor from the reboiler is 4.0 wt%. Calculate the following:

(a) The kilograms of overhead product and bottoms produced per hour.
(b) The kilograms of reflux per hour.
(c) The kilograms of liquid entering the reboiler and the reboiler vapor per hour.
(d) The kilograms of steam and liters of cooling water used per hour.

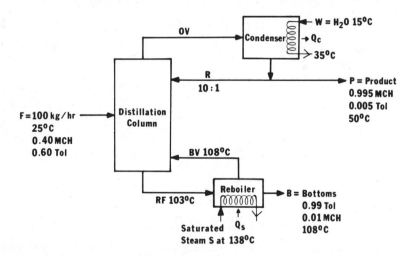

Solution

Step 1. Choose a basis of 100 kg of feed (i.e., the feed for one hour of operation).

Step 2. Set up equations for material balances on the overall process.

Overall total material balance:

$$F = P + B \qquad (4.9)$$

$$100 \text{ kg} = P + B \qquad (4.10)$$

Overall methylcyclohexane balance. Letting X represent the weight fraction of MCH in the respective streams the MCH balance is:

$$(F)(X_F) = (P)(X_p) + (B)(X_B) \qquad (4.11)$$

$$(100)(0.40) = (P)(0.995) + (B)(0.01) \qquad (4.12)$$

Substituting Eq. 4.10 for B in Eq. 4.12 leads to

$$(100)(0.40) = P(0.995) + (100 - P)(0.01) \qquad (4.13)$$

solving for P and B leads to an answer to question (*a*)

$$P = \text{overhead product} = 39.6 \text{ kg/hr}$$

$$B = \text{bottoms product} = 60.4 \text{ kg/hr}$$

Step 3. Set up material balances around the condenser: Since the reflux ratio is $10:1$ we can write the equality

$$R = (10)(P) = (10)(39.6) = 396 \text{ kg/hr} \qquad (4.14)$$

This is the answer to question (*b*). We can also calculate the total overhead vapor rate OV

$$OV = R + P = 369 + 39.6 = 435.6 \text{ kg/hr} \qquad (4.15)$$

Step 4. Set up material balances around the reboiler.

Total balance $RF = BV + B$ $\qquad (4.16)$

MCH *balance*: ($x = $ fraction of MCH)

$$(RF)(x_{RF}) + (BV)(x_{BV}) + (B)(x_B) \qquad (4.17)$$

substituting the known values into Eq. 4.16 and 4.17 leads to:

$$RF = BV + 60.4 \text{ kg} \qquad (4.18)$$

$$(RF)(x_{RF}) = (60.4 \text{ kg})(0.01) + (BV)(0.04) \qquad (4.19)$$

In looking at Eq. 4.18 and 4.19 it is apparent that we have three unknowns and only two independent equations. We can write additional equations around the distillation column, but these will not resolve the problem since we would still be left with one unknown more than the number of independent equations. It is at this stage in the solution of the problem that energy balances can be used effectively.

Step 5. Set up an overall energy balance around the entire process. Let the reference temperature be 25°C. Assume the solutions are ideal so that the thermodynamic properties (enthalpies and heat capacities) are additive. No work or potential or kinetic energy change is involved in this problem. Thus:

$$Q = \sum \Delta H_{\text{output}} - \sum \Delta H_{\text{input}} \qquad (4.20)$$

$$Q = Q_{\text{steam}} + Q_{\text{condenser}} = \Delta H_{\text{product}} + \Delta H_{\text{bottoms}} - \Delta H_{\text{feed}} \qquad (4.21)$$

$$Q_s + Q_c = (P)(\Delta H_{fP}^\circ) + P \int_{25^\circ}^{50^\circ} Cp_p \, dT + B(\Delta H_{fB}^\circ) + B \int_{25^\circ}^{108^\circ} Cp_B \, dT$$

$$- F(\Delta H_{fF}^\circ) - F \int_{25^\circ}^{25^\circ} Cp_F \, dT \qquad (4.22)$$

Since $F = P + B$ the terms $(P)(\Delta H_{fP}^\circ) + B(\Delta H_{fB}^\circ) - F(\Delta H_{fF}^\circ)$ cancel out. Also the term $F \int_{25^\circ}^{25^\circ} Cp_F \, dT$ is equal to zero. Therefore Eq. 4.22 reduces to:

$$Q_s + Q_c = P \int_{25^\circ}^{50^\circ} Cp_p \, dT + B \int_{25^\circ}^{108^\circ} Cp_B \, dT \qquad (4.23)$$

To solve this equation we need the heat capacities for the product and bottoms stream. These values can be calculated from the heat capacities of the individual components. Heat capacities obtained from Appendix C follow:

Methylcyclohexane. B.p. $= 101.1°C$; mol wt $= 98.1$; liquid mean $Cp = 49.08$ cal/(g·mol)(°C) in the temperature range 25 to 115°C. $\Delta H_{\text{vap}} = 7.58$ kcal/g·mole $= 77.3$ kcal/kg

Toluene. B.p. $= 110.8°C$; mol wt $= 92.1$; liquid mean $Cp = 39.43$ cal/ (g·mole)(°C) in the temperature range 25 to 115°C. $\Delta H_{\text{vap}} = 8.0$ kcal/ g·mole $= 86.9$ kcal/kg

The product has a composition corresponding to 0.995 MCH and 0.005 Tol on a weight fraction basis. The mean Cp of MCH in weight units is 0.500 kcal/(kg)(°C), and for Tol it is 0.428 kcal/(kg)(°C). Therefore the mean Cp for the product stream is:

$$Cp_p = (0.995)[0.500 \text{ kcal/(kg)(°C)}] + (0.005)[0.428 \text{ kcal/(kg)(°C)}]$$

$$= 0.499 \text{ kcal/(kg)(°C)}$$

The bottoms have a composition corresponding to 0.99 Tol and 0.01 MCH. Therefore the mean Cp for the bottoms stream is

$$Cp_B = (0.99)[0.428 \text{ kcal/(kg)(°C)}] + (0.01)[0.500 \text{ kcal/(kg)(°C)}]$$

$$= 0.429 \text{ kcal/(kg)(°C)}$$

Putting these values for Cp_P and Cp_B and the values of P and B calculated earlier into Eq. 4.23 leads to

$$Q_s + Q_c = 39.6 \text{ kg/hr} \int_{25°}^{50°} [0.499 \text{ kcal/(kg)}(°\text{C})](dT)$$

$$+ 60.4 \text{ kg/hr} \int_{25°}^{108°} [0.429 \text{ kcal/(kg)}(°\text{C})](dT) \qquad (4.24)$$

Since Eq. 4.24 still has two unknown terms in it, we need to establish more independent equations.

Step 6. Set up an energy balance on the condenser. This time we will let the reference temperature be 50°C. This choice simplifies the calculation because this is the temperature of the R and P streams, and therefore their enthalpies will be zero with respect to the reference temperature. Let the condenser be the system and the water the surroundings.

$$Q_{\text{system}} = Q_c = \sum \Delta H_{\text{output}} - \sum \Delta H_{\text{input}} \qquad (4.25)$$

$\sum \Delta H_{\text{output}} = \Delta H_p + \Delta H_R = 0$ since the reference temperature is 50°C and both R and P are at 50°C.

$$\sum \Delta H_{\text{input}} = \Delta H_{OV} \qquad (4.26)$$

Assuming that the temperature of stream OV is the same as the boiling point of MCH (101.1°C) allows us to write Eq. 4.26 as

$$\Delta H_{OV} = OV \Bigg[(x_{\text{MCH}}) \int_{50°}^{101.1°} (Cp_{\text{MCH}}) dT + (x_{\text{MCH}})(\Delta H_{\text{vap MCH}})$$

$$+ (x_{\text{Tol}}) \int_{50°}^{101.1°} (Cp_{\text{Tol}})(dT) + (x_{\text{Tol}})(\Delta H_{\text{vap Tol}}) \Bigg] \qquad (4.27)$$

where (x_{MCH}) and (x_{Tol}) are the weight fractions of MCH and Tol. Substituting for the known values in Eq. 4.27 leads to

$$\Delta H_{OV} = (435.6 \text{ kg/hr}) \Bigg[(0.995) \int_{50°}^{101.1°} (0.500 \text{ kcal/kg}°\text{C})(dT)$$

$$+ (0.995)(77.3 \text{ kcal/kg})$$

$$+ (0.005) \int_{50°}^{101.1°} (0.428 \text{ kcal/kg}°\text{C})(dT)$$

$$+ (0.005)(86.9 \text{ kcal/kg}) \Bigg] \qquad (4.28)$$

Solving Eq. 4.28 for ΔH_{OV} gives

$$\Delta H_{OV} = 44{,}812 \text{ kcal/hr}$$

Therefore,

$$Q \text{ system} = Q_c = -\sum \Delta H_{input} = -\sum \Delta H_{OV} = -44{,}812 \text{ kcal/hr}$$

and since Q system $= -Q$ surrounding where Q surroundings $= Q$ water.

$$Q \text{ water} = \Delta H \text{ water} = (2)(Cp_{H_2O})(t_2 - t_1) \tag{4.29}$$

Substituting known values into Eq. 4.29 leads to

$$-(-44{,}812 \text{ kcal/hr}) = (W)[1 \text{ kcal/(kg)(°C)}](35° - 15°)$$

$$W = 2241 \text{ kg/hr}$$

This is the amount of cooling water used per hour and answers part of question (d).

Step 7. Since we now know Q_c we can go back to Eq. 4.24 and solve for Q_s, the amount of steam used.

$$Q_s = (39.6 \text{ kg/hr})[0.499 \text{ kcal/(kg)(°C)}](25°C)$$

$$+ \ 60.4 \text{ kg/hr } [0.429 \text{ kcal/(kg)(°C)}](83°C) - (-44{,}812 \text{ kcal/hr})$$

$$Q_s = 47{,}457 \text{ kcal/hr}$$

From the steam tables in Reference 1 ΔH_{vap} at 138°C is 512.9 kcal/kg. Assume the steam leaves at its saturation temperature and is not subcooled. Then

$$\text{kg steam used/hr} = \frac{47{,}457 \text{ kcal/hr}}{512.9 \text{ kcal/kg}}$$

$$= 92.5 \text{ kg/hr}$$

This is the second half of the answer to question (d).

Step 8. Set up an energy balance around the reboiler:

$$Q_{steam} = \sum \Delta H_{output} - \sum \Delta H_{input} \tag{4.30}$$

or,

$$Q_s = B(\Delta H_B) + BV(\Delta H_{BV}) - (RF)(\Delta H_{RF}) \tag{4.31}$$

We know $Q_s = 47{,}457$ kcal/hr. We do not know RF or the value of ΔH_{RF}. However, the heat capacity of stream RF would be about that of toluene. The value of BV is still unknown, but all the other values (ΔH_{BV}, B, ΔH_B and ΔH_{RF}) are known or can be calculated. Thus, if we combine the energy balance around the reboiler with the overall material balance around the reboiler, we can find both *RF* and *BV*.

Energy balance. Choose a reference temperature of 108°C and Eq. 4.31 becomes:

$$47{,}457 \text{ kcal/hr} = (B) \int_{108°}^{108°} Cp_B \cdot dT + (BV)(\Delta H_{\text{vap } BV})$$

$$+ BV \int_{108°}^{108°} Cp_{BV} \cdot dT - RF \int_{108°}^{103°} Cp_{RF} \cdot dT \quad (4.32)$$

Equation 4.32 reduces to

$$47{,}457 \text{ kcal/hr} = (BV)[(0.99)(86.9 \text{ kcal/kg}) + (0.01)(77.3 \text{ kcal/kg})]$$

$$- RF[0.428 \text{ kcal/(kg)(°C)}](-5°C) \quad (4.33)$$

The material balance around the reboiler was given in Eq. 4.18.

$$RF = BV + 60.4 \text{ kg} \quad (4.18)$$

Eqs. 4.18 and 4.33 can now be solved for *BV* and *RF* giving:

$$BV = \text{reboiler vapor per hour} = 532 \text{ kg}$$

$$RF = \text{kilograms liquid entering reboiler} = 592 \text{ kg}$$

These last two values represent the answer to question (*b*).

4.3 SUMMARY

From the information presented in this chapter you should have learned how an energy balance is expressed, both in words and in mathematical form. You should be able to apply the energy balance equation to any steady-state system. In so doing, you should be able to state what each term is in light of the conditions of the problem, know what assumptions can and cannot be made, and know what terms can be omitted. In certain situations, it will be necessary to use material and energy balances in combination. You should be able to recognize and analyze such problems and be able to write down the appropriate balances in an accomplished manner.

In applying the energy balance, you should be able to locate enthalpy or heat capacity data for most common compounds and, if such data are not available, be able to make suitable approximations. You should be familiar

with the principles of thermochemistry, and especially how to use reference
temperatures.

4.4   REFERENCES

1. R. C. Weast, *Handbook of Chemistry and Physics*, 56th ed., Chemical Rubber Publishing Co.,
Cleveland, Ohio, 1975, pp. E18–E23.

Supplementary References

Hall, N. A., and W. E. Ibele, *Engineering Thermodynamics*, Prentice-Hall, Englewood Cliffs,
New Jersey, 1960.

Henley, E. J., and H. Bilber, *Chemical Engineering Calculations*, McGraw-Hill, New York, 1959.

Himmelblau, D. M., *Basic Principles and Calculations in Chemical Engineering*, 2nd ed., Prentice-
Hall, Englewood Cliffs, New Jersey, 1967.

Hougen, O. A., K. M. Watson, and R. A. Rogatz, *Chemical Process Principles*, Part II, 2nd ed.,
Wiley, New York, 1959.

Schmidt, A. X., and H. L. List, *Material and Energy Balances*, Prentice-Hall, Englewood Cliffs,
New Jersey, 1962.

4.5   PROBLEMS

1. By use of heat capacity data for liquid water, calculate (*a*) the enthalpy
    of liquid water at 89°C and 1 atm relative to 40°C and 1 atm and (*b*) liquid
    water at 210°F relative to 100°F and 1 atm.

2. One gram mole of air is heated from 25°C to 500°C. Calculate the change
    in enthalpy.

3. Use of the steam tables:
    (*a*)   What is the enthalpy change needed to change 5 kg of liquid water
          at 20°C to steam at 1 atm and 125°C?
    (*b*)   What is the enthalpy change needed to heat 2 kg of water at 50 psia
          and 30°C to steam at 150°C and 50 psia?
    (*c*)   What is the enthalpy change needed to heat 5 lb of water at 50 psia
          and 45°F to steam at 300°F and 50 psia?

4. Benzene is cooled in the heat exchanger shown below. How much heat
    must be removed by the cooling water per hour?

BENZENE 75°C ⟶   ⟶ BENZENE 25°C
1000 kg/hr

COOLING WATER

5. Calculate the amount of work that the pump must do in order to transfer the crude oil through the system shown below.

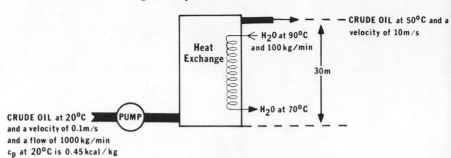

6. A chemist for a gas company finds that a new natural gas source analyzes as $91\% \ CH_4$, $4\% \ C_2H_6$, $2\% \ C_3H_8$, and $3\% \ N_2$. How much heat can be obtained from the combustion of $1000 \ ft^3$ of this gas measured at 1 atm and $60°F$? Assume that the combustion is complete.

7. Finely divided zinc sulfide is oxidized in a continuous adiabatic reactor according to the following reaction:

$$ZnS + 1.5\,O_2 = ZnO + SO_2$$

Air $(21\% \ O_2, 79\% \ N_2)$ is fed to the reactor at a rate of 145% of the theoretical amount required by the reaction. The solids in the reactor are sufficiently mixed by the air flow so that the reaction mass may be considered to reach a uniform constant adiabatic reaction temperature. The equilibrium constant at this elevated temperature is large, rates are high, and the solid product to be withdrawn is estimated at 96 mole $\% \ ZnO$, the rest is ZnS. The zinc sulfide and air enter the reactor at $40°C$, while the solid product and off-gases leave at the temperature of the adiabatic reactor.

(a) Calculate the adiabatic reactor temperature.

(b) The hot off-gases in such processes are used to generate steam in a waste heat boiler. If the off-gases are cooled to $150°C$ in the boiler, how much heat can theoretically be recovered per mole of ZnS feed?

8. In a small battery plant, $98\% \ H_2SO_4$ at $25°C$ is mixed with water at $20°C$ to make a $30\% \ H_2SO_4$ solution. The 30% solution is made up in 250-gallon batches. If the maximum allowed temperature of the final solution is $35°C$, how much heat must be removed per batch?

9. You are asked by your boss to prepare a material and energy balance for a small laboratory scale proces in which maleic anhydride is to be synthesized by the vapor phase air oxidation of benzene. The reaction is to be conducted in a tube reactor over a vanadium pentoxide catalyst

at 400°C and 1 atm of pressure. Air and benzene are fed at the correct stiochiometric ratio for the oxidation reaction. The conversion of benzene per pass is 50%, with a maleic anhydride yield of 80%. Assume that the remaining portion of the converted benzene goes to carbon dioxide. Thus, the reaction products can be assumed to consist of a three-component liquid mixture of maleic anhydride, unconverted benzene, and water of reaction, plus a gaseous mixture containing nitrogen (from the air feed), unreacted oxygen, and carbon dioxide. Consider that all of the reaction products are obtained at a temperature of 30°C and that all the reactants are initially at 25°C. Calculate a material and energy balance for the production of 1 kg of maleic anhydride per hour. (Note: Your balance should indicate the amount of heat required to bring the reactants to the reaction temperature, the amount of heat addition or removal required to keep the reaction temperature constant and the amount of heat removal required to bring the products to a temperature of 30°C.)

10. Using data from the literature, design a small pilot-plant-size process for producing 5 kg of aspirin per hour, by the reaction of salicylic acid with acetic anhydride. (Note: Your design should include a flow diagram, plus a complete material and energy balance.)

5
CHEMICAL TRANSPORT

A CHEMIST'S VIEW OF FLUID FLOW

When a chemical reaction is carried out, reagents must be brought together in a reaction vessel and the products must be removed. The method by which the reactants are added and the products removed depends primarily upon the physical state of the materials, the reaction conditions, and the quantities of materials involved. Typically in the laboratory, where batch-type reactions are performed, chemicals are generally transferred in the liquid state by physically pouring from one container to another. Likewise, gases are transferred through small-bore tubing by means of induced pressure differentials. Occasionally, solids may be added directly to or removed from a reaction vessel, but in general most of the chemical transfers will be of a fluid, either a liquid or gas.

Most chemical transfers in a chemical plant are also of materials in the fluid state. This is primarily by design, for it is easier and more economical to transfer large quantities of liquids and gases than solids. Thus a typical chemical plant consists of, in addition to a collection of processing equipment, an assembly of pumps, blowers, compressors, and a network of piping to transport fluids from one location to another. In general, chemists are not directly involved in the design and construction of such large scale fluid transfer systems. These problems are generally handled by engineers. However chemists often do become involved in fluid flow problems on a smaller scale.

In designing, constructing, and operating bench-top or pilot-plant-size continuous process equipment, the chemist will encounter numerous problems dealing with the flow of fluids. In the first place, in common with the chemical engineer, he is concerned with the transport of fluids from one location to another through pipes or tubing; this requires determination of the pressure drops in the system, and hence of the size pump required for pumping, selection of the most suitable pumps, and measurement of the flow rates. Further, in those processes where heat transfer or mass transfer to a flowing fluid occurs, the nature of the flow may have a profound effect on the transfer coefficient for the process. Therefore, industrial chemists should possess an understanding of the basic principles and relationships that govern the flow of liquids and gases in chemical processes.

We take the expression "fluid flow" to mean the movement of materials such as liquids, gases, or dispersed solids through certain bounded regions. Energy balances and material balances, along with the laws of fluid friction, constitute the basis of the principles of fluid flow. Application of these fundamentals provides methods for determining relationships between rates of flow and pressure drops in a given system. These relationships may be used to determine power requirements or flow rates for different types of equipment.

5.1   TYPES OF STEADY FLOW

When a fluid passes through a pipe at a steady mass rate of flow, the mass of fluid entering one end of the pipe in unit time must equal the mass of fluid leaving the other end of the pipe in the same unit time. As discussed in Chapter 3, this is merely a statement of the law of conservation of matter expressed for the conditions of no accumulation or depletion. Similarly, on the basis of unit time, the mass of fluid passing any total cross-sectional area of the pipe equal the mass of fluid flowing past any other total cross-sectional area of the pipe.

The flow of a fluid through a pipe can be divided into two general classes, streamline flow or turbulent flow, depending upon the type of path followed by the individual particles of the fluid. When the flow of all the fluid particles is essentially along lines parallel to the axis of the pipe, the flow is called streamline (also sometimes called viscous or laminar). When the course followed by the individual particles of the fluid deviates greatly from a straight line so that vortices and eddies are formed in the fluid, the flow is called turbulent.

The distinction between streamline flow and turbulent flow can be shown clearly by means of a simple experiment. The experiment is carried out by injecting a small stream of colored liquid into a fluid flowing inside a glass tube. If the fluid is moving at a sufficiently low velocity, the colored liquid will flow through the system in a straight line. Under these conditions, when there is no mixing of the two fluids, streamline flow exists. As the velocity of the liquid is increased little eddies and whirls will begin to appear. If the velocity is increased to even greater values complete mixing will occur, and under these conditions turbulent flow is said to exist.

Since significant differences exist between the characteristics of streamline and turbulent flow, it is essential when dealing with fluid flow problems that the type of flow be identified. Normally, this can be accomplished by use of a dimensionless parameter (note: dimensionless means that there are no physical units for the particular symbol or group) known as the Reynolds number.

Reynolds Number. The flow, whether streamline or turbulent, has been shown experimentally to depend on the inside diameter of the tube D, the velocity of the flowing fluid v, the density of the fluid ρ and the viscosity of the fluid μ. The numerical value of a dimensionless grouping of these four variables known as the Reynolds number serves to indicate whether the flow is streamline or turbulent. The expression for this grouping is given below:

$$\text{Reynolds number} = \text{Re} = \frac{Dv\rho}{\mu} \tag{5.1}$$

In the English system the units that must be used to give a dimensionless grouping are: pipe diameter in feet; velocity in feet per second; density in pounds per cubic foot; and viscosity in pounds per foot-second. If the metric system is used the units are: pipe diameter in centimeters; velocity in centimeters per second; density in grams per cubic centimeter; and viscosity in grams per centimeter-second. When the Reynolds number exceeds 2100 turbulent flow may exist, while streamline flow exists at Reynolds numbers less than about 2100. An example calculation using the Reynolds number expression is given in section 5.3.

Streamline flow may occur at Reynolds numbers higher than 2100 if the condition is obtained by gradually increasing the Reynolds number toward and then past the value of 2100. However, if the flow is originally turbulent, it will stay turbulent until the Reynolds number drops below about 2100.

5.2 THE ROLE OF VISCOSITY IN FLUID FLOW

The viscosity of a fluid is a property of the material by virtue of which it resists shearing forces. Fluids having low viscosities, such as water, offer less resistance to a shearing force than fluids having high viscosities, such as oils. A thin-bladed knife will easily cut through water, whereas much more force would be required to force the knife through a heavy oil at the same speed. The force exerted on the knife as it cuts through the fluid is a shearing force, and the difference in resistance offered by the two fluids is due to the difference in their viscosities.

The concept of viscosity is particularly important in considering the flow of fluids since the magnitude of the viscosity affects the resistance to flow offered by the fluid. A fluid having a low viscosity flows more freely than a fluid having a high viscosity.

The expression for viscosity may be obtained by considering two small parallel layers of fluid, each with an area A and a differential distance dx apart. Under these conditions, a certain shearing force must be exerted on the top layer to cause it to move parallel to the other layer at a relative differential velocity of dv. It has been observed experimentally that this force is directly proportional to the difference in velocities dv and to the area A of the layers and is inversely proportional to the distance dx between the layers. Equation 5.2 summarizes these observations.

$$\text{Force} = (\mu)(A)\left(\frac{dv}{dx}\right) \tag{5.2}$$

The proportionality constant is represented by μ and is defined as the viscosity.

For gases and most liquids the value of μ is constant if the temperature and pressure are fixed. Fluids of this type are called Newtonian fluids. The viscosity of such fluids can be calculated by rearranging Eq. 5.2 to give:

$$\text{Viscosity} = \mu = \frac{(\text{force})}{A/dv/dx} \tag{5.3}$$

As can be seen from Eq. 5.3 viscosity has the dimensions of mass/(length) · (time). In English units absolute viscosity has the dimensions of pounds/ (feet)(second). The absolute viscosity in metric units is called the poise and has the dimensions of grams/(centimeters)(seconds). Thus the poise is equivalent to 0.0672 English units lb/(ft) · (s) of viscosity. It will be found in industrial chemistry practice that absolute viscosity is most frequently quoted in units of centipoise. A centipoise is equivalent to 0.01 poise. Therefore, to convert centipoises to English units, multiply by 0.000672.

The viscosity of air at room temperature is approximately 0.02 centipoise, while the viscosity of water at room temperature is about 1 centipoise. Oils have viscosities ranging from 10 to 5000 centipoises, depending on the temperature and type of oil.

5.3 VELOCITY DISTRIBUTION IN PIPES

Because of the resistance encountered by a flowing fluid at the pipe-wall surface, the fluid particles at the wall surface may be considered to have no net forward velocity. This means that there is no slippage at the wall and that the velocity of the fluid is zero at the wall surface. The particles farther from the wall are less affected by this frictional resistance, and the maximum velocity of the fluid particles occurs at the center of the pipe. Figure 5.1 illustrates a cross-sectional view of the velocity distribution of a fluid flowing in a long straight pipe. As can be seen, the maximum velocity occurs at the center of the pipe, with the velocity gradually decreasing to zero at the wall.

The average linear velocity of flow through a pipe is taken as the flow, expressed as volume per unit time, divided by the cross-sectional area of the pipe. For example, if a fluid is flowing at the rate of 6 ft³/min through a pipe having an inside diameter of 2 in., the average linear velocity of the fluid is

$$\frac{(6 \text{ ft}^3/\text{min})(1 \text{ min}/60 \text{ s})}{(3.14)(1 \text{ in.})^2(\text{ft}/12 \text{ in.})^2} = 4.59 \text{ ft/s}$$

If this liquid has a density of 30 lb/ft³ and a viscosity of 0.002 lb/(s)(ft), the Reynolds number for this flowing liquid would be

$$\text{Re} = \frac{Dv\rho}{\mu} = \frac{(2 \text{ in.})(1 \text{ ft}/12 \text{ in.})(4.59 \text{ ft/s})(30 \text{ lb}/(s)(\text{ft})^3)}{(0.002) \text{ lb}/(s)(\text{ft}))}$$

$$\text{Re} = 11,470$$

Since the Reynolds number is greater than 2100, the flow in this example may be considered as turbulent. As an exercise, you should prove to yourself that the Reynolds number is indeed dimensionless by converting all of the parameters in the previous example to metric units and then calculating the Reynolds number.

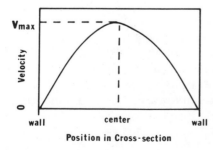

Figure 5.1 Velocity distribution of a fluid flowing in a straight pipe.

5.4 FLOW MEASUREMENTS

The measurement of flow rates in a continuous process is essential since the measured values enable the process to be examined and controlled. However the required accuracy of the measurements depends on the nature of the operation and the ultimate use of the data. Unnecessary accuracy raises the cost of equipment and complicates work, while insufficient accuracy minimizes the usefulness of the results. The optimum accuracy is usually a matter of judgment.

The flow rates to be metered in the laboratory are usually much smaller than those used in pilot plant work and certainly significantly smaller than the flow rates found in industrial plants. Some of the measuring devices discussed below can be used to measure flow rates in all three ranges, whereas others may only be used in one or two of the flow ranges. It should also be recognized that in an industrial plant, the measuring device frequently must supply a signal that can be transmitted to some remote site where the measurement is read. In a laboratory or pilot plant setup, it is often more practical to read the measuring device directly. Quite often in the pilot plant and industrial plant, the flow measuring device is connected to some type of flow control mechanism. A more extensive discussion about flow control is given in Chapter 9. The following section deals only with a brief discussion of some of the more common flow measurement devices.

Flow measurement is generally accomplished by a device that falls into one of the following classifications: constriction devices, velocity devices, or displacement devices.

5.4.1 Constriction Devices

In this type of flowmeter, a primary device creates a change in fluid velocity that is sensed as a differential pressure. The differential pressure depends on the velocity (velocity squared) and density of the flowing fluid. The measuring system consists of the primary element (the restriction) that produces a velocity change and a secondary element that measures the resulting differential pressure and relates it to flow. Some of the more common constriction devices are as follows.

Venturi Tube.   A venturi tube is shown in Fig. 5.2. A venturi tube is a piece of pipe in which the diameter has been decreased for a short distance in order to increase the speed of a fluid, which will increase its pressure. The magnitude of the pressure drop is related to the rate of flow. Venturi tubes are difficult to manufacture and are generally limited to applications where viscosity

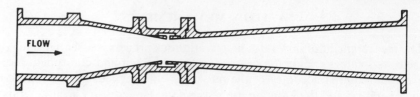

Figure 5.2 Sectional view of a venturi tube.

is high, where there is considerable suspended solids, or maximum accuracy is required.

Orifice Plate. This type of flowmeter is essentially a flat plate with a hole very accurately cut into it as shown in Fig. 5.3. The restriction to flow causes a pressure drop that will be related to the rate of flow in the pipe. As a given orifice can be used for only a relatively small range of flow, flanges are used to hold the plate in position and allow orifices with different sized holes to be inserted easily.

The disadvantage of the orifice plate is that it cannot be used with solids and that it has a higher pressure drop than some of the other constriction devices. It is, however, the most popular primary flow element because it can be manufactured cheaply, and generally the higher pressure drop is not a serious problem.

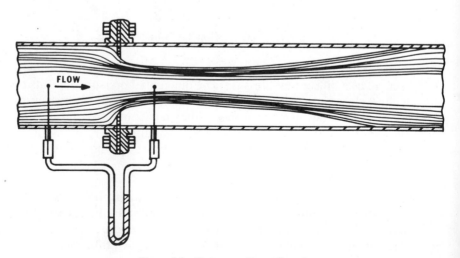

Figure 5.3 Cutaway of an orifice plate.

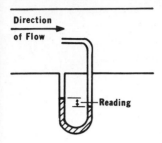

Figure 5.4 Schematic diagram of a simple pitot tube.

Pitot Tube. Figure 5.4 shows the general appearance of a simple pitot tube. It consists of one tube with its opening normal to the direction of flow and a second tube in which the opening is parallel to the flow.

The velocity of flow is calculated from the difference between the pressure at the opening parallel to the flow, registering the static pressure, and the pressure in the impact tube, called the stagnation or impact pressure. This device is satisfactory in many gas streams or clear liquids, but in general it must be calibrated in use by some other means and is not very accurate.

Rotometer. The rotometer is illustrated in Fig. 5.5. In each of the three meters discussed above, the area of the constriction was constant and the pressure drop varied with flow rate; in the rotometer, the pressure drop remains nearly constant and the area of the constriction varies. The fluid (gas or liquid) flows vertically upward through the tapered rotometer tube, and the float comes to equilibrium at a point where the annular flow area is such that the velocity increase has produced the necessary pressure difference. A higher flow rate causes the float to rise to a point where the annular area is larger. Flow rates in the range of 3 ml/min to 200 liters/min for gases and between 0.07 ml and 50 liters/min for liquids can be measured very conveniently by use of a rotometer.

In general, constriction type flowmeters are the most economical of all the flow rate devices. They perform well with both liquids and gases. Their major limitations are difficulty in reading low flow rates, a square-root relationship in readout, and inability to cope with certain types of fluids (e.g., slurries).

5.4.2 Velocity Devices

The most popular flowmeter in this classification is the magnetic flowmeter. As shown in Fig. 5.6, a magnetic flowmeter consists of a nonmagnetic insulated pipe in a magnetic field with a pair of electrodes in contact with the

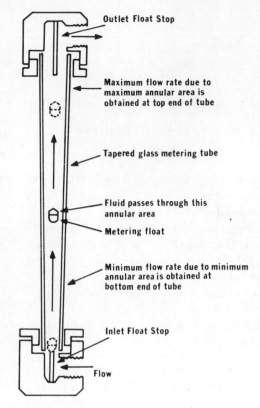

Outlet Float Stop

Maximum flow rate due to maximum annular area is obtained at top end of tube

Tapered glass metering tube

Fluid passes through this annular area

Metering float

Minimum flow rate due to minimum annular area is obtained at bottom end of tube

Inlet Float Stop

Flow

Figure 5.5 Schematic diagram of a roto-meter.

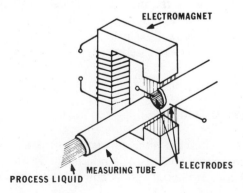

ELECTROMAGNET

PROCESS LIQUID MEASURING TUBE ELECTRODES

Figure 5.6 A magnetic flowmeter.

liquid within the pipe. As liquid flows through it acts as a moving conductor, cutting the magnetic lines of force and inducing a voltage between the electrodes. The magnitude of this induced voltage is directly related to liquid velocity. The magnetic flow transmitter generally consists of a stainless steel tube with an insulating liner, surrounded by two field coils (to generate the magnetic field). The chart readout is a straight-line relationship because the relationship between electrode voltage and flow rate is linear.

The magnetic flowmeter can be used with any liquid that has a specified minimum conductivity. This prevents its use with hydrocarbons or other nonconductive liquids, or with gases even when they are ionized. However, it works well with most liquids, particularly those that contain even a minute quantity of water. Liquids that contain suspended solids, such as slurries, are easily handled. Magnetic flowmeter outputs are often multiplied by density to measure mass flow.

5.4.3 Displacement Devices

Positive displacement meters measure discrete quantities of the flowing fluid and indicate flow in terms of an integrated or totalized flow volume. Two of the more popular devices that fall into this classification are the turbine flowmeter and the wet-test meter.

Turbine Flowmeter.   This device, as shown in Fig. 5.7, consists of a section of metal pipe, a multibladed rotor mounted in the center of the straight-through passage, and a magnetic pickup coil mounted outside the fluid passage. A shaft held in place by fixed radian vanes supports the rotor assembly. As flow through the transmitter spins the rotor, movement of each rotor blade past the face of the pickup coil changes the total flux through the coil, thereby inducing a pulse. The number of these pulses is proportional to total flow and can be employed directly in digital flow counting, or continuous in-line blending systems. The turbine meter is adaptable to a variety of fluids, but flowing materials containing solids require careful selection of a meter designed for this type of service.

Wet-Test Meter.   A schematic of a wet-test meter is shown in Fig. 5.8. In this measuring device a gas fills a rotating segment, and an equal volume of gas is expelled from another segment. The number of cycles is counted and recorded on a series of dials. This type of meter usually appears very bulky for the quantities they are measuring. This is because the linear velocity of a gas in a pipe is normally very high compared to that of a liquid, and the large volume is needed so that the speed of the moving parts can be reduced and wear minimized.

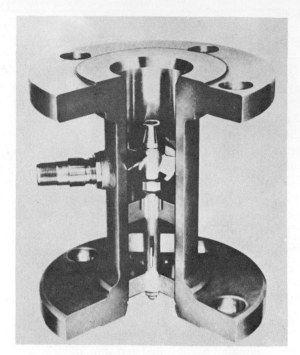

Figure 5.7 Turbine flowmeter. Photograph courtesy of the Foxboro Co.

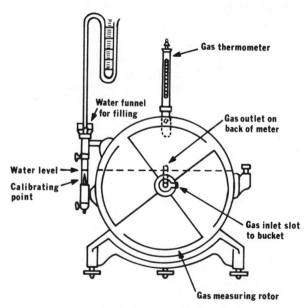

Figure 5.8 A wet-test meter.

5.5 ENERGY REQUIREMENTS IN FLUID FLOW

In the chemical industries fluids may have to be pumped long distances from storage to reactor units to separation units and back to storage. Associated with such transfers will often be a substantial loss in pressure both in the pipeline and in the actual units themselves. Consequently, calculations dealing with power requirements for pumping, with designs for the most suitable flow system and with estimates of the most economical sized pipe, constitute a most important stage in the design of a chemical plant. Normally, chemists are not involved in such large-scale design procedures. However, chemists often run into similar problems on a smaller scale when designing and assembling bench and pilot plant scale processes. Therefore, it is desirable for the industrial chemist to possess a basic understanding about the energy requirements involved in transferring a fluid from one point to another.

As discussed in Chapter 4, a fluid flowing through any type of conduit, such as a pipe, contains energy in three fundamental forms: potential, kinetic, and internal energy. Consequently, the energy balance expression developed in Chapter 4 (Eq. 4.2) can be applied to a flowing fluid. However, as discussed in Section 5.3, the point velocity of a fluid flowing through a pipe varies over the cross-sectional area of the pipe. Therefore, we need to correct the kinetic energy term in Eq. 4.2 for this occurrence.

When the velocity distribution is approximately uniform across the pipe, as in turbulent flow, only small errors are introduced in Eq. 4.2 if the kinetic energy term is expressed as $mv^2/2g$ where v is the average linear velocity. When the flow is streamline, the velocity distribution over the cross-sectional area of the pipe is parabolic, and the kinetic energy term may be expressed as mv^2/g. Since the average linear velocity v of a fluid flowing in a conduit can be determined easily, it is customary to use the following expression for the kinetic energy term:

$$\frac{mv_i^2}{2\alpha g}$$

where $\alpha = 1.0$ if the flow is turbulent and $\alpha = 0.5$ if the flow is streamline. Substituting into Eq. 4.2 gives the following common form for the total energy balance used in fluid flow problems:

$$mx_1 + \frac{mv_1^2}{2\alpha_1 g} + mP_1V_1 + mE_1 + Q = mx_2 + \frac{mv_2^2}{2\alpha_2 g} + mP_2V_2 + mE_2 + W$$

$$(5.4)$$

It should be remembered that this equation takes all types of energy changes into consideration, including surface effects, chemical effects, and frictional effects. It should also be remembered that the $E + PV$ terms can be combined into one enthalpy term.

5.6 APPLICATION OF THE TOTAL ENERGY BALANCE EQUATION TO FLUID FLOW SYSTEMS

Example 5.1

One pound-mole of dry nitrogen per second is introduced into a horizontal pipe. The nitrogen is at a temperature of 200°F and a pressure of 200 lb/in² and is flowing at an average linear velocity of 20 ft/s. The nitrogen leaves the pipe at an absolute pressure of 100 lb/in² and a temperature of 100°F. The average linear velocity of the nitrogen leaving the pipe is 100 ft/s. The mean heat capacity of nitrogen in this temperature range may be taken to be 7.0 Btu/(lb · mole)(°F). It may also be assumed that turbulent flow exists at all points in the pipe. Calculate the amount of heat lost from the walls of the pipe as British thermal units per second.

Solution

Step 1.   Choose a basis of 1 lb mole of dry nitrogen ≡ to 1 s of operation.

Step 2.   To solve this problem we must solve for the Q term in Eq. 5.4. Therefore the next step is to determine if enough information is given in the problem to evaluate all of the other energy terms. Since the pipe is horizontal, $x_1 = x_2$, and these terms disappear from the total energy balance.

Also since the system is neither doing any work nor is any work being done on the system, $w = 0$. Our energy balance equation reduces to:

$$Q = mE_2 + mP_2V_2 - mE_1 - mP_1V_1 + \frac{mv_2^2}{2\alpha_2 g} - \frac{mv_1^2}{2\alpha_1 g}$$

Step 3.   Let us group the $m(E + PV)$ terms and calculate the change in enthalpy as $m(H_2 - H_1) = m[(E_2 + P_2V_2) - (E_1 + P_1V_1)]$. We can calculate this enthalpy change as $m(H_2 - H_1) = (mCpm)(T_2 - T_1)$, where $m = 1$ lb · mole because the units of Cp are in pound-moles.

$$\therefore \quad m(H_2 - H_1) = (1 \text{ lb} \cdot \text{mole})\left(\frac{7 \text{ Btu}}{(\text{lb} \cdot \text{mole})(°F)}\right)(100°F - 200°F)$$

$$= -700 \text{ Btu}$$

Step 4. Solve for the kinetic energy terms. Since the flow is turbulent $\alpha = 1$. Also, 1 lb · mole of nitrogen weighs 28 lb. Therefore m is 28 lb.

$$m\left(\frac{v_2^2}{2g} - \frac{v_1^2}{2g}\right) = (28\text{ lb})\left(\frac{(100\text{ ft/s})^2 - (20\text{ ft/s})^2}{(2)(32.17\text{ ft/s}^2)}\right)$$

$$= 4.18 \times 10^3\text{ ft} \cdot \text{lb}$$

Since there are 778 ft · lb/Btu, this reduces to

$$\frac{(4.18 \times 10^3\text{ ft} \cdot \text{lb})}{(778\text{ ft} \cdot \text{lb/Btu})} = 5.4\text{ Btu}$$

Step 5. We can now solve for Q.

$$Q = -700\text{ Btu} + 5.4\text{ Btu} = -694.6\text{ Btu}$$

Since the basis was for 1 s, this is the amount of heat that is lost (i.e., the negative sign indicates loss from the system) from the walls of the pipe per second.

5.7 THE TOTAL MECHANICAL-ENERGY BALANCE EQUATION

Friction losses occur in all systems in which materials are flowing. When the friction losses are regarded as occurring within the flowing material, they are included in the E terms of Eq. 5.4. However, it has been found that friction losses are much easier to deal with if the total energy balance expression is written in a form involving only mechanical energies. This can be accomplished fairly easily, for a fluid flowing through a pipe will ordinarily not undergo a chemical change, and minor energy changes such as surface effects can usually be neglected. Making these assumptions allows one to set up a mechanical-energy balance expression for a system containing a fluid flowing at a steady rate.

If a unit weight of fluid entering a steady-flow system is assumed as a basis, then the total mechanical-energy balance between the entrance and exit points of the system may be expressed as shown in Eq. 5.5.

$$x_1 + \frac{v_1^2}{2\alpha_1 g} + P_1 V_1 + \int_1^2 P\,dV = x_2 + \frac{v_2^2}{2\alpha_2 g} + P_2 V_2 + \sum F + W \quad (5.5)$$

The dimensions of the individual energy terms in Eq. 5.5 are energy per unit weight of fluid, such as foot-pounds per pound mass (which reduces to the equivalent units of feet), or kilogram—meters per kilogram (or meters), or gram—centimeters per gram (or centimeters). In this form the terms in

Eq. 5.5 have the dimensions of length as of a column of fluid. For this reason these terms are frequently referred to as "head," such as feet, meters, or centimeters of the fluid. This common terminology, with pressure head for the term PV, velocity head for the term $v^2/2\alpha g$, and static head for x_1, must be applied with caution as it is frequently and erroneously applied to the terms of Eq. 5.5.

As mentioned above, the flow of a fluid through any system is always accompanied by friction. The friction causes loss in mechanical energy that could have done work but was lost as a result of irreversible changes occurring in the flowing fluid. The term $\sum F$ in Eq. 5.5 represents the loss in mechanical energy caused by frictional effects and is best defined as the amount of mechanical energy necessary to balance the output energy against the input energy in a total mechanical-energy balance. As we show in the next section, a knowledge of these frictional losses cannot be accurately determined except by actual experiment.

Mechanical energy may enter and leave a system in the form of kinetic energy $v^2/2\alpha g$ and potential energy x. Mechanical energy in the form of work W may be done on or by the system. The fluid entering the system adds mechanical energy because of the pushing force PV exerted by the adjacent fluid, and in a similar manner mechanical energy is delivered to the surroundings at the point where the fluid leaves the system.

When the system involves a compressible fluid such as a gas, there is usually a change in volume per unit mass as the fluid flows from one point to another. Such a volume change will occur when there are changes in temperature and pressure. When the volume of a fluid changes, a certain amount of work must be supplied or withdrawn. This work of expansion or contraction between points 1 and 2 in a system is expressed by the term $\int_1^2 P \, dV$ in Eq. 5.5. Many systems that the chemist and chemical engineer deal with involve noncompressible fluids, and therefore the $\int P \, dV$ is zero.

Before we get into the application of the total mechanical-energy balance, it is necessary that we learn more about how to handle the friction term in Eq. 5.5. The next section gives a brief discussion on this subject.

5.8 FRICTION LOSSES

5.8.1 Friction Loss in Straight Lengths of Pipe

When a fluid flows through a straight piece of pipe the amount of energy lost due to friction depends on the properties of the flowing fluid and the size of the pipe system. For the case of steady flow through long straight pipes of uniform diameter, the parameters that will affect the amount of frictional losses are the velocity at which the fluid is flowing, the density of the fluid, the viscosity of the fluid, the diameter of the pipe, the length of the pipe, and the degree of roughness of the inside surface of the pipe.

The mechanical energy lost as friction, represented by the letter F, is equal to the pressure drop over the system caused by friction, divided by the density of the flowing fluid. This equality is represented in Eq. 5.6, where

$$F = \frac{\Delta P_f}{\rho} \tag{5.6}$$

ΔP_f represents the pressure drop due to friction and ρ the density. In engineering practice the units of ΔP_f are pounds per square foot and in the metric system by kilograms per square meter. The units of ρ are pounds per cubic foot or kilograms per cubic meter. The value of P_f can be obtained by use of the Fanning expression given in Eq. 5.7.

$$\Delta P_f = \frac{2fv^2 L\rho}{gD} \tag{5.7}$$

In this equation f, often called the Fanning factor, represents the friction factor, v the average velocity of the flowing fluid, L the length of the pipe, g the gravitational constant, and D the diameter of the pipe. Eq. 5.7 is applicable only to systems in which the fluid density, viscosity, and linear velocity are constant.

The friction factor f is unitless and is based on experimental data and has been found to be a function of the Reynolds number and the relative roughness of the inside surface of the pipe. It is known that a rough pipe leads to a larger friction factor for a given Reynolds number than a smooth pipe does. If a rough pipe is smoothed, the friction factor is reduced. When further smoothing brings about no further reduction in friction factor for a given Reynolds number, the pipe is said to be hydraulically smooth.

Figure 5.9 shows several idealized kinds of roughness. The height of a single unit of roughness is denoted by k and is called the roughness parameter.

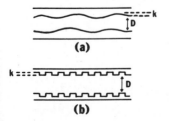

Figure 5.9   Types of roughness.

From dimensional analysis, f is a function of both Re and the relative roughness k/D, where D is the diameter of the pipe. For a given kind of roughness (e.g., that shown in Fig. 5.9a), it can be expected that a different curve of f versus Re would be found for each magnitude of the relative roughness and also that, for other types of roughness such as shown in Fig. 5.9b, a different family of curves of Re versus f would be found for each type of roughness. Experiments on artifically roughened pipe have confirmed these expectations. It has also been found that all clean new commercial pipes seen to have the same type of roughness and that each material of construction has its own characteristic roughness parameter. Old and corroded pipes are often very rough, and the character of the roughness differs from that of a clean pipe.

Roughness has no appreciable effect on the friction factor for laminar flow unless k is so large that the measurement of the diameter becomes uncertain.

For ease and convenience the frictional characteristics of round pipe, both smooth and rough, are summarized by the friction-factor chart, Fig. 5.10, which is a plot of log f versus log Re. This plot is sometimes referred to as Moody's Diagram.

Since roughness has no appreciable effect on the friction factor for laminar

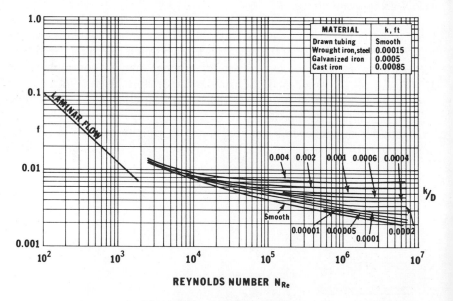

Figure 5.10 Friction-factor chart.

flow, a formula for f as a function of Reynolds number exists, and is shown in Eq. 5.8.

$$f = \frac{16}{\text{Re}} \tag{5.8}$$

A log–log plot of this equation is a straight line with a slope of -1. This plot line is shown in Fig. 5.10 for Reynolds numbers less than 2100.

For turbulent flow the lowest line represents the friction factor for a smooth pipe. The other curved lines in the turbulent-flow range represent the friction factors for various types of commercial pipe, each of which is characterized by a different value of k. The parameters for several common metals are given in Fig. 5.10. Clean wrought-iron or steel pipe, for example, has a k value of 1.5×10^{-4}, regardless of the diameter of the pipe. Drawn copper and brass pipe may be considered to be smooth.

To use the chart in Fig. 5.10 the Reynolds number must first be calculated from the rate of flow, the density, and viscosity of the fluid and the diameter of the pipe. Using the Reynolds number as the abscissa, the friction factor is read as the ordinate and is used in Eqs. 5.6 and 5.7 to calculate the desired friction loss. The friction loss so calculated, when combined with other forms of friction within the flow system, may then be substituted for the $\sum F$ term in Eq. 5.5.

5.8.2 Friction Loss from Changes in Velocity

Whenever the velocity of a fluid is changed, either in direction or magnitude, by a change in the direction or size of the pipe, friction in addition to the skin friction from flow through the straight pipe is generated. In most situations these frictional effects cannot be calculated precisely, and it is necessary to rely on empirical data. Often it is possible to estimate friction of this kind in specific cases from a knowledge of the losses in known arrangements.

The loss of mechanical energy as friction due to the sudden enlargement of the cross-sectional areas of the pipe in which the fluid is flowing can be calculated by use of Eq. 5.9.

$$F_e = \frac{(v_1 - v_2)^2}{2\alpha g} \tag{5.9}$$

The term v_1 is the average linear velocity of the fluid in the section preceding the enlargement, and v_2 is the average linear velocity of the fluid after the enlargement. The units of F_e are the equivalent of foot-pounds force per pound mass when the velocity is expressed in feet per second. When the flow is turbulent α should be taken as 1, and for laminar flow it is 0.5.

The loss of mechanical energy as friction due to sudden contraction of the cross-sectional area of the flow system under consideration can be calculated by the following equation:

$$F_c = \frac{K_c v_2^2}{2\alpha g} \tag{5.10}$$

The term v_2 is the downstream velocity, or the velocity in the pipe after the contraction has occurred, and K_c is called the contraction-loss coefficient. Experimentally, for laminar flow, $K_c < 0.1$, and the contraction loss F_c is negligible. For turbulent flow K_c is given by the empirical equation

$$K_c = (0.4)\left(1 - \frac{Sb}{Sa}\right) \tag{5.11}$$

where Sa and Sb are the cross-sectional areas of the upstream and downstream pipes, respectively.

5.8.3 Friction Loss Due to Pipe Fittings

Fittings and valves disturb the normal flow lines and cause friction. In short lines with many fittings the loss from the fittings may be greater than that from the pipe itself. The friction loss F_f from fittings is found from Eq. 5.12.

$$F_f = \frac{K_f \bar{v}_a^2}{2g} \tag{5.12}$$

The term K_f is the loss factor for the fitting, and $\bar{v}_a^2$ is the average velocity in the pipe leading to the fitting. Factor K_f is found by experiment and differs for each type of fitting. A short list of factors is given in Table 5.1.

Table 5.1   Loss Coefficients for Standard Threaded Pipe Fittings

Fitting	K_f
Globe valve, wide open	10.0
Gate valve, wide open	0.2
Gate valve, half open	5.6
Tee	1.8
Elbow, 90°	0.9
Elbow, 45°	0.4

5.8.4   Some Practical Aspects of Frictional Effects

Frictional effects are extremely important in flow processes. In many cases friction may be the main cause for resistance to the flow of a fluid through a given system. Consider the common example of water passing through a pipe. If no frictional effects were present, it would be possible to use pipes of very small diameters for all flow rates. Under these conditions the pumping power costs for forcing 100,000 gal of water per hour through a pipe with a diameter of $\frac{1}{8}$ in. would be the same as the power costs for forcing the same amount of water through a pipe of equal length having a diameter of 2 ft. However, frictional effects are present, and they must be taken into consideration when dealing with any real flow process.

When a fluid is to be pumped between two points, the diameter of the pipeline should be chosen so that the overall cost of operation is a minimum. The smaller the diameter, the lower is the initial cost of the line but the greater is the cost of pumping; an economic balance must therefore be achieved.

In deciding on a lower limit for the diameter of a pipe, a useful guideline to follow is that, in general, liquids of approximately the same density and viscosity as that of water should not flow in steel pipes at linear velocities higher than 4 or 5 ft/sec. For vapors such as steam at pressures in the range of 20–50 psi, linear velocity in steel pipes should not be greater than 50 to 60 ft/s.

5.9   CONVERSION OF UNITS IN THE MECHANICAL-ENERGY BALANCE EXPRESSION

As pointed out earlier, the conventional units for the individual energy terms in energy balance Eq. 5.5 are foot-pounds force per pound mass. However, it is often desirable to convert these units to pressure-drop units or to standard power units, such as the pressure drop due to friction in a steady-flow system, which can be calculated from the friction term $\sum F$. For example, suppose that the value of $\sum F$ has been calculated in the normal manner to be 150 ft·lb force/lb mass for water with an average density of 62.1 lb mass/ft^3. The pressure drop due to friction is then equal to the value of $\sum F$ multiplied by the average liquid density, or

$$\text{pressure drop} = \Delta P = (150 \text{ ft} \cdot \text{lb/lb})(62.1 \text{ lb/ft}^3)(1 \text{ ft}^2/144 \text{ in}^2)$$
$$= 64.6 \text{ lb/in}^2$$

Many practical flow problems require the determination of the size of motor necessary to pump a liquid at a set rate through a given system. The power requirements of the motor can be determined by the use of the total mechanical-energy balance. For example, suppose that the application of the total mechanical-energy balance over a system has indicated that the magnitude of the term w is -125 ft $\cdot$ lb/lb (the negative sign indicates that work is being supplied from an external source). In addition, suppose that the fluid in the system was a hydrocarbon mixture with a density of 40 lb/ft^3 flowing at a rate of 100 gal/min, and that the pump and motor operate with an overall efficiency of 50%. Then the horsepower supplied by the pump motor that is necessary to handle this flow may be calculated in the following manner.

From Appendix A we find that 7.48 gal = 1 ft^3, and 1 hp = 550 ft $\cdot$ lb force/s. Now the foot-pound force per second delivered to the fluid as work is equal to w multiplied by the mass flow.

$$(125 \text{ ft} \cdot \text{lb/lb})(40 \text{ lb/ft}^3)(100 \text{ gal/min})(1 \text{ ft}^3/7.48 \text{ gal})(1 \text{ min/60 s})$$

$$= 1114 \text{ ft} \cdot \text{lb/s}$$

The horsepower delivered to the fluid is

$$(1114 \text{ ft} \cdot \text{lb/s}) \cdot (1 \text{ hp/550 ft} \cdot \text{lb/s}) = 2.02 \text{ hp}$$

Since the equipment is only 50% efficient, the total horsepower delivered by the motor, or the power requirement of the motor is

$$2.02 \text{ hp/0.50} = 4.04 \text{ hp}$$

5.10 APPLICATION OF THE TOTAL MECHANICAL-ENERGY EQUATION

Example 5.2

A pump draws a solution with a density of 114.6 lb/ft^3 from a storage tank of large cross section through a 3-in. (inside diameter) pipe. The velocity in the suction line is 4 ft/s. The pump discharges through a 2-in. pipe to an overhead tank. The end of the discharge line is 75 ft above the level of the solution in the feed tank. Friction losses in the entire system are 20 ft $\cdot$ lb/lb mass of solution. What pressure must the pump develop in pounds per square inch to deliver this flow? What theoretical horsepower (i.e., 100% efficient) would the pump be required to have in this system? (Note: The flow may be taken to be turbulent.)

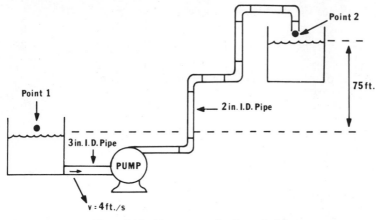

Figure 5.11 Flow system for Example 5.2.

Solution

Step 1. Choose a basis of 1 lb mass of flowing fluid. In applying Eq. 5.5 to this system the following steps have been taken:

$$x_1 + \frac{v_1^2}{2\alpha_1 g} + P_1 V_1 + \int_1^2 P \, dv = x_2 + \frac{v_2^2}{2\alpha_2 g} + P_2 V_2 + \sum F + W$$

Step 2. If the reference plane passes through point 1 then
$$x_1 = 0 \quad \text{and} \quad x_2 = 75 \text{ ft} \cdot \text{lb/lb}.$$

Step 3. Point 1 and point 2 are both at atmospheric pressure; $P_1 V_1 = P_2 V_2$ and they cancel from the energy balance.

Step 4. Since the surface area at point 1 is large, the velocity of the surface is approximately 0; therefore, $v_1^2/2\alpha_1 g = 0$. The velocity at point 2 can be calculated from the volume flow at the bottom tank exit.

$$\text{(Velocity)}_1 (\text{cross section area pipe entrance})$$
$$= \text{(velocity)}_2 (\text{cross section area pipe)}_2$$

$$(4 \text{ ft/s})(3.14)(0.125 \text{ ft})^2 = (v_2)(3.14)(0.0833 \text{ ft})^2$$

$$v_2 = \frac{(4 \text{ ft/s})(3.14)(0.125 \text{ ft})^2}{(3.14)(0.0833 \text{ ft})^2} = 9.01 \text{ ft/s}$$

hence,

$$\frac{v_2^2}{2\alpha g} = \frac{(9.01 \text{ ft/s})^2}{(2)(1)(32.17 \text{ ft} \cdot \text{lb/(s}^2)(\text{lb})} = 1.26 \text{ ft} \cdot \text{lb/lb}$$

Step 5. Since the fluid is incompressible,

$$\int_1^2 P \, dV = 0$$

Step 6. From the information given in the problem,

$$\sum F = 20 \text{ ft} \cdot \text{lb/lb}$$

Step 7. Putting the values calculated above into the energy balance equation gives:

$$0 + 0 + 0 + 0 = 75 \text{ ft} \cdot \text{lb/lb} + 1.26 \text{ ft} \cdot \text{lb/lb} + 0 + 20 \text{ ft} \cdot \text{lb/lb} + W$$

$$W = -96.26 \text{ ft} \cdot \text{lb/lb}$$

This is the work done on the system per pound of fluid. To calculate the work per unit time, we need to know the mass flow.

Step 8. The mass flow per second is calculated from the velocity, cross-sectional area, and the density.

$$\text{Volume flow} = (4 \text{ ft/s})(3.14)(0.125 \text{ ft})^2 = 0.196 \text{ ft}^3/\text{s}$$

$$\text{Density} = 114.6 \text{ lb/ft}^3$$

$$\text{Mass flow} = (114.6 \text{ lb/ft}^3)(0.196 \text{ ft}^3/\text{s}) = 22.46 \text{ lb/s}$$

Step 9. The work on a time basis is

$$W = (-96.26 \text{ ft} \cdot \text{lb/lb})(22.46 \text{ lb/s}) = -2162 \text{ ft} \cdot \text{lb/s}$$

Step 10. Since 1 hp is equal to 550 ft · lb/s, the theoretical horsepower of the pump is

$$\text{theoretical horsepower} = \frac{2162 \text{ ft} \cdot \text{lb/s}}{550 \text{ ft} \cdot \text{lb/(s)(hp)}} = 3.93 \text{ hp}$$

Step 11. To calculate the pressure that the pump must develop, you must remember that work = (force) · (distance) = (pressure) · (volume). This means that the work a pump does is equal to the product of its working pressure and the volume of liquid pumped.

In this problem the pressure is calculated as follows:

$$W = P \cdot V$$

$$P = \frac{W}{V} = \frac{2162 \text{ ft} \cdot \text{lb/s}}{0.196 \text{ ft}^3/\text{s}} = 11030.6 \text{ lb/ft}^2$$

or

$$\text{pump pressure} = (11030.6\ \text{lb/ft}^2)(1\ \text{ft}^2/144\ \text{in}^2) = 76.6\ \text{lb/in}^2$$

Example 5.3

In the system shown below water is pumped from a reservoir to a constant-level storage tank. Calculate the cost of operating the pump per hour, if the efficiency of the pump assembly is 40%. The water flow in the system is to be maintained at 100 gal/min. The temperature of the water is 60°F. The material of construction for the pipe is steel. Assume that the cost for electrical energy is 3.0 cents/kW-hr.

Solution

Step 1. Choose as a basis 1 lb of flowing water and then apply the total mechanical-energy balance equation between points 1 and 2. Some properties of water that we need to know are:

density of water at 60°F = 62.3 lb/ft^3
viscosity of water at 60°F = 1.12 cP

or

$$(1.12\ \text{cP})(6.72 \times 10^{-4}\text{lb/s} \cdot \text{ft} \cdot \text{cP}) = 7.52 \times 10^{-4}\text{lb/s} \cdot \text{ft}$$

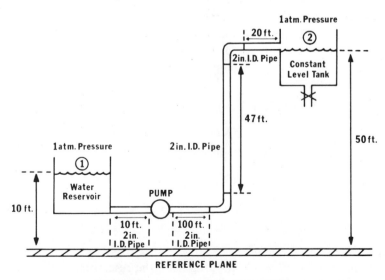

Figure 5.12 Flow system for Example 5.3.

Step 2. Solve for the velocity of the water in the 2-in. pipe.

$$\text{Vel} = (100 \text{ gal/min})(1 \text{ min/60 s})(1 \text{ ft}^3/7.48 \text{ gal})\left(\frac{1}{1 \text{ in.}}\right)^2\left(\frac{1}{3.14}\right)(144 \text{ in}^2/\text{ft}^2)$$

$$= 10.22 \text{ ft/s}$$

Reynolds number in 2-in. pipe is

$$\text{Re} = \frac{(D)(v)(\rho)}{(\mu)}$$

$$\text{Re} = (2 \text{ in.})\left(\frac{1 \text{ ft}}{12 \text{ in.}}\right)(10.22 \text{ ft/s})(62.3 \text{ lb/ft}^3)\left(\frac{1}{7.52 \times 10^{-4} \text{ lb/s} \cdot \text{ft}}\right)$$

$$= 141,114$$

Thus the flow is turbulent.

Step 3. Solve for the friction terms in the flow system. The total friction term $\sum F$ will consist of friction due to the total length of 2-in. pipe, to two 90° elbow fittings, to a contraction from the reservoir to the 2-in. pipe, and to an expansion from the 2-in. pipe to the constant-level tank. Therefore, the $\sum F$ term can be written as $\sum F = \sum F_{\text{pipe}} + \sum F_{\text{contraction}} + \sum F_{\text{expansion}} + \sum F_{\text{fittings}}$ where $F(\text{pipe}) = P_f/\rho = 2 \cdot f \cdot v^2 \cdot L/g \cdot D$. The friction factor f can be obtained from Fig. 5.10. The value of k for steel pipe is 0.00015 ft, and k/D for the 2-in. pipe is 0.0009. Using this value and a Re value of 141,114, the friction factor from Fig. 5.10 is 0.0051. Since the volume flow is constant throughout the system, the velocity of the water is the same in all sections of 2-in. pipe. Therefore the total length of 2-in. pipe can be lumped together to calculate the pipe friction term.

$$L = 10 \text{ ft} + 100 \text{ ft} + 47 \text{ ft} + 20 \text{ ft} = 177 \text{ ft}$$

$$\therefore \quad F(\text{pipe}) = \frac{(2)(0.0051)(10.22 \text{ ft/s})^2(177 \text{ ft})}{(32.17 \text{ ft} \cdot \text{lb/s}^2 \cdot \text{lb}) \cdot (2/12 \text{ ft})} = 35.17 \text{ ft} \cdot \text{lb/lb}$$

The term F_c due to contraction is calculated by use of Eq. 5.10, that is

$$F_c = K_c \cdot \left(\frac{v_2^2}{2\alpha g}\right)$$

where $K_c = (0.4)(1 - Sb/Sa)$.

In calculating the F_c value in going from the reservoir to the 2-in. pipe, the ratio of cross-sectional area Sb/Sa can be taken to be zero since the area of the reservoir Sa is so much greater than the area of the 2-in. pipe. There-fore, K_c will be equal to 0.4, and F_c will be equal to

$$F_c = 0.4 \left[\frac{(10.22 \text{ ft/s})^2}{(2)(1)(32.17 \text{ ft} \cdot \text{lb/s}^2 \cdot \text{lb})} \right] = 0.65 \text{ ft} \cdot \text{lb/lb}$$

The term F_e due to enlargement is calculated from Eq. 5.9,

$$F_e = \frac{(v_1 - v_2)^2}{2\alpha g}$$

At point 2, the surface is so large that the velocity can be taken to be zero. Since the flow in the pipe is turbulent, $\alpha = 1$.

$$F_e = \frac{(10.22 \text{ ft/s})^2 - (0)^2}{(2)(1)(32.17 \text{ ft} \cdot \text{lb/s}^2 \cdot \text{lb})} = 1.62 \text{ ft} \cdot \text{lb/lb}$$

The term F_f can be calculated from Eq. 5.12

$$F_f = \frac{(K_f)(\bar{v}a^2)}{2g}$$

From Table 5.1, K_f for a 90° elbow is 0.9. Hence

$$F_f = \frac{(0.9)(10.22 \text{ ft/s})^2}{(2)(32.17 \text{ ft} \cdot \text{lb/s}^2 \cdot \text{lb})} = 1.46 \text{ ft} \cdot \text{lb/lb}$$

Since there are two elbows in the system

$$\sum F_f = (2)(F_f) = (2)(1.46) = 2.92 \text{ ft} \cdot \text{lb/lb}$$

The total friction term $\sum F$ can now be calculated as follows:

$$\sum F = 35.17 + 0.65 + 1.62 + 2.92 = 40.36 \text{ ft} \cdot \text{lb/lb}$$

Step 4. Since both points 1 and 2 are at atmospheric pressure, the *PV* term is constant, and they cancel from the energy balance expression.

Step 5. Since the surface areas of the vessels at point 1 and 2 are so large with respect to the 2-in. pipe, the velocity of the liquid at the points can be taken to be zero. Therefore the $v^2/2\alpha g$ terms will be zero.

Step 6. Since water is a noncompressible liquid, $\int_1^2 P \, dV = 0$.

Step 7. Since point 1 is 10 ft and point 2 is 50 ft above the reference plane, $x_1 = 10 \text{ ft} \cdot \text{lb/lb}$ and $x_2 = 50 \text{ ft} \cdot \text{lb/lb}$.

Step 8.   The total mechanical-energy equation can now be written as

$$10 \text{ ft} \cdot \text{lb/lb} + 0 + 0 + 0 = 50 \text{ ft} \cdot \text{lb/lb} + 0 + 0 + 40.36 \text{ ft} \cdot \text{lb/lb} + w$$

$$w = -50 - 40.36 + 10 = -80.36 \text{ ft} \cdot \text{lb/lb}$$

Step 9.   The mass flow in the system is

$$\text{mass flow} = (100 \text{ gal/min})(1 \text{ min/60 s})(1 \text{ ft}^3/1 \text{ gal})(62.3 \text{ lb/ft}^3)$$

$$= 103.83 \text{ lb/s}$$

The work on a time basis is

$$w = (-80.36 \text{ ft} \cdot \text{lb/lb})(103.83 \text{ lb/s}) = -8344 \text{ ft} \cdot \text{lb/s}$$

This would be the work that the pump must supply if it was 100% efficient. Since the pump is only 40% efficient, it must supply:

$$w = (8344 \text{ ft} \cdot \text{lb/s})(1/0.4) = 20,860 \text{ ft} \cdot \text{lb/s}$$

Step 10.   We can now calculate the cost per hour of operation by using the conversion 1 kw = 738 ft · lb/s.

$$\text{Cost/hr} = (20,860 \text{ ft} \cdot \text{lb/s})(1 \text{ kw/738 ft} \cdot \text{lb/s})(3 \text{ cents/kw} \cdot \text{hr})$$

$$= 84.8 \text{ cents/hr}$$

5.11   FLUID FLOW EQUIPMENT

As indicated previously, the transportation of fluids in chemical processes is an extremely important operation. This section deals briefly with the equipment used in moving fluids, as the several previous sections dealt with the mechanics of fluid flow and pipe resistance.

5.11.1   Pipe

The first requisite in transporting a fluid is a channel through which the flow may take place. In the chemical industry this channel is normally satisfied by some form of pipe. The pipe used may be made of steel, steel alloys, cast iron, glass, glass- or plastic-lined steel, lead, plastic, aluminum, or copper. It is not uncommon to find several of these materials used in the same plant. The pipe chosen is generally the most economical of those materials that will be resistant to the corrosive effect of the chemicals being handled. Pipe of the proper size and wall thickness is then selected on the basis of temperature, pressure, and maximum allowable pressure drop.

The most commonly used pipe is made of carbon steel. Steel pipe was originally classed in three thicknesses for different operating pressures: standard, extra strong, and double-extra-strong. These three classes are now obsolete, and thicknesses follow a set formula, expressed as the "schedule number" as established by the American Standards Association. Ten schedule numbers are in current use: 10, 20, 30, 40, 60, 80, 100, 120, 140, and 160.

For all pipe sizes below 12 in. in diameter, schedule 40 is most often used, and schedule 80 is the next most popular. As an example of how these two differ, a 1-in. diameter low-carbon steel pipe in both grades would have an outside diameter of 1.315 in. In a schedule 40 grade the wall is 0.133 in. thick, whereas in schedule 80 the wall is 0.179 in. thick. Most chemical plants try to limit the number of schedules used in order to avoid confusion and excessive inventories. In small sizes, one inch and below, most plants do not use any wall thinner than schedule 40.

Stainless steel pipe is popular in the chemical industry because of its general resistance to corrosion. Stainless steel is classified by type number. Of the 300 series, which shows good general resistance to corrosion, the most popular is type 316 with type 304 being next in order. Type 304 is more economical than type 316, but the general resistance to corrosion is less. The 300 series has high chromium and nickel content. The major difference between the two is that type 316 has 16 to 18 % chromium, 10 to 14 % nickel, and 2 to 3 % molybdenum, whereas type 304 has 18 to 20 % chromium, 8 to 10 % nickel, and 2 % manganese.

Since glass resists the corrosive effects of almost all chemicals, glass pipe is often used in the chemical industry. It also has the advantage of permitting one to see what is going on in the pipe, which is often desirable, especially in certain pilot plant operations. However, the use of glass pipe does have some disadvantages, such as being subject to breakage and the relative expense. Therefore, steel pipe lined with glass, Kel-F, or Teflon is often used in the place of glass pipe. The lined pipe has all of the advantages of glass pipe except transparency. The lined steel pipe is much stronger than the glass pipe; and in pipe of large diameter, this becomes very significant.

5.11.2 Fittings

Threaded pipe is most commonly encountered in industry when small diameters are used (e.g., less than 6 in.). Sections of threaded pipe are joined together by fittings such as illustrated in Fig. 5.13. Couplings join successive straight lengths of pipe with no change in direction or size. When the size is to be reduced or enlarged, a reducing coupling is used. When the direction is to be changed, an elbow is used. If more than two branches of piping are

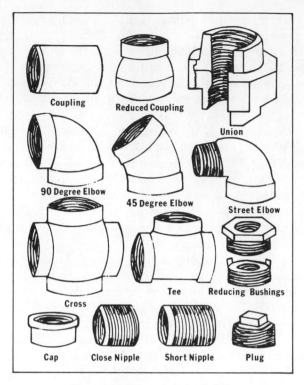

Figure 5.13 Threaded pipe fittings.

to be connected at the same point, tees and crosses are used. If a pipe must be broken at intervals for maintenance then a union is used as a connector. The two halves of the union are tightened to the pipe sections independently, and the final connection made by tightening the bonnet of the union. A reducing bushing is used to make a size reduction at a tapped connection. A simultaneous change in direction and connection to a tapped outlet may be made by a street elbow having male threads on one end and female threads on the other. The end of a pipe is closed with either a cap or a plug.

Carbon steel pipe up to 2 in. in diameter is generally threaded. From 2 in. to about 10 in. in diameter, it may be threaded or welded, but over 10 in. it is generally welded. Fittings for welding are available in the same shapes as shown in Fig. 5.13 for threaded pipe fittings.

Stainless steel fittings similar to the carbon steel fittings mentioned are available and are generally obtained in the same type of stainless steel as the pipe on which they are to be used. However, as it is difficult to thread stainless steel pipe and obtain a tight connection, it is generally welded.

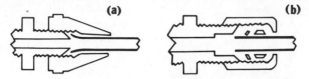

Figure 5.14 (*a*) Sectional view of a flared fitting; (*b*) sectional view of a compression fitting.

Compression or flared fittings are generally used with copper and stainless steel tubing. A schematic drawing of these two types is shown in Fig. 5.14. The fittings are available in tubing-to-tubing unions, elbows, tees, and so on, and are also available in pipe-to-tubing fittings and connectors with both male and female pipe ends.

5.11.3 Valves

The flow of fluid in a pipe is controlled by a valve or series of valves. In the chemical industry a number of different types of valves are used. Some of the more popular types are discussed below.

Gate valves are shown in Fig. 5.15. Gate valves have a gate, disc, or wedge

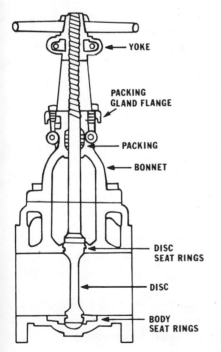

Figure 5.15 Sectional view of a gate valve.

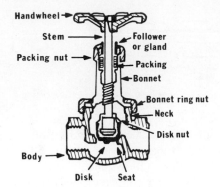

Figure 5.16 Sectional view of a globe valve.

that can be raised or lowered. When partially open this type of valve exhibits a crescent-shaped opening for flow that changes in area extremely rapidly with slight adjustment of the valve handle, thus making this type of valve rather undesirable for partial-flow control, although quite suitable for ordinary open-and-shut control.

The globe valve, so-called because of the bulbous shape of the valve body, is shown in Fig. 5.16. In this type of valve the direction of flow is changed so that it runs essentially parallel to the seating surfaces. This reduces erosion of the seating surfaces; but by changing direction of the flow several times, the pressure drop through the valve is considerable. An excellent partial-flow control valve is obtained if the globe valve is constructed with a tapered metallic plug that fits into a conical seat, the plug and the seat being of different taper to furnish a line contact for the seal. When this valve is made with a slender tapering needle seating in a small orifice drilled in the valve body, it is called a needle valve. Such a valve furnishes precise control and is excellent where a small stream is desired and a large drop in pressure is not disadvantageous.

The butterfly valve is most economical for use with large pipes. In this type of valve the opening gate rotates, rather than moving vertically.

5.11.4 Pumps

One of the most common and important pieces of equipment in a chemical plant is the pump. It is a machine that does work on a flowing fluid. The quantity of work is represented by the symbol w in the energy balance quation.

The majority of all pumps may be classified as centrifugal, reciprocating, and rotary.

Centrifugal pumps arc the most commonly used because of lower cost,

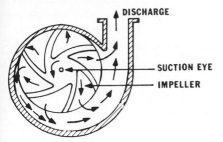

Figure 5.17   Sectional view of a centrifugal pump.

uniform flow, low maintenance cost, and ability to handle some undissolved solids. If liquids are to be pumped to great heights or to pressures higher than 100 psig other types of pumps become more economical.

Basically, the centrifugal pump consists of a rotating impeller inside a case. The liquid enters the center, or eye, of the impeller and the blades of the impeller throw the liquid out, giving the liquid a centrifugal force as shown in Fig. 5.17.

A typical centrifugal pump piping arrangement is shown in Fig. 5.18. The line from the feed vessel should be as short as possible. Normally, the valve on the suction side of the pump is a gate valve, and the valve on the discharge side a globe valve. The flow is regulated by opening or closing the globe valve. The pressure that the pump is putting out will be indicated by the pressure gage. The check valve is placed in the line to prevent liquid from backing up through the pump when the pump is turned off or accidentally stops running. The vent valve (generally a needle valve) is used to release any gases that might be vapor locking the pump and also to allow for the easy removal of samples.

The volume flow pumping capacity of a centrifugal pump can be altered by using different sizes and types of impeller and motors of different horsepower. For each pump the manufacturer provides performance curves that

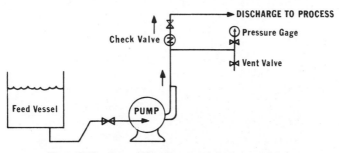

Figure 5.18   A centrifugal pump piping arrangement.

indicate the discharge pressure to be expected for a given rate of pumping, size of motor, and diameter of impeller. These curves can be used to determine if the specific pump in question can satisfactorily perform the duty for which it is intended.

Reciprocating pumps develop a higher pressure by the direct action of a piston or plunger on the fluid confined in a cylinder forcing the high-pressure fluid through the discharge valves. This type of pump is excellent for developing extremely high head pressures but is also relatively expensive if large flows are desired. It is also satisfactory for delivering a fixed amount of liquid in a given time in spite of varying pressure and temperature changes. However, since reciprocating pumps require both suction and discharge valves, this type of pump cannot be used on very viscous liquids or on liquids containing undissolved solids.

A typical piping arrangement when using a reciprocating pump is shown in Fig. 5.19. The rate of flow from the discharge line can be regulated by using the discharge line valve or by opening and closing the bypass valve. Since it is difficult to leak back through a reciprocating pump, it is better to use the bypass valve to reduce output rather than throttle back on the discharge valve. This is the reason for inserting a safety relief valve, as this type of pump will develop high pressures.

Output of a reciprocating pump may also be controlled by varying the speed of the drive, or by adjusting the setting of the connecting rod, thus adjusting the length of stroke of the piston. A single piston pump delivers a pulsating flow, but a two-piston and three-piston pump will deliver a smoother flow. A surge dampener pot may also be used to smooth the flow.

Rotary pumps include the entire field of pumps that combine rotating movement of the working parts with positive displacement. The rotating parts move in relation to the casing so as to create a space that first enlarges,

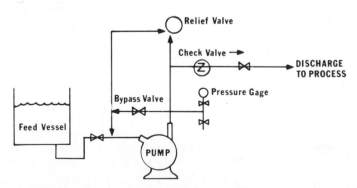

Figure 5.19   Typical reciprocating pump piping arrangement.

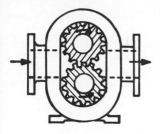

Figure 5.20   Sectional view of a gear pump.

drawing in the fluid in the suction line; is sealed; and then reduces in volume, forcing the fluid through the discharge port at a higher pressure. This sequence is illustrated by the gear pump shown in Fig. 5.20.

Rotary pumps may be divided into five main types according to the character of the rotating parts: gear, lobe, screw, vane, or cam. The gear pump is probably the most commonly used type. The gear pump is excellent for very viscous liquids, but it cannot handle liquids that contain hard or abrasive solids. Because of internal slippage, the gear pump is limited to about 150 psig on light liquids, but it may go up to 1000 psig on very viscous liquids. Generally, the rate of flow is controlled by a bypass valve or by varying the speed of rotation of the gears. The piping should be similar to that for the reciprocating pump (Fig. 5.19).

5.12   SUMMARY

This chapter was intended to acquaint you with the basic concepts in fluid flow and the associated equipment used to transfer liquids. In applying these concepts to prôblems in industry, you should be able to design a functional pipe layout for chemical processes on the bench and pilot plant scale level. You should also be able to choose the proper type of pump design and calculate the size required for the problem at hand.

5.13   SUPPLEMENTARY REFERENCES

Barna, P. S., *Fluid Mechanics for Engineers*, Butterworths, New York, 1951.

Brown, G. G., *Unit Operations*, Wiley, New York, 1950.

Considine, D. M., *Process Industries and Controls Handbook*, McGraw-Hill, New York, 1957.

Coulson, J. M., and J. F. Richardson, *Chemical Engineering*, Vol. 1, Macmillan, New York, 1964.

McCabe, W. L., and J. C. Smith, *Unit Operations of Chemical Engineering*, 2nd ed., McGraw-Hill, New York, 1967.

Schmidt, R. G., *Practical Manual of Chemical Plant Equipment*, Chemical Publishing Co., New York, 1967.

Stepanoff, A. J., *Centrifugal and Axial Flow Pumps*, Wiley, New York, 1957.

Streeter, V. L., *Fluid Mechanics*, McGraw-Hill, New York, 1958.

5.14 PROBLEMS

1. In the equipment shown in the figure below, a pump draws a solution, specific gravity = 1.2, from a storage tank through a 3-in. diameter steel pipe. The velocity in the 3-in. line is 4 ft/s. The pump discharges through

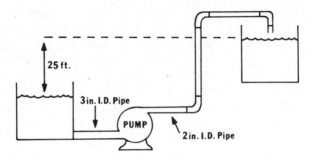

   a 2-in. diameter steel pipe to an overhead tank. The end of the discharge pipe is 25 ft above the level of the solution in the feed tank. Friction losses in the entire piping system are 15 ft · lb/lb. The efficiency of the pump is 50%.

   (*a*) What pressure must the pump develop in pounds force per square inch?

   (*b*) What horsepower must the pump have?

2. Ethylene glycol at 50°C is to be pumped from a large pilot plant operation to a storage tank at a rate of 100 gal/min. The transfer line from the pilot plant to the storage tank is 2-in. commercial steel pipe and is 1000 ft long. The fittings used in the line are as follows: ten 90° standard elbows, four tees (side run), and four gate valves. The end of the discharge line is 15 ft above the pipe entrance at the pilot plant. Calculate the theoretical horsepower of the centrifugal pump and the pump pressure used for pumping the ethylene glycol.

3. While working as chief chemist for a small company, you are asked to help in the design of a new process for recovering bromine from sea water. One of the pumps in this process is to have a sea water pumping capacity of 30,000 gal/min. The sea water has a specific gravity of 1.03 and is at a

temperature of 70°F. The water is pumped through a line 1000 ft long to a tower whose top is 50 ft above sea level.

    (*a*) What size pipe would you recommend for pumping the sea water to the tower? (Hint: Look back in this chapter for the maximum recommended pumping speeds of liquids.)

    (*b*) What pressure would the pump be required to develop?

    (*c*) With an overall efficiency of 65%, what should be the horsepower of the pump's electric motor?

4. Nitrogen is flowing with a velocity of 150 m/s at 20°C and an absolute pressure of 10 atm in a horizontal pipe. The nitrogen passes through an orifice. At a point slightly downstream from the orifice, the linear velocity of the nitrogen is 250 m/s and the absolute pressure is 7 atm. Considering nitrogen as an ideal gas, what is the temperature at this point? The mean heat capacity at constant pressure for nitrogen over the temperature range involved can be taken as 5.98 Cal/(g · mole) (°C). The flow can be assumed to be turbulent throughout the entire system.

5. A special liquid hydrocarbon mixture is to be used in a liquid extraction tower. The preliminary design of the unit requires the mixture to be pumped from a storage tank that is under 25 psia pressure. The liquid level in the tank is 5 ft above the floor. The mixture is to be pumped from the storage vessel through 100 ft of 2 in.-inside-diameter steel pipe with six 90° elbows into the top of the extraction tower, 25 ft above the floor level. The operating pressure in the tower is to be 75 psia. The rate of hydrocarbon flow through the tower is estimated to be 25 gal/min. The viscosity of the mixture is 18 cP, and its density is 55.1 lb/ft^3. Assuming the pump assembly operates with an overall efficiency of 52%, what horsepower input will be required for the motor? If the cost of electricity is 3 cents/kW-hr, what is the power cost per hour for operating this extraction tower?

6. As a member of a pollution control group, you have been assigned the responsibility for obtaining solvent extraction data on the separation of copper from a waste water stream. Your bench scale study using a batch extraction technique shows that a 5% solution of 8-quinoline in benzene results in a 98% extraction of copper from the waste stream when the volume ratio of benzene to waste is 1:50. You are now prepared to test the extraction on a continuous basis by using countercurrent contact. The scale of your continuous process has been set at an extraction rate of 1 gal of waste solution per minute. Draw a flow diagram for the process. Assign a set of practical specifications to all components in the fluid flow part of the process (e.g., pipe sizes, pipe lengths, materials of construction, type of fittings, pump types, pump sizes, relief valves, etc.).

Estimate the total material cost to construct the fluid flow part of this process, excluding the extraction tower. (Note: Price data for pumps, fittings, pipe, etc. can be obtained from industrial equipment suppliers' catalogs. A list of equipment manufacturers can be found in *Lab Guide*, a yearly publication of the American Chemical Society.)

7. What type of materials of construction would you specify for a fluid flow system in which the following liquids were being handed?

   (*a*) Aqueous HF at 100°C

   (*b*) Concentrated nitric acid at 50°C

   (*c*) Concentrated hydrochloric acid at 30°C

   (*d*) Molten *p*-dichlorobenzene

   (*e*) A gaseous chlorine stream

   (*f*) Glacial acetic acid at 40°C

   (*g*) 100% H_2SO_4 at 30°C

8. Specify a flow measuring device that would be suitable for use in each of the following systems:

   (*a*) A gas stream flowing at 50 l/min, in a 1.0-cm i.d. stainless steel tube

   (*b*) A water stream flowing at 1 l/min. in a 1.5-cm i.d. copper tube

   (*c*) A slurry flowing at 3 l/min. in a 2.0-cm i.d. glass tube

   (*d*) Molten polyethylene flowing in a 2-in. schedule 40 steel pipe

9. R. A. Nash [*Chem. Tech.*, April, 1976, p. 240] describes some of the problems that arise when a bench-scale procedure for formulating a product is scaled-up to pilot plant size. As indicated in Nash's article, many of the scale-up problems in product formation involve material transfer phenomena. Using the modern vanishing cream formulation given by Nash, describe the material transfer problems you would expect to encounter in going from a bench size cream formulation process to a pilot plant size process. Indicate how you would solve these problems.

10. Ethylene glycol at 25°C is draining by gravity from the bottom of a tank. The depth of liquid above the drawoff connection in the tank is 15 ft. The drawoff line is made of 2-in. schedule 40 pipe and is 100 ft. long. The line contains three elbows, one union, one gate valve, and one globe valve. The ethylene glycol discharges into an open container where the discharge point is 25 feet below the drawoff connection of the tank. What will be the ethylene glycol discharge rate in gallons per hour?

6

HEAT TRANSFER

WHICH WAY DOES IT GO?

There are three major chemical operations that involve the movement of heat:

1. Heat transfer requirements of chemical reactants. Chemical reactions are usually carried out at temperatures greater than the surroundings. This is often done to increase the rate of the reaction, or to shift the equilibrium in favor of the production of more product. Heat must therefore be supplied to the reactants in order to raise them to the reaction temperature.

For certain reactions (e.g., where one of the reactants may be thermally unstable in the feed mixture at the temperature of the surroundings), it may be necessary to cool the reactants to a temperature below the surroundings. In such a case it is necessary to remove heat from the feed stream.

2. Heat transfer during the course of the reaction. Most chemical reactions are either exothermic or endothermic. Consequently, in order to maintain control of the temperature of the reaction mixture, heat must either be removed or added to the chemical reactor.

3. Heat transfer requirements of chemical products. When the products of a reaction exit from the reactor, two basic needs for heat transfer may exist. One need may be to cool the products to the temperature of the surroundings before they leave the plant. The other need involves the recognition that the desired product of a chemical reaction often leaves the reactor mixed with unconverted reactants, by-products, and so on. In such cases the product stream requires some type of separation operation to be performed on it. All of the major separation processes, such as distillation, crystallization, absorption, involve the addition or removal of energy in the form of heat.

The average chemistry student may recognize that heat transfer is an important operation in industrial chemical processes, but most believe that its method of application is the domain of study for chemical engineers. However, this is not true. There are three primary reasons why industrial chemists need to have more than just a superficial understanding of heat transfer concepts.

1. The first reason deals with safety factors. Chemists often do not appreciate the importance of heat transfer principles because they have become accustomed to running reactions in 100 to 500 ml flasks, where the surface area of the reaction vessel in relation to the mass of the reaction mixture is large. In such cases the heat of reaction is liberated through the walls of the container without any difficulty. However, suppose that it becomes necessary to scale up the reaction so that a 22-liter reaction vessel is required. As we shall see in later discussions, the heat transfer surface area in relation to reaction mixture mass is reduced significantly during scale-up. Consequently, serious problems could result if provisions for satisfactory heat removal are not made.

Very often in such scale-up situations, because the rate of heat transfer is quite different from that normally experienced by chemists in typical laboratory situations, the unsuspecting chemist is not prepared for the tremendous buildup in temperature in the reaction vessel. The buildup in temperature also helps to compound the heat transfer problem because of its effect on the rate of the reaction. As indicated in Chapter 7, a rule of thumb often used is that the rate of a reaction doubles for every 10°C increase in temperature. Now we can understand how an exothermic reaction can become a runaway case during scale-up. If sufficient heat removal equipment is not provided for the temperature of the reaction mixture increases, thereby increasing the reaction rate—with a resulting increase in the amount of heat generated. This additional heat increases the temperature, and this increases the reaction rate to an even higher level. Such a sequence of events leads to loss of temperature control, and a runaway reaction results. In a confined container runaway reactions often lead to explosions and fires. Therefore it is essential that chemists know when and how to utilize the proper type and amount of heat transfer equipment.

2. Chemists working with materials that are subject to thermal decomposition should have a basic understanding about heat transfer concepts. Very often, improper use of heat transfer equipment when operating chemical reactors, distillation units, and the like can lead to localized regions of high temperature. If the temperature at these "hot spots" is great enough to cause thermal decomposition of the materials in contact with them, then the main product stream may become contaminated with undesirable decompo-

sition products. Therefore it is important that chemists know how to add and remove heat from chemical process equipment without creating undesirable temperature extremes.

3. The third factor, and one that is becoming increasingly more important, involves the cost of heat energy. In the chemical industry heat is produced primarily by burning either natural gas, heavy fuel oil, or coal. While electrical energy is the primary source of heat used by chemists in laboratory experiments, it finds only limited use in chemical process plants.

The relative costs of heat energy vary widely from place to place because of differences in transportation costs due to varying distances between use and sources of supply. However, one common feature about all heat costs is that they have been increasing very rapidly in recent years. This is primarily due to the fourfold increase in the price of crude oil during the last five years. Table 6.1 compares typical fuel costs for 1971 and 1976. The costs are given for Gulf Coast locations. Fuel costs for the Northeast would be higher.

In looking at the data in Table 6.1, it becomes apparent why chemical companies have become very concerned about rising heat costs. Many companies have set up extensive programs designed to locate and eliminate practices, procedures, and equipment in their chemical operations that waste heat energy.[1] Such programs in some companies have resulted in as much as a 20% reduction in heat energy requirements. In addition to eliminating practices that waste heat, companies are now very interested in developing chemical processes that are low energy consumers. Therefore chemists working in the area of process development must be acutely aware of heat transfer concepts and the effect they have on the cost of producing a chemical product.

Before leaving this section, you should note that Table 6.1 shows why electrical energy is not used industrially for heating unless the need for it is

Table 6.1   Typical Fuel Costs of Heat (Louisiana-Texas Gulf Coast)

Fuel	Unit Fuel Costs		Fuel Costs per Million Btu of Heat	
	1971	1976	1971	1976
Natural gas	$0.20/1000 ft^3	$1.44/1000 ft^3	$0.20	$ 1.44
Coal	$8.00/ton	$25.00/ton	$0.32	$ 1.00
Bunker C. fuel oil	$2.00/bbl	$8.00/bbl	$0.30	$ 1.20
Electricity	0.7¢/kw-hr	3.0¢/kw-hr	$2.36	$10.13

critical. For example, heat from electrical energy is eight times more expensive than from Bunker C fuel oil. Electrical heating is used only when its unique advantages justify its costs, such as when it is necessary to develop high temperatures to attain a high degree of control or when it is necessary to keep products of combustion out of the furnace system.

6.1   THE NATURE OF HEAT FLOW

The concept of heat is dependent upon its transfer from a body at a high temperature to another at a lower temperature, and, accordingly, heat may be defined as the energy which is so transferred. The net flow of heat is always in the direction of the temperature decrease. Heat is transferred by three basic methods: conduction, convection, and radiation.

Conduction is the transfer of heat from one part of a body to another part of the same body, or between two bodies in physical contact without significant displacement of the particles of the body. The mechanism of heat flow is one where the momentum of individual molecules is transported along the temperature gradient.

Convection is the transfer of heat from one point to another within a fluid, or between a fluid and a solid or another fluid, by the movement or mixing of the fluids involved. Since convection is a macroscopic phenomenon, it can occur only when forces act on the particle or stream of fluid and maintain its motion against the forces of friction. The identification of convection with heat flow is a matter of convenience because, in practice, it is difficult to separate convection from true conduction when they both are occurring at the same time.

Radiation is a term given to the transfer of heat by the emission of electromagnetic radiation from one material and its absorption by another material. In general, radiation becomes important at high temperatures.

In most systems the actual transfer of energy as heat is accomplished by more than one of these three modes of transfer. However, conduction-convection and radiation can be studied separately and their separate effects added together in cases where more than one is important.

6.2   HEAT TRANSFER BY CONDUCTION

6.2.1   Basic Equation

The instantaneous rate of heat transfer through a homogeneous body by conduction is directly proportional to the temperature-difference driving force across the body and to the cross-sectional area of the body at right angles to the direction of heat flow. The rate of heat transfer is inversely proportional to the thickness of the body along the length of the path through which heat flows. These proportionalities can be expressed mathematically in what is known as Fourier's law:

$$\frac{dQ}{dt} = -kA\left(\frac{dT}{dx}\right) \tag{6.1}$$

where dQ/dt = rate of heat flow
$\quad\quad T$ = temperature
$\quad\quad x$ = length of heat flow path
$\quad\quad A$ = area of surface through which heat is transferred
$\quad\quad k$ = proportionality constant

The negative sign reflects the physical fact that heat flow occurs from hot to cold regions and therefore the sign of the gradient is opposite that of the heat flow. In using Eq. 6.1 it must be clearly understood that the area A is that of a surface perpendicular to the flow of heat, and distance x is the length of path measured perpendicularly to area A.

The proportionality factor k is defined as the thermal conductivity of the material through which the heat is flowing. Fourier's law states that k is independent of the temperature gradient but not necessarily of temperature itself. This law has been validated by experiment for many substances. In comparing the values of k given in Table 6.2, it can be seen that k is not a strong function of temperature, and for small temperature ranges k may be considered constant.

In American engineering units the units of k are (Btu/hr · ft · °F) and in the centimeter-gram-second system the units are (cal/s · cm · °C). In general, then, the thermal conductivity is a measure of the number of British thermal units of heat conducted in 1 hr through 1 ft^2 of area measured perpendicular to the direction of heat flow when the gradient of temperature in the direction of heat flow is 1°F per ft. A comparison of k for different substances in Table 6.2 shows that insulating materials like glass fiber have small k values of the order of 0.01, whereas good heat conductors such as copper metal has a k equal to 220 at 100°F.

6.2.2 Steady Flow of Heat in Homogeneous Bodies

In most industrial chemical processes heat is transferred from one point to another under steady conditions of temperature difference, length of heat-flow path, and cross-sectional area. This state is known as the condition of steady heat flow. For example, a flat furnace wall initially at equilibrium

Table 6.2   Thermal Conductivities k for Various Substances[a]

Substance	k,Btu/ hr·ft·°F		
	0°F	100°F	200°F
Metals: (k is in general range 10-200)			
Copper	224	220	218
Aluminum	114	117	119
Mild steel	26.6	26.2	26.0
Stainless steel (304)	8.1	8.55	9.0
Non-metals: (k is in general range 0.01-0.2)			
Asbestos (36 lb/ft^3)	0.082	0.097	0.110
Glass fiber (9 lb/ft^3)	0.010	0.011	0.020
Glass fiber (3 lb/ft^3)	0.013	0.015	0.025
Fused silica - alumina (110 lb/ft^3)	0.24	--	0.27
Fire brick (125 lb/ft^3)	0.47	--	0.62
Liquids: (k is in general range 0.01-1)			
Water	0.298	0.350	0.402
Benzene	0.10	0.09	0.08
Acetone	0.108	0.103	0.0975
Gases: (k is in general range 0.001-0.1)			
Air	0.0108	0.0141	0.0162
Methane	0.0165	0.0205	0.025

[a]Values obtained from Reference 2.

with the air, at 75°F, is shown in Fig. 6.1. The initial state of temperature equilibrium is represented by line I. Assume now that one side of the wall is suddenly exposed to hot gases at 1000°F. If there is negligible resistance to heat flow between the gas and the wall, the temperature at the gas side of the wall immediately rises to 1000°F, and heat flow begins. After the elapse of some time t, the temperature distribution may look like that represented by curve II. Finally, if the wall is kept in contact with hot gas and cool air, the temperature distribution shown by line III is obtained. This temperature distribution remains unchanged as long as the furnace is kept in active use. In this case, the difference between the temperature at the hot gas and air walls of the furnace remains the same, and a condition of steady flow of heat exists.

For common cases of steady heat flow, Eq. 6.1. may be expressed as

$$\frac{dQ}{dt} = \text{constant} = \frac{Q}{t} = \frac{k_m A \Delta T}{x} \tag{6.2}$$

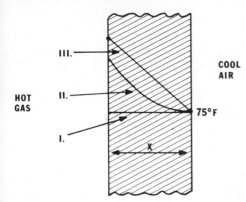

Figure 6.1 Temperature distributions in a flat furnace wall. I, Before exposure to heat; II, during heating at time t; III, at steady heat flow.

where Q represents the total amount of heat transferred in time t. The temperature difference driving force, taken as the higher temperature minus the lower temperature, is indicated by ΔT, and k_m is the average thermal conductivity of the material through which the heat is transferred. Since the thermal conductivity of most homogeneous substances varies almost linearly with temperature, k_m is evaluated at the average temperature of the material. The application of Eq. 6.2 is illustrated in Example 6.1.

Example 6.1

A layer of glass fiber (3 lb/ft^3) 3-in. thick is used as thermal insulation in the walls of a laboratory drying oven. The surface temperature of the cold side of the insulation is 100°F, and that of the warm side is 200°F. The total wall area of the oven is 24 ft^2. What is the rate of heat flow through the walls of the oven in British thermal units per hour?

Solution

The arithmetic mean thermal conductivity k_m of the glass fiber between 100 and 200°F is

$$k_m = \frac{0.015 + 0.025}{2} = 0.020 \text{ Btu/hr} \cdot \text{ft} \cdot °\text{F}$$

Also,

$$A = 24 \text{ ft}^2; \qquad \Delta T = 200 - 100 = 100°\text{F}; \qquad x = \tfrac{3}{12} = 0.25 \text{ ft}$$

Substituting these values into Eq. 6.2 gives

$$\frac{Q}{t} = \frac{(0.020 \text{ Btu})}{\text{hr} \cdot \text{ft} \cdot °\text{F}} \cdot (24 \text{ ft}^2) \cdot \frac{(100°\text{F})}{(0.25 \text{ ft})} = 129 \text{ Btu/hr}$$

6.2.3   Series Resistance to Flow of Heat

As described in the preceding paragraphs, when heat flows through a homogeneous body by conduction, there is a certain resistance to the flow determined by the cross-sectional area, the thermal conductivity, and thickness of the material. In order to establish a certain rate of heat flow, this resistance must be overcome by setting up a certain temperature-differential driving force.

Now, if one solid is placed in series with another solid so that heat must pass through both of them, the resistance to the flow of heat will be greater than it would be for one of the solids alone. For example, suppose that three solids of equal cross-sectional area are placed in series as shown in Fig. 6.2. When a steady flow of heat exists, exactly the same amount of heat per unit time must be transferred through each of the solids. This can be expressed mathematically as:

$$\left(\frac{Q}{t}\right)_1 = \left(\frac{Q}{t}\right)_2 = \left(\frac{Q}{t}\right)_3 = \left(\frac{Q}{t}\right)_{total} \tag{6.3}$$

The rate of heat transfer through solid 1, according to Eq. 6.2, must be

$$\left(\frac{Q}{t}\right)_1 = \left(\frac{k_m A \, \Delta T}{x}\right)_1 = \left(\frac{k_m A}{x}\right)_1 \Delta T_1 \tag{6.4}$$

Also,

$$\left(\frac{Q}{t}\right)_2 = \left(\frac{k_m A}{x}\right)_2 \Delta T_2 \tag{6.5}$$

$$\left(\frac{Q}{t}\right)_3 = \left(\frac{k_m A}{x}\right)_3 \Delta T_3 \tag{6.6}$$

The total temperature difference across the three solids must be the sum of the individual temperature differences across each of the solids, or

$$\Delta T_{total} = \Delta T_1 + \Delta T_2 + \Delta T_3 \tag{6.7}$$

Combining Eqs. 6.4 to 6.7 gives

$$\Delta T_{total} = \frac{(Q/t)_1}{(k_m A/x)_1} + \frac{(Q/t)_2}{(k_m A/x)_2} + \frac{(Q/t)_3}{(k_m A/x)_3} \tag{6.8}$$

Since Eq. 6.3 shows that $(Q/t)_1 = (Q/t)_2$ and so on, we can write Eq. 6.8 as follows:

$$\left(\frac{Q}{t}\right)_{total} = \frac{\Delta T_{total}}{[1/(k_m A/x)_1 + 1/(k_m A/x)_2 + 1/(k_m A/x)_3]} \tag{6.9}$$

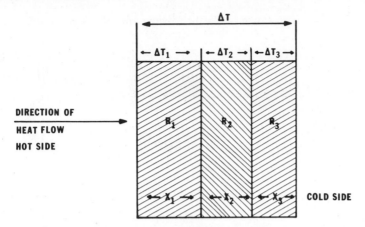

Figure 6.2   Thermal resistances in series.

Since the term $(x/k_m A)$ represents the resistance encountered by heat flowing through the material, the symbol R is often used to represent the grouping of these parameters. Therefore Eq. 6.9 can be written as,

$$\left(\frac{Q}{t}\right)_{total} = \frac{\Delta T_{total}}{[R_1 + R_2 + R_3]} \tag{6.10}$$

where $R_1 = (x/k_m A)_1$, and so on. Equation 6.10 shows that in heat flow through a series of layers, the overall resistance equals the sum of the individual resistances. Example 6.2 illustrates the use of Eq. 6.10.

Example 6.2

The flat wall of a laboratory scale catalyst regeneration furnace is constructed of a 6.0-in. layer of fire brick, backed by a 10.0-in. layer of asbestos fiber. The surface temperature in the furnace is 1500°F, and the surface temperature of the outer wall is 150°F. The total wall area for this furnace is 50 ft². (a) How much heat is lost per hour from the furnace? (b) What is the temperature of the interface between the brick and the asbestos?

Solution

(a) The rate of heat loss is given by Eq. 6.10

$$\frac{Q}{t} = \frac{\Delta T_{total}}{R_B + R_A}$$

where R_B is the resistance of the fire brick and R_A is the resistance of the asbestos.

$$R_B = \left(\frac{x}{k_M \cdot A}\right)_{\text{brick}} = \left(\frac{6/12 \text{ ft}}{(0.62 \text{ Btu/hr} \cdot \text{ft} \cdot {}^\circ\text{F})(50 \text{ ft}^2)}\right)$$

$$R_B = 0.016(\text{hr})({}^\circ\text{F})/\text{Btu}$$

$$R_A = \left(\frac{x}{k_m \cdot A}\right)_{\text{asbestos}} = \left(\frac{10/12 \text{ ft}}{(0.110 \text{ Btu/hr} \cdot \text{ft} \cdot {}^\circ\text{F})(50 \text{ ft}^2)}\right)$$

$$R_A = 0.151 \; (\text{hr})({}^\circ\text{F})/\text{Btu}$$

The values of k_m were selected at the highest temperature given in Table 6.2.

$$\frac{Q}{t} = \left(\frac{(1500{}^\circ\text{F} - 150{}^\circ\text{F})}{(0.016 \; (\text{hr})({}^\circ\text{F})/\text{Btu} + 0.151 \; (\text{hr})({}^\circ\text{F})/\text{Btu})}\right) = 7988 \text{ Btu/hr}$$

(b)   The temperature of the interface can be calculated by recognizing that the total rate of heat conduction through the wall is equal to the rate of heat conduction through either the asbestos or brick layer. Therefore,

$$\left(\frac{Q}{t}\right)_{\text{total}} = \left(\frac{Q}{t}\right)_{\text{asbestos}} = \frac{\Delta T_{\text{asbestos}}}{R_{\text{asbestos}}}$$

$$\Delta T_{\text{asbestos}} = \left(\frac{Q}{t}\right)(R) = (7988 \text{ Btu/hr})(0.151 \; (\text{hr})({}^\circ\text{F})/\text{Btu})$$

$$= 1206{}^\circ\text{F}$$

$$\Delta T_{\text{asbestos}} = T_{\text{interface}} - T_{\text{outside}} = T_{\text{interface}} - 150{}^\circ\text{F}$$

$$T_{\text{interface}} = 1206{}^\circ\text{F} + 150{}^\circ\text{F} = 1356{}^\circ\text{F}$$

6.2.4   Conduction of Heat through Cylindrical Vessels

Very often it is necessary to transfer heat to a vessel where the surface area varies. For example, a tubular reactor or a distillation tower exhibit a changing surface area along the lines of heat transfer. To illustrate the concept involved in transferring heat in such vessels, let us consider the hollow tubular reactor represented by Fig. 6.3. The inside radius of the tube is r_1, the outside radius is r_2, and the length of the tube is L. The thermal conductivity of the material of which the tubular reactor is made is k. The temperature of the outside surface is T_2, and that of the inside surface is T_1.

For the purpose of calculating the rate of heat flow from the reactor, consider a very thin cylinder, concentric with the main cylinder, of radius r,

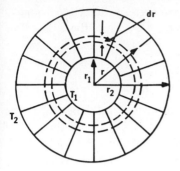

Figure 6.3 Flow of heat through a tubular reactor.

where r is between r_1 and r_2. The thickness of the wall of this cylinder is dr. Applying Eq. 6.2 to this set of conditions gives

$$\frac{Q}{t} = -k_m \frac{(2\pi r L)}{dr} dT \qquad (6.11)$$

where $(2\pi r L)$ is the area perpendicular to heat flow and the x of Eq. 6.2 is equal to dr and ΔT is equal to dT. Rearranging Eq. 6.11 and integrating between the appropriate limits gives

$$\int_{r_1}^{r_2} \frac{dr}{r} = \frac{2\pi L k_m}{(Q/t)} \int_{T_2}^{T_1} dT$$

or

$$\frac{Q}{t} = \frac{(k_m)(2\pi L)(T_1 - T_2)}{\ln(r_2/r_1)} \qquad (6.12)$$

Equation 6.12 can be used to calculate the heat flow from a cylindrical shaped vessel, but it is often expressed in the more convenient form

$$\frac{Q}{t} = \frac{(k_m)(\bar{A}_L)(T_1 - T_2)}{r_2 - r_1} \qquad (6.13)$$

The term $\bar{A}_L$ is calculated as

$$\bar{A}_L = \frac{2\pi L(r_2 - r_1)}{\ln(r_2/r_1)} \qquad (6.14)$$

$\bar{A}_L$ can be thought of as the area of a cylinder of length L and radius $\bar{r}_L$, where

$$\bar{r}_L = \frac{r_2 - r_1}{\ln(r_2/r_1)} \qquad (6.15)$$

The term $\bar{A}_L$ is known as the logarithmic mean area of the cylinder and $\bar{r}_L$ is known as the logarithmic mean radius. The application of these heat flow expressions is illustrated in Example 6.3.

Example 6.3

A 1-in. o.d. tube reactor is insulated with a 3-in. layer of asbestos. If the temperature of the outer surface of the insulation is 100°F and the rate of heat loss per foot of reactor length is 50 Btu/hr, what must the temperature in the inside of the reactor be? (For asbestos in this temperature range k_m is 0.105 Btu/hr · ft · °F.)

Solution

Assume that the temperature in the inside of the reactor is the same as the temperature of the reactor-insulation interface. Therefore only the resistance of the asbestos layer need be considered, and the log mean area of this layer per foot of reactor is given by Eq. 6.14.

$$\bar{A}_L = \frac{(2 \cdot \pi)(1 \text{ ft})(3.5/12 \text{ ft} - 0.5/12 \text{ ft})}{\ln[(3.5/12)/(0.5/12)]} = 0.807 \text{ ft}^2$$

where the outside radius $r_2 = 3.5$ in and inside radius $r_1 = 0.5$ in. Substituting this value of $\bar{A}_L$ into Eq. 6.13 gives

$$(T_1 - T_2) = \frac{(Q/t)(r_2 - r_1)}{(k_m)(A_L)} = \frac{(50 \text{ Btu/hr})(0.292 \text{ ft} - 0.042 \text{ ft})}{(0.105 \text{ Btu/hr} \cdot \text{ft} \cdot °\text{F}) \cdot (0.807 \text{ ft}^2)}$$

$$(T_1 - T_2) = 147.5°\text{F}$$

Therefore

$$T_1 = 147.5°\text{F} + T_2 = 147.5°\text{F} + 100°\text{F}$$

$$T_1 = 247.5°\text{F}$$

6.3 HEAT TRANSFER BY CONVECTION

As indicated in Section 6.1, heat may be transferred from one substance to another by the physical process of mixing a hot substance with a cold substance. When heat is transferred in this manner it is known as convection. However, in practically all cases a certain amount of heat is transferred by conduction simultaneously with convection heat transfer. Since it is not practical to differentiate between convection and conduction when they are both occurring at the same point, a simplified approach has been developed that takes both types of transfer into consideration. This approach utilizes the concept of the film heat transfer coefficient.

6.3.1   Concept of the Film

When a fluid is flowing past a stationary surface, such as shown in Fig. 6.4, a thin film of the fluid is postulated as existing between the flowing fluid and the stationary surface. It is assumed that all of the resistance to transmission of heat between the flowing fluid and the wall containing the fluid is due to the film at the stationary surface. If a plot is prepared as shown in Fig. 6.4 with temperature as the ordinate and distance perpendicular to the wall as the abscissa, then the concept of the film resistance becomes evident. In the figure the metal wall of the tube separates the hot fluid on the left from the cold fluid on the right. The change in temperature with distance is shown by the line $T_1 T'_1 T'_2 T_2$. The temperature profile is thus divided into five regions.

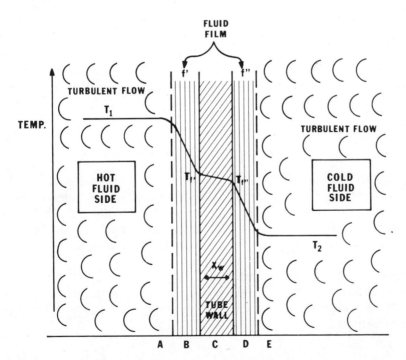

Figure 6.4   Temperature profile in a convection-conduction heat transfer system.

Region A. Convective heat transfer occurs because of the motion of the elements of the fluid.

Region B. This region encloses the fluid film where flow velocity is too slow to provide convection heat transfer. Consequently, heat transfer is by conduction in this region.

Region C. This region consists of the solid material of the tube wall through which heat is transferred by conduction.

Regions D and E. These regions for the colder fluid correspond to B and A, respectively, for the fluid inside the tube.

From the temperature profile in Fig. 6.4, it can be seen that the largest temperature changes occur in the two film regions. This is because conduction in these zones provides a slow rate of heat transfer. Heat transfer problems such as this one are best solved by the use of individual heat-transfer coefficients for the two fluid streams. The separate coefficients are then combined with the wall resistance to obtain an overall coefficient of resistance.

The individual heat-transfer film coefficient denoted by h is defined by Eq. 6.16.

$$h = \frac{(Q/t)}{A_f \cdot \Delta T_f} \qquad (6.16)$$

where A_f is the film area and ΔT_f is the temperature drop across the film. When the heat flow rate Q/t is expressed in British thermal units per hour, A_f in squarefeet, and ΔT_f in degrees Fahrenheit, then the units of h are Btu/(hr)(ft^2)(°F).

6.3.2   Overall Coefficient

In order to calculate the rate of heat transfer through the wall shown in Fig. 6.4, it is necessary to combine the individual film coefficients and the resistance of the wall into one overall resistance coefficient. The overall coefficient can be calculated by combining the film concept developed in the preceding paragraphs with the appropriate expressions for heat transfer through resistances in series developed in Section 6.2.

When heat flows from the hot fluid in Fig. 6.4 to the cold fluid, the same amount of heat per unit area and per unit time must be transferred through film f', the wall, and the film f''. The total amount of heat transferred can be expressed as

$$\frac{Q}{t} = U \cdot A \cdot \Delta T_{\text{total}} = h'A_f'T_f' = \frac{kA_w \, \Delta T_w}{x_w} = h''A_f''T_f'' \qquad (6.17)$$

where U = the overall heat transfer coefficient
$\quad A$ = the base area chosen for the evaluation of U
$\quad \Delta T_{total}$ = the difference in T between the hot fluid and cold fluid
$\quad h'$ = the hot-side film coefficient
$\quad A'_f$ = the area of the hot-side film, f'
$\quad \Delta T'_f$ = the temperature drop across the film, f'
$\quad k$ = the thermal conductivity of the wall material
$\quad A_w$ = the wall area
$\quad \Delta T_w$ = the drop in T across the wall
$\quad x_w$ = the wall thickness
$\quad h''$ = the cold-side film coefficient
$\quad A''_f$ = the area of the cold-side film, f''
$\quad \Delta T''_f$ = the temperature drop across the film, f''

The total temperature difference must equal the sum of the temperature differences across each resistance.

$$\Delta T_{total} = \Delta T'_f + \Delta T_w + \Delta T''_f \tag{6.18}$$

Combining Eqs. 6. 17 and 6.18 gives

$$\Delta T_{total} = \left(\frac{Q}{t}\right)\left(\frac{1}{h'A'_f} + \frac{x_w}{kA_w} + \frac{1}{h''A''_f}\right) \tag{6.19}$$

but ΔT_{total} is also given by

$$\Delta T_{total} = \left(\frac{Q}{t}\right)\left(\frac{1}{UA}\right) \tag{6.20}$$

Therefore,

$$\frac{1}{U} = A\left(\frac{1}{h'A'_f} + \frac{x_w}{kA_w} + \frac{1}{h''A''_f}\right) \tag{6.21}$$

If the transfer of heat from a hot fluid to a cold fluid, as depicted in Fig. 6.4, is through a flat wall then all the areas are equal and Eq. 6.21 reduces to

$$\frac{1}{U} = \frac{1}{h'} + \frac{x_w}{k} + \frac{1}{h''} \tag{6.22}$$

If, however, the hot fluid is flowing through the inside of a tube and the cold fluid is flowing over the outside of the tube then $A'_f \neq A_w \neq A''_f$. In this case A'_f would be the area of the inside wall of the tube, A_w would be the log mean wall average as calculated by Eq. 6.14, and A''_f would be the area of the outside tube wall. The area A upon which U is based can be either A'_f, A_w, or A''_f so long as the same area is used in the application of Eq. 6.17. For example, if the overall coefficient U in a tube heat exchanger is based on the inside tube area A'_f, then Eq. 6.21 becomes

$$\frac{1}{U} = A'_f \left(\frac{1}{h'A'_f} + \frac{x_w}{kA_w} + \frac{1}{h''A''_f} \right) \tag{6.23}$$

which reduces to

$$\frac{1}{U} = \left(\frac{1}{h'} + \frac{x_w A'_f}{kA_w} + \frac{A'_f}{h''A''_f} \right). \tag{6.24}$$

When U is calculated by use of Eq. 6.24 then A'_f must be used for the area A in Eq. 6.17. The overall heat transfer coefficient U could equally as well be calculated with the base area being A_w or A''_f. In such a case, the appropriate value of A_w or A''_f would be used in Eq. 6.17.

In actual service heat transfer surfaces such as depicted in Fig. 6.4 do not remain clean. Dirt, scale, and other solid deposits form on one or both sides of the tubes. The presence of such materials provides additional resistance to heat flow and therefore reduces the overall coefficient. The effect of such deposits is taken into account by adding a term $1/Ah_d$ to the term in parentheses in Eq. 6.21 for each scale deposit. Thus if scale is present on both the inside and outside surface of the tube, Eq. 6.21 becomes

$$\frac{1}{U} = A \left(\frac{1}{h'A'_f} + \frac{x_w}{kA_w} + \frac{1}{h''A''_f} + \frac{1}{A'_f h'_d} + \frac{1}{A''_f h''_d} \right) \tag{6.22}$$

where h'_d and h''_d are the fouling factors for the scale deposits on the inside and outside tube surfaces, respectively.

6.3.3 Magnitude of Heat-Transfer Coefficient and Fouling Factors

The values of the coefficients h and h_d vary greatly, depending on the fluids involved, their flow velocities, the wall material, and so on. Some typical ranges are given in Table 6.3.

Example 6.4 illustrates the calculation of the overall heat transfer coefficient based upon the use of the film coefficient theories developed in the preceding discussion.

Example 6.4

Benzene flowing in the inner pipe of a double-pipe exchanger is cooled with water flowing in the jacket. The inner pipe is made from Schedule 40 1-in. steel pipe. The thermal conductivity of steel may be taken to be 26.2 Btu/

Table 6.3   Range of Values for h and h_d

Type of Process	Btu/(ft^2) (hr) (°F)	
	h	h_d
Steam condensing dropwise	4000-18,000	
Steam condensing filmwise	800-3500	
Organic vapors condensing	150- 450	1200-1800
Water-heating or cooling	20-2000	800-1000
Oils-heating or cooling	5- 250	
Air-heating or cooling	0.1- 10	

hr · ft · °F. The individual coefficients and fouling factors in Btu/(ft^2)(hr)(°F) are:

Benzene side coefficient: $h' = 210$
Inside fouling factor: $h'_d = 1250$
Water side coefficient: $h'' = 350$
Outside fouling factor: $h''_d = 860$

What is the overall heat transfer coefficient, based on the outside area of the inner pipe?

Solution

From Section 5.11.1 in Chapter 5, the diameters and wall thickness of Schedule 40 1-in. pipe are:

Outside diameter $= D'' = 1.315$ in.
Inside diameter $= D' = 1.049$ in.
Wall thickness $= x_w = 0.133$ in.

Using these values we can calculate the various areas for a 1-ft section of pipe.

$$A'_f = \pi D' \cdot L = (3.14)\left(\frac{1.049 \text{ ft}}{12}\right)(1 \text{ ft}) = 0.274 \text{ ft}^2$$

$$A''_f = \pi D \cdot L = (3.14)\left(\frac{1.315 \text{ ft}}{12}\right)(1 \text{ ft}) = 0.344 \text{ ft}^2$$

The log mean wall area $\bar{A}_w$ can be calculated by use of Eq. 6.14

$$\bar{A}_w = \frac{\pi L(D'' - D')}{\ln(D''/D')} = \frac{(3.14)(1)(1.35/12 - 1.049/12)}{\ln[(1.35/12)/(1.049/12)]}$$

$$= 0.312 \text{ ft}^2$$

Substituting these values into Eq. 6.22 with the condition that U is based on the outside area (i.e., $A = A''$) gives:

$$\frac{1}{U} = 0.344\left[\frac{1}{(210)(0.274)} + \frac{(0.133/12)}{(26.2)(0.312)} + \frac{1}{(350)(0.344)}\right.$$

$$\left. + \frac{1}{(0.274)(1250)} + \frac{1}{(0.344)(860)}\right]$$

$$\frac{1}{U} = 0.01147$$

$$U = 87.2 \text{ Btu/(ft}^2)(\text{hr})(^\circ\text{F})$$

6.4 THE PRACTICE OF HEAT EXCHANGE

As discussed in the beginning of this chapter, the successful operation of almost all chemical processes involves the transfer of heat. Such transfers are normally accomplished by use of heat exchangers.

Heat exchangers exist in many different forms, but typically involve two fluids passing on opposite sides of a conducting wall through which heat flows, as shown in Fig. 6.5. The exchanger in Fig. 6.5a is operating in a cocurrent mode, that is both streams are flowing in the same direction. In Fig. 6.5b the streams flow in opposite directions, and the mode of operation is referred to as countercurrent.

The simple heat exchanger design depicted in Fig. 6.5 is often modified quite extensively, depending on the nature of the streams involved. For example, both cocurrent and countercurrent operation are involved in the double-pass shell-and-tube heat exchanger shown in Fig. 6.6. In this exchanger one fluid passes down through one-half of the tube and then back through the other half, while the other fluid passes on the outside of the tubes. The amount of heat exchanged in the equipment depicted in Fig. 6.5 and 6.6 depends on the flow rates, temperature differences, and thermal properties of the fluids, as well as the surface area of the exchangers and in particular whether the mode of operation is cocurrent or countercurrent.

Before considering the differences between cocurrent and countercurrent heat transfer, it is necessary to establish a set of equations that describe the

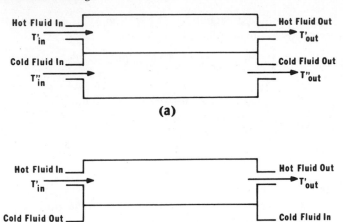

Figure 6.5 (*a*) A heat exchanger operating cocurrently; (*b*) a heat exchanger operating countercurrently.

performance of a heat exchanger. First, we should recognize that heat is being transferred by both conduction and convection processes. Therefore heat transfer equations developed in the preceding section are applicable. Consequently the total rate at which heat is transferred in the heat exchanger is given by Eq. 6.17, that is

$$\frac{Q}{t} = U \cdot A \cdot \Delta T_{\text{total}}$$

Since the difference in temperature between the hot stream and cold stream, ΔT_{total}, varies along the length of the exchanger, an average ΔT must be used.

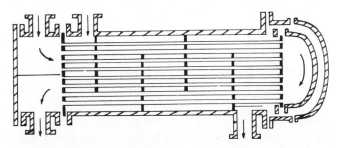

Figure 6.6 A double-pass shell-and-tube heat exchanger.

Using the countercurrent exchanger in Fig. 6.5b as an example, we can calculate the arithmetic mean ΔT as follows:

$$\Delta T_{AM} = \frac{(T'_{in} - T''_{out}) + (T'_{out} - T''_{in})}{2} \tag{6.23}$$

Where T'_{in} = temperature of the hot stream in
$\quad T'_{out}$ = temperature of the hot stream out
$\quad T''_{in}$ = temperature of the cold stream in
$\quad T''_{out}$ = temperature of the cold stream out

When $(T'_{in} - T''_{out})$ is greater than twice $(T'_{out} - T''_{in})$ then the log mean temperature must be used. The log mean temperature expression is very similar to the log mean radius equation developed in Section 6.2 and is given as follows for a countercurrent heat exchanger

$$\Delta T_{LM} = \frac{(T'_{in} - T''_{out}) - (T'_{out} - T''_{in})}{2.3 \log[(T'_{in} - T''_{out})/(T'_{out} - T''_{in})]} \tag{6.24}$$

The value of ΔT calculated from either Eq. 6.23 or 6.24 is used in Eq. 6.17 to solve for the rate of heat transfer in the exchanger.

According to the concepts of energy balance developed in Chapter 4, the heat going into the exchanger must be equal to the heat leaving. Normally the exterior of a heat exchanger is well insulated to minimize heat loss to the surroundings, thereby allowing heat to flow only between the two streams involved. When such an exchanger is operating in the steady-state, the loss of heat by the hot stream exactly equals the heat gained by the cold stream, and we can write,

$$Q = (M_{hot})(Cp_{hot})(T'_{in} - T'_{out}) = (M_{cold})(Cp_{cold})(T''_{out} - T''_{in}) \tag{6.25}$$

where M_{hot} = mass of the hot stream per unit time
$\quad Cp_{hot}$ = the mean heat capacity of the hot stream over the temperature range involved
$\quad M_{cold}$ = mass of the cold stream per unit time
$\quad Cp_{cold}$ = the mean heat capacity of the cold stream over the temperature range involved

We now have a basis for comparing the operation of a countercurrent to a cocurrent exchanger as illustrated in Example 6.5. This example shows that for a fixed overall heat transfer coefficient, the exchanger area required to transfer a given amount of heat between two streams in countercurrent flow is smaller than in cocurrent flow. In addition, it should be noticed that in the cocurrent exchange the temperatures of the fluids approach each other as the hot fluid transfers heat to the cold fluid. Regardless of the size of the heat exchanger in this mode of operation, the hot and cold fluids are driven toward an intermediate temperature. Because of this characteristic, cocurrent heat exchangers are rarely used in industrial chemical processes. On the other hand, in countercurrent operation the cold fluid temperature is driven toward the highest hot fluid temperature, and the hot fluid temperature is driven toward the coldest cold fluid temperature.

Example 6.5

Suppose that it is necessary to cool a hot stream of isooctane product that exits from a distillation column in an alkylation process. The stream exits at 175°F and must be cooled to 110°F by heat exchange with water available at 60°F. The flow rate of each stream and the respective mean heat capacities are as follows:

	Isooctane Stream	Water Stream
Flow rate (lb/s)	200	100
Heat capacity [Btu/(lb)(°F)]	0.33	1.0

Assuming that the overall coefficient of heat transfer is the same, compare the area requirements of countercurrent and cocurrent heat exchangers necessary to accomplish the cooling required in this process.

Solution

Step 1. Calculate the heat lost by the hot stream from either exchanger. Using Eq. 6.25 we find

$$Q = (M_{hot})(Cp_{hot})(T'_{in} - T'_{out})$$
$$= (200 \text{ lb/s})[(0.33 \text{ Btu/(lb)(°F)}](175° - 110°)$$
$$Q = 4290 \text{ Btu/s}$$

Step 2. Calculate the temperature of the leaving water stream. Again, using Eq. 6.25 we find,

$$Q = 4290 \text{ Btu/s} = (M_{cold})(Cp_{cold})(T''_{out} - T''_{in})$$
$$= (100 \text{ lb/s})[1 \text{ Btu/(lb)(°F)}(T'' - 60)]$$
$$T''_{out} = \frac{4290}{100} + 60 = 102.9°F$$

Step 3. The log mean temperature driving forces can be calculated by using Eq. 6.24.

Countercurrent	Cocurrent

$T'_{in} - T''_{out} = 175 - 102.9 = 72.1°F$ $T'_{in} - T''_{in} = 175 - 60 = 115°F$

$T'_{out} - T''_{in} = 110 - 60 = 50.0°F$ $T'_{out} - T''_{out} = 110 - 102.5 = 7.5°F$

$$\Delta T_{LM_{cc}} = \frac{(72.1 - 50.0)}{2.3 \log(72.1/50.0)}$$

$$\Delta T_{LM_{c}} = \frac{115 - 7.5}{2.3 \log(115/7.5)}$$

$$= 59.7$$

$$= 39.8$$

Step 4. The area of the respective exchangers is given by Eq. 6.17.

$$A = \frac{(Q/t)}{U \cdot \Delta T_{LM}}$$

Therefore

$$\frac{\text{area for countercurrent flow (cc)}}{\text{area for cocurrent flow (c)}} = \frac{(Q/t)_{cc}/(U_{cc} \cdot \Delta T_{LM_{cc}})}{(Q/t)_{c}/(U_{c} \cdot \Delta T_{LM_{c}})}$$

Since the same amount of heat is transferred, and U is taken to be the same in both modes of operation, this equation reduces to,

$$\frac{A_{cc}}{A_{c}} = \frac{\Delta T_{LM_{c}}}{\Delta T_{LM_{cc}}} = \frac{\Delta T_{LM_{c}}}{\Delta T_{LM_{cc}}} = \frac{39.8}{59.7} = 0.67$$

Consequently the cooling required in this process can be accomplished with a countercurrent exchanger that has only 67 % of the area of a cocurrent exchanger.

6.5 HEAT EXCHANGE CONCEPTS IN SCALE-UP

Chemistry students, during their laboratory training stage, rarely encounter situations where their experimental design is dictated by heat transfer limitations. This is so because laboratory experiments are typically performed in vessels that possess a high surface to volume ratio. For example, the surface to volume ratio of a 1-liter spherical flask is 0.48. In such a vessel adequate heating and cooling can most often be accomplished through the surface area of the vessel, thereby eliminating the need for additional heat exchange surface. However, if a 100-liter spherical flask is used, the surface to volume ratio is reduced to 0.10, and now additional heat exchange surface may be required.

Failure to adjust for the change in surface area to volume ratio in scaling up a reaction process can lead to runaway reactions, explosions, and so on. For example, while the surface area of a 1-liter vessel may provide sufficient heat transfer area to prevent an exothermic reaction from reaching dangerous temperature levels, a 100-liter vessel may not. Example 6.6 illustrates such a situation.

Example 6.6

Consider the exothermic condensation reaction.

$$2A \longrightarrow B \quad \Delta H = -79 \text{ Btu}$$

The reaction proceeds at a rate of formation of B of 0.6 mole/1 · hr at 117°F. When the reaction mixture is at this temperature, and the surrounding room temperature is 77°F, compare the rate of heat transfer with the rate of heat generation for a 2-liter and a 100-liter spherical reaction vessel. Assume that the overall heat transfer coefficient for pyrex glass vessels remains constant at 3 Btu/(hr)(ft^2)(°F) and that the reaction mixture fills the vessel.

Reaction in 2-liter vessel:

$$\text{volume} = \frac{4}{3}\pi r^3 = 2000 \text{ cm}^3$$

$$r = 7.82 \text{ cm}$$

$$\text{area} = 4\pi r^2 = (4)(3.14)(7.82)^2 = 768 \text{ cm}^2$$

or

$$\text{area in ft}^2 = (768 \text{ cm}^2)(1.076 \times 10^{-3} \text{ ft}^2/\text{cm}^2) = 0.826 \text{ ft}^2$$

The rate of heat transfer in this vessel under the conditions cited is

$$\frac{Q}{t} = U \cdot A \cdot \Delta T = [(3 \text{ Btu}/(\text{hr})(\text{ft}^2)(°F)] \cdot (0.826 \text{ ft}^2) \cdot (117°F - 77°F)$$

$$\frac{Q}{t} = 99.1 \text{ Btu/hr}$$

The rate of heat generation in this vessel is

$$\text{heat generation} = (\text{rate of reaction})(\text{heat of reaction})$$

$$\text{heat generation} = (0.6 \text{ mole B/hr} \cdot \text{liter of reaction mixture})$$

$$(2\text{-liter reaction mix})(79 \text{ Btu/mole B}) = 94.8 \text{ Btu/hr}$$

Since the rate of heat generation, 94.8 Btu/hr is less than the rate of heat transfer, 99.1 Btu/hr, no additional heat transfer surface is required in order to prevent an increase in temperature of the reaction mixture.

Reaction in 100-liter vessel:

$$\text{volume} = \frac{4}{3\pi r^3} = 100{,}000 \text{ cm}^3$$

$$r = 28.8 \text{ cm}$$

$$\text{area} = (4\pi r^2) = (4)(3.14)(28.8 \text{ cm})^2 = 1.042 \times 10^4 \text{ cm}^2$$

or

$$\text{area} = 11.2 \text{ ft}^2$$

The rate of heat transfer is

$$\frac{Q}{t} = [3 \text{ Btu/(hr)(ft}^2)(°F)](11.2 \text{ ft}^2)(117° - 77°)$$

$$= 1344 \text{ Btu/hr}$$

The rate of heat generation for 100 liters of reaction mixture is

$$\text{heat generation} = (0.6 \text{ mole/(hr)(liter)})(100 \text{ liters})(79 \text{ Btu/mole})$$

$$= 4740 \text{ Btu/hr}$$

Obviously the rate at which heat is being generated (4740 Btu/hr) is much greater than the rate of heat transfer (1344 Btu/hr). If a temperature buildup is to be prevented, additional heat transfer surface must be supplied so that the rate of heat transfer will be greater than or equal to the rate of heat generation.

In analyzing this problem it becomes apparent how the heat generation-transfer rates get out of balance during scale-up. In going from the 2-liter flask to the 100-liter flask the heat transfer surface and, consequently the rate of heat transfer, increased by a factor of 13.6. However, the rate at which heat is generated increased by a factor of 50, thus creating an imbalance.

6.6 HEAT EXCHANGE ARRANGEMENTS

In analyzing the scale-up of the exothermic reaction in Example 6.6, it was determined that additional heat transfer surface is required in order to keep the temperature of the reaction mixture under control. The increase can be effected in several different ways. For example, if it is not intended to reserve the heat for other uses, a stream of cooling water might be passed through a bundle of tubes immersed in the reaction mixture. The amount of surface area and the rate of water addition could be calculated by use of Eqs. 6.17 and 6.25. Another method for providing the proper amount of cooling is to circulate some of the reaction mixture through a heat exchanger maintained outside of the reaction vessel. Again, water could be used as the cooling fluid.

Normally, in bench type and pilot plant scale operation, recovery of heat for economic reasons is not done. However, for a full-scale plant the recovery of heat from an exothermic reaction of the type just considered may be economically feasible. This might be accomplished by using the feed stream of another process as the cooling fluid, thereby using the heat of the exothermic reaction to bring the temperature of this stream to the proper value required for the process.

In supplying heat to a reactor in which an endothermic reaction is occurring, the same types of heat exchanger arrangements can be employed. Only now, a hot exchange fluid must be circulated through the exchanger. This fluid might be steam under pressure, a hot oil stream, or some other heated fluid such as hot gases. Instead of heating the reaction mixture, one can sometimes supply the heat required for the endothermic reaction by heating the reactant stream to a temperature greater than that required for the reaction. As the reactants cool, they give off the heat required for the endothermic reaction.

6.7 SUMMARY

The material in this chapter is intended to serve only as an introduction to the study of heat transfer concepts. We hope that it has made you aware of the type and complexity of problems industrial chemists may encounter. If you are interested in or require a more extensive treatment of heat transfer phenomena, then one of the references cited at the end of this chapter should be consulted.

6.8 REFERENCES

1. J. E. Biles, *Combustion* **48**, 17 (1976).
2. G. G. Brown, *Unit Operations*, Wiley, New York, 1950, p. 584.

Supplementary References

Blackadder, D. A., and R. M. Nedderman, *A Handbook of Unit Operations*, Academic, New York, 1971.

Clarke, L., *Manual for Process Engineering Calculations*, McGraw-Hill, New York, 1947.

Coulson, J. M., and J. F. Richardson, *Chemical Engineering*, Vol. 1, 2nd ed., Pergamon, New York, 1964.

Holland, F. A., R. M. Moores, F. A. Watson, and J. K. Wilkinson, *Heat Transfer*, American Elsevier, New York, 1970.

Kays, W. M., and A. L. London, *Compact Heat Exchangers*, 2nd ed., McGraw-Hill, New York, 1964.

McCabe, W. L., and J. C. Smith, *Unit Operations of Chemical Engineering*, 2nd ed., McGraw-Hill, New York, 1967.

6.9 PROBLEMS

1. (a) A furnace wall is constructed of firebrick, 6 in. thick. The temperature at the inside of the wall is 1500°F, and the temperature of the outside of the wall is 150°F. If the mean thermal conductivity of the brick under these conditions is 0.17 Btu/hr · ft · °F, what is the rate of heat loss through 10 ft^2 of wall surface?

    (b) If the coefficient of heat transfer between the outside furnace wall and the air is 2.2 Btu/(hr)(°F)(ft^2) and the air temperature is 70°F when the inside temperature of the furnace is 1500°F, what is the rate of heat loss through 10 ft^2 of wall surface? What is the temperature of the outside furnace wall under these conditions?

    (c) If the furnace wall in (b) was constructed of an outside layer of 2-in. thick firebrick and an inside layer of chrome brick 4 in. thick [$k = 0.80$ Btu/hr · ft · °F], answer the questions asked in part (b). In addition, calculate the temperature between the two layers of brick.

2. A furnace wall is constructed of firebrick, insulating brick, and common brick, each 4 in. thick. The temperature of the inside wall surface is 950°C, while the outside temperature is 30°C. The resistance between the joints of the different types of brick may be neglected. The conductivities in Btu/hr · ft · °F are; firebrick 0.7, insulating brick 0.046, and common brick 0.4.

    (a) If the furnace described above has 500 ft^2 of wall area, how much heat is lost during a 24-hr period by conduction through the walls?

    (b) Calculate the temperature at the center of the insulating brick.

3. A standard 1-in. schedule 40 steel pipe carries saturated steam at 275°F. The pipe is insulated with a 1.5-in. layer of 85 % magnesia pipe covering, and outside this magnesia there is a 3-in. layer of fiber glass insulation.

The outside temperature of the glass insulation is 85°F.

(a)   Calculate the heat loss per hour from 100 ft of pipe.

(b)   Calculate the temperatures at the boundaries between the steel and magnesia and between the magnesia and fiber glass.

4. Aniline is to be cooled from 95°C to 60°C in a double-pipe heat exchanger having a total length of 100 ft. The exchanger consists of a 1.25-in. schedule 40 steel pipe inside a 2-in. schedule 40 steel pipe. For cooling, a stream of toluene at a temperature of 25°C and a rate of 8500 lb/hr passes through the 2-in. pipe. The aniline flow rate is 10,000 lb/hr. If flow is counter-current what is the toluene outlet temperature; the logarithmic mean temperature difference; and the overall heat transfer coefficient?

5. Suppose you were going to run an exothermic reaction in a spherical vessel. The reaction reaches a steady state and gives off 5000 Btu of heat per hour. The temperature of the reaction mixture is 200°F, and the temperature outside of the vessel is 80°F. What size (volume in liters) spherical vessel must be used in order to dissipate all of the heat of reaction through the walls of the vessel so that the reaction mixture does not rise above 200°F? You may use an overall heat-transfer coefficient U for this liquid-to-gas heat exchange of 3 Btu/(hr)(ft^2)(°F). The value of U is based on the inside area of the reaction vessel.

6. Water flows through a 2-in. standard schedule 40 steel pipe. The pipe is jacketed with steam at 230°F. The conditions are such that the overall heat-transfer coefficient U is 250 Btu/(hr)(ft^2 of outside pipe area) (°F). The length of the pipe is 10 ft. If the temperature of the water is 80°F at the pipe entrance and 180°F at the pipe exit, calculate the average amount of heat gained by the water per hour for this exchange system.

7. You have been assigned a project to develop a more efficient heat-conducting black paint for use on solar collection panels. The solar collection panels are to be used to heat home hot water systems. For the purpose of testing various paint formulations, consider that you have set up a single collection panel on the roof of your laboratory. The hot water exit from the collection panel is connected by means of a 0.5-in. i.d. copper tubing to an 80 gal storage vessel. The storage vessel is located in your laboratory and is cylindrically shaped with a radius of 1 ft. The length of copper tubing between the collector plate and the storage vessel is 40 ft, with half of the tubing located outside of the laboratory and the other half inside. Suggest an insulation and its thickness for insulating the hot water line and the storage vessel. With your insulation, what would be the rate of heat loss from the entire system, when the hot water temperature is 145°F, the laboratory temperature is 75°F, and the temperature outside of the laboratory is 85°F?

8. Suggest several possible methods and materials that might be used to improve the heat-conducting ability of the paint to be tested on the solar panels in the project described in problem 7.

9. Calculate the overall heat transfer coefficients based on both inside and outside areas for benzene at atmospheric pressure condensing on the inside of a Liebig condenser with water at 20°C flowing cocurrent at 5 cm/s through the outer jacket. The condenser is made of Pyrex and is 50 cm long. The inner tube has an i.d. of 1.5 cm, while the outer tube has an i.d. of 3.5 cm. The condenser is made of 2-mm-thick glass.

10. A cylindrical shaped reaction vessel, 4-ft i.d. and 7-ft high and containing an aqueous solution, is to be maintained at a temperature of 80°C by means of a steam heating coil. The tank is constructed of 0.25-in.-thick 316 stainless steel. The ambient temperature outside the reaction vessel is 25°C. Natural gas purchased at a unit fuel price given in Table 6.1 is used to fire the steam boiler. The combined boiler-heating coil efficiency can be assumed to be 50%. Using prevailing price and quality (i.e., thermal conductivity) information that is common to your geographical location, calculate how many days the reactor must be operated under the specified conditions in order to pay for a 2-in.-thick layer of polyurethane insulation. How many days of operation would be required to pay for a 4-in-thick layer? How many days of operation would be required to pay for a 2-in.-thick layer if the boiler was fired with coal at 50% efficiency?

7

KINETICS

EXPERIMENTAL DATA, RATE CONSTANTS, AND REACTOR DESIGN

The successful development and operation of processes in which chemical reactions occur requires knowledge of the conditions under which the reactions attain commercially feasible rates. These conditions are important to the industrial chemist, for they determine the final flow design, equipment requirements, and economics of the process. For example, the residence time, temperature, and catalyst requirements determine the type and dimensions of the reactor required. Operating conditions also determine the product stream composition and therefore dictate the type and size of separation and recovery equipment required. The operating conditions also may set feed stream composition and purity requirements because of problems associated with side reactions and catalyst poisons.

The area of study dealing with the rate by which one chemical species is converted to another is referred to as chemical kinetics. You have probably heard it stated in some of your previous chemistry courses that the ultimate goal of chemical kinetics is to develop a fundamental rate equation that fits the rate data and is consistent with observations on the reaction mechanism. The rate in this case is the mass, expressed in moles, of a product produced or reactant consumed per unit time and unit volume. The mechanism is the sequence of individual events whose overall result produces the observed reaction. However, it is important to realize that it is not necessary to know the mechanism of a reaction in order to develop a chemical process. What is necessary, so that the chemical reaction vessel may be designed, built, and operated successfully, is a satisfactory rate equation. Although a knowledge of the mechanism is of great value in extending the rate data beyond the original experiments, it is not a prerequisite to the development of a commercial chemical process.

Many of the reactions that students encounter in their study of chemical kinetics are carried out in homogeneous systems. However, many industrial processes are based upon heterogeneous reactions. Such systems often involve the passage of a reactant stream over a solid catalyst surface. Therefore it is necessary that industrial chemists be capable of performing kinetic experiments on both homogeneous and heterogeneous systems.

The purpose of this chapter is to present an insight into how kinetic studies in both homogeneous and heterogeneous systems are performed and how such studies form an important part of the overall development of a chemical process. For the sake of convenience, homogeneous systems are discussed first and then heterogeneous catalyzed systems are considered.

7.1 HOMOGENEOUS REACTIONS

7.1.1 Separation of Kinetic and Physical Parameters

It is important in rate studies to make sure that the recorded data pertain to the rate of transformation of atoms and molecules from one structural form to another and that they not be complicated with terms involving physical processes such as heat and mass transfer. For example, when more than one reagent is involved in a homogeneous reaction such as the liquid phase chlorination of benzene or the gas phase oxidation of methane, the participating molecules must first collide before they can combine. In practice, this requires mixing of the reagents, either before or after heating to the proper temperature. The conversion of reactants to products in any given reactor, consequently, is determined by the slowest of three steps, namely, mixing, heat transfer, or kinetics of the reaction itself.

It is common practice to operate a laboratory experiment in such a way as to minimize the significance of physical phenomena. For instance, a laboratory reactor may be operated at near-isothermal conditions, eliminating heat-transfer and temperature gradient considerations. Such a method of operation may be uneconomical in a commercial scale system. However, once the kinetics are established in the laboratory, the results can be extrapolated to the practical situation in which mass and heat transfer phenomena are important.

7.1.2 Methods of Mass Transfer in Chemical Reaction Vessels

When chemical reaction rates are relatively slow, reagent mixing can be completed well before the reaction has proceeded very far. Consequently, the design and dimensions of the reaction vessel will be determined by the rate of the chemical reaction. On the other hand, when chemical reaction rates are fast, then mixing of the reactants becomes the rate-limiting step that fixes the design and dimensions of the reaction vessel. Interestingly enough, in such a mixing-controlled situation information on the reaction rate is not necessary since the chemical reaction occurs so rapidly after mixing that the reaction time and space requirements for the chemical reaction are minimal.

Vessels used to carry out chemical reactions have a great variety of sizes, shapes, and operating conditions. One such vessel that is in common use and is most familiar to the chemist is the small flask or beaker used in the laboratory for liquid-phase reactions. At the other extreme in size are the large cylindrical reaction vessels used in the petroleum industry which may be up to 40 feet in diameter. In addition to the difference in size and shape of reaction vessels, the mode of operation may also differ. For example, in a typical laboratory operation, a charge of reactants is added to a beaker or flask, and the reaction mixture is brought to the desired temperature. The contents of the vessel are held at this condition for a predetermined time, and then the reaction mixture is removed. When the reaction vessel is operated in this manner, it is referred to as a batch reactor. The operation of a batch reactor is characterized by the variation in extent of reaction and properties of the reaction mixture with time.

A commercial scale batch reactor most often consists of a kettle or tank as shown in Fig. 7.1. However, a batch reactor may also be a closed loop of tubing provided with a circulating pump. A batch reactor may be closed, except for a vent, to prevent loss of material and danger to the operating personnel. The vent system may be fitted with a reflux condenser to prevent loss of vaporized liquids, or with a relief valve if the reaction is carried out under pressure.

Instead of adding the total charge of reactants at one time to the reaction vessel they might be fed continuously in conjunction with continuous removal of products. When a reaction vessel is operated in this manner, it is referred to as a continuous-flow reactor.

There are a number of design possibilities for a continuous-flow reactor. For example, it may consist of a tank or kettle, much like a batch reactor, with provisions for continuously adding reactants and withdrawing product as shown in Fig. 7.2. When adequate mixing is provided, the composition and temperature of the reaction mass will tend to be the same in all parts of the reactor and equal to that of the exit stream. A vessel operated in this mode is referred to as a continuous stirred-tank reactor (CSTR).

As shown in future sections, CSTR type flow reactors are best suited for carrying out relatively slow reactions.

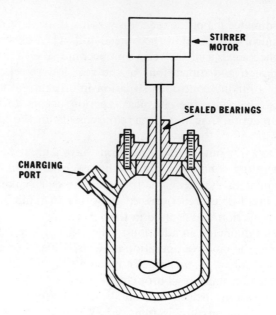

Figure 7.1 A batch reactor.

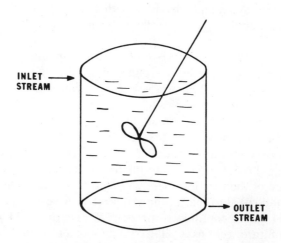

Figure 7.2 A CSTR reactor.

Figure 7.3   A tube reactor.

Continuous-flow reactors may also be tubular in design as shown in Fig. 7.3. In such reactors the length is generally large with respect to the tube diameter, and the velocity in the direction of flow is sufficient to retard back mixing; that is, it is possible to approach plug-flow performance. A tubular reactor operated in this manner is referred to as a plug-flow reactor (PFR), and it is assumed that all elements of the reaction mass have an equal velocity in the reactor.

It should be recognized that the two classifications, batch or continuous and tank or tube, are independent. For example, a laboratory flask, which might be classified as a tank reactor, can be made into a continuous-flow type by adding tubes for continuous addition of reactants and withdrawal of product.

In practice there are some general relations between the physical nature of the reaction mixture and the type of reactor used. For example, homogeneous gas-phase reactions are normally carried out in continuous tube-type flow reactors rather than a batch or CSTR reactor. For liquid-phase and liquid-solid-phase heterogeneous reactions, both CSTR and tube reactors are used. Batch-operated tank reactors are often used for small-scale production and when flexible operating conditions (i.e., temperature and pressure) are required. Such systems frequently involve costly reactants and products, as in the pharmaceutical industry.

The kinetics involved in carrying out reactions in batch and flow reactors are discussed in later sections. However, before taking up these new concepts, it is necessary that we review the basic rate law expressions that are fundamental to all kinetic studies.

7.1.3   Fundamental Rate Expressions

When miscible reagents are mixed in a reactor, a homogeneous gaseous or liquid solution forms in which a chemical reaction occurs. The rate at which the homogeneous reaction occurs in either the gaseous or liquid phase is conveniently expressed in terms of the appearance or disappearance of an arbitrarily chosen key component. For the reaction shown in Eq. 7.1, the rate of production r of

$$aA + bB \longrightarrow xX + zZ \tag{7.1}$$

the key component X in moles per unit time and per unit volume is expressed as follows:

$$r_x = \frac{1}{V} \cdot \frac{dN_x}{dt} = \text{moles/(volume)(time)} \tag{7.2}$$

where V is the volume of the reaction mixture, N_x represents the number of moles of product X in this volume, and t is time. If N refers to moles of reactant so that dN/dt is negative, a minus sign is commonly used in front of the derivative so that the rate is always positive. Any component in reaction 7.1 may be chosen as the key component, with rates of appearance and disappearance of the reaction partners related by the stoichiometric coefficients as follows:

$$\frac{1}{x} \cdot \frac{dN_X}{V dt} = \frac{1}{z} \cdot \frac{dN_Z}{V dt} = -\frac{1}{a} \cdot \frac{dN_A}{V dt} = -\frac{1}{b} \cdot \frac{dN_B}{V dt} \tag{7.3}$$

The course of a reaction is normally measured by the change in concentration of a reactant or product. However, volume changes may also occur, such as when the reaction results in a net change in the total number of moles in the reaction mixture. Then concentration changes arise from a change in volume as well as from the reaction. The influence of volume change can be accounted for by expressing N as $N_x = C_x V$, where C_x represents the concentration of X. Thus Eq. 7.2 expressed in this form becomes:

$$r_x = \frac{1}{V} \cdot \frac{d(C_x \cdot V)}{dt} \tag{7.4}$$

If the volume, or density, of the reaction mixture is constant, then Eq. 7.4 reduces to the common form

$$r_x = \frac{dC_x}{dt} \tag{7.5}$$

Care must be exercised in deciding when Eq. 7.5 is applicable. For example, in a flow reactor used for a gaseous reaction with a change in the number of moles, it is not correct. However, it would be correct for all gas-phase reactions in a tank-type reactor since the gaseous reaction mixture fills the entire vessel so that the volume is constant. For many liquid-phase systems, density changes during the reaction are small, and Eq. 7.5 is valid. The use of Eqs. 7.1 through 7.5 will become clear as we consider various kinds of reactions and reactors.

Generally, the rate of a reaction r is a function of several variables, including the instantaneous concentrations of the various species involved in the process, temperature, nature of solvent, the presence of a catalyst, and the ionic strength. The elucidation of the form of this function, known as the rate law, is often the primary goal of chemists working in the area of experimental reaction kinetics.

When the reaction is homogeneous and all variables are held constant except reactant concentration, it is generally found experimentally that the rate of a reaction is proportional to the α power of the concentration of one of the reactants A to the β power of the concentration of reactant B, and so on, so that the rate law, expressed in terms of the formation of X, takes the form

$$r_x = \frac{dC_x}{dt} = k(C_A)^{\alpha}(C_B)^{\beta} \cdots \tag{7.6}$$

The exponents α, β, and so on, are generally whole numbers. The constant of proportionality k is the rate constant, and its dimensions are (concentration)$^{n-1} \cdot$ (time)$^{-1}$. The number $n = \alpha + \beta + \cdots$ is known as the overall order of the reaction. It can also be said that the above reaction is of the α order with respect to A, the β order with respect to B, and so on.

It is important to realize that by no means can all reactions be spoken of as having an order. In many cases the relationship between the rate and the concentrations is much more complicated than that represented by Eq. 7.6; frequently, for example, concentrations appear also in the denominator of the rate expression. Such complex rate equations arise when the reaction occurs by a complex mechanism.

There is no necessary correlation between order and the stoichiometric coefficients in the reaction equation; that is, it is not required that $\alpha = a$ and $\beta = b$ in Eq. 7.6. For example, the stoichiometry of the ammonia-synthesis reaction is

$$2\,N_2 + 3\,H_2 \rightleftharpoons 2\,NH_3$$

but in the presence of several different types of catalyst systems, the rate equation has been found to be first order in nitrogen and zero order in hydrogen.

Kinetic studies of many different kinds of reactions have shown that formation of the final products from the original reactants usually occurs in a series of relatively simple steps. This observation provides an explanation for the difference between order and stoichiometric coefficients. This difference comes about because the rates of the individual steps in the reaction are usually different, and the rate of the overall reaction is determined primarily by the slowest of these steps.

The sequence of the individual steps that accounts for the formation of the products is the mechanism of the reaction. If the mechanism is known, it is usually possible to evaluate a rate equation such as 7.6 and, hence, the order of the reaction. However, it is generally not possible to infer a mechanism solely from the rate data.

The rate constant k in Eq. 7.6 includes the effects of all variables other than concentration on the rate of the reaction. The most important variable that affects k is temperature, but other factors may also be significant. For example, a homogeneous reaction may proceed only in the presence of a catalyst. In such a case k may depend upon the concentration and nature of the catalytic substances. The physical makeup and geometry of the reaction vessel may affect the value of k. For example, a reaction may be primarily homogeneous but have appreciable surface effects. In such cases k will vary with the nature and extent of the reactor surface in contact with the reaction mixture.

The dependency of k on temperature for an elementary process is given by the general integrated form of the Arrhenius equation

$$k = Ae^{-E_a/RT} \tag{7.7}$$

in which A is the temperature-independent constant of integration; usually termed the "frequency factor," and E_a is referred to as the activation energy, since it represents the energy level that the reactants must exceed before they will react. Combining Eqs. 7.6 and 7.7 yields

$$r = Ae^{-E_a/RT}(C_A)^\alpha(C_B)^\beta \cdots \tag{7.8}$$

This equation provides a description of the rate in terms of the measurable variables, concentration, and temperature. It is limited to elementary steps in a reaction because the Arrhenius equation is so restricted. However, the exponential effect of temperature most often accurately represents experimental rate data for an overall reaction, even though the activation energy is not clearly defined and may be considered a combination of activation energy values for several of the elementary steps.

All the rate expressions discussed so far apply only to reactions that go substantially to completion at equilibrium, that is, to reactions for which the equilibrium constant K is large. Suppose the reaction represented in Eq. 7.1 is not unidirectional but reaches a state of equilibrium as represented in Eq. 7.9.

$$aA + bB \rightleftharpoons xX + zZ \tag{7.9}$$

For such a reaction the net rate of production of X may take the following form

$$r_X = k_F(C_A)^\alpha(C_B)^\beta - k_R(C_X)^\gamma(C_z)^v \tag{7.10}$$

The term k_F is the rate constant for the forward reaction and k_R for the reverse reaction. Thus the rate of formation of X is given by the first term in Eq. 7.10 and the rate of removal by the second term. When equilibrium is reached the rate of production of X is zero and Eq. 7.10 can be written as

$$k_F(C_A)^\alpha(C_B)^\beta - k_R(C_X)^\gamma(C_z)^\upsilon = 0 \qquad (7.11)$$

However, the ratio of the forward and reverse rate constants will be equal to the equilibrium constant only if the forward and reverse reactions are elementary processes, that is, only if α, β, γ, and υ equal a, b, x, and z.

7.1.4 Generation and Treatment of Rate Data

The purpose of this section is to give a very brief review of some of the laboratory methods used in obtaining and treating rate data. For a more extensive discussion on this subject, the student should consult one of the references given at the end of this chapter.[1-4]

From the preceding discussion it follows that the form of the rate expression and the value of the rate constant for a reaction cannot be predicted or calculated and therefore must be determined experimentally. The most commonly used experimental method for homogeneous reactions involves adding known quantities of the reactants to a batch reactor arranged to operate isothermally at constant volume. The reactants are mixed thoroughly, and the change in concentration of the key component with time is measured. The data are then compared with various rate equations to find the one giving the best agreement. The comparison can be made by either the integration or differential method.

The integration method involves a comparison of the predicted and observed compositions of the reaction mixture as a function of time. To use this approach it is necessary to integrate the rate expression (e.g., Eq. 7.6) to give concentration as a function of time. To illustrate the approach that must be taken in the use of this method, let us assume that the reaction given in Eq. 7.1 is irreversible (i.e., goes essentially to completion) and that we wish to test our data to see if the rate of the reaction is first order with respect to reactant A. Writing Eq. 7.6 in a form that satisfies these assumptions gives

$$r = -\frac{dC_A}{dt} = k(C_A) \qquad (7.12)$$

If the initial condition is $(C_A) = (C_A)_0$ and $(C_A)_t$ represents the concentration of A at time t, then integration leads to

$$\ln \frac{(C_A)_0}{(C_A)_t} = k \cdot t \qquad (7.13)$$

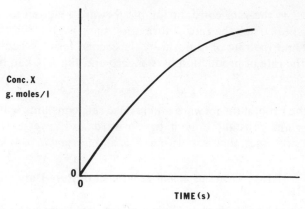

Figure 7.4 Concentration versus time data for the reaction $aA + bB \rightarrow xX + zZ$.

A plot of $\ln\left[(C_A)_0/(C_A)_t\right]$ versus t should be a straight line with slope equal to k. If such a plot is not a straight line then the rate of the reaction is not first order with respect to reactant A at the conditions under which the data were obtained, and a different order equation must be tested.

The differential method involves a comparison of predicted and observed rates obtained by differentiating the experimental data. For example, suppose the concentration versus time data shown in Fig. 7.4 were obtained for reaction 7.1. The slope of this curve at any point is equal to the rate of reaction since

$$r_x = \frac{dC_x}{dt}$$

Thus, a collection of values of r at different times can be obtained from a curve such as Fig. 7.4. Now if the reaction is first order in A, then the rate is given by Eq. 7.14.

$$r_x = dC_x = k_1 \cdot (C_A) \tag{7.14}$$

Equation 7.14 may also be written as

$$\log r_x = \log k_1 + \log (C_A) \tag{7.15}$$

Similarly, if the reaction is second order in A, then

$$r_x = \frac{dC_x}{dt} = k_2 \cdot (C_A)^2 \tag{7.16}$$

which may be written as

$$\log r_x = \log k_2 + 2 \log (C_A) \tag{7.17}$$

For the first-order case when the log r is plotted against log (C_A) then a straight line with a slope of 1.0 should be obtained. For the second-order case the result should be a straight line with a slope of 2.0 in accordance with Eq. 7.17.

Either of the two methods discussed can be used in the analysis of experimental data. However, the integration method sometimes tends to mask variations in rate constant calculations; and in such instances, the differential method should be used.

Even though we have only discussed the analysis of data for first and second order equations, the procedure for handling other orders is similar. Orders higher than 2 are not very common; however, fractional orders do exist when the reaction represents a sequence of several elementary steps.

For those reactions that proceed by a complex series of stages, the analytical treatment of rate data as described in the preceding discussion becomes extremely difficult. In such cases experiments are often undertaken to study the individual steps separately. However the rate equation for these individual steps may be very complicated and not susceptible to treatment by simple analytical techniques. In such cases it is necessary to resort to computer aided calculations. For a more complete discussion on the use of computers in chemical kinetics calculations, you should consult one of the references listed at the end of this chapter.[4,5] However, basically the procedure involves the use of the computer to perform numerical integration of the rate equations where the reaction rate equations, say for Eq. 7.1, may be written as

$$\frac{dC_A}{dt} = f_1(C_A, C_B, \ldots, C_i, \ldots) \tag{7.18}$$

$$\frac{dC_B}{dt} = f_2(C_A, C_B, \ldots, C_i, \ldots) \tag{7.19}$$

where $C_A, C_B, \ldots, C_i, \ldots$ are the concentrations of the reactants. For a small change of time δt each equation may be written in the form

$$\delta C_A = f_1(C_A, C_B, \ldots, C_i, \ldots)\,\delta t \tag{7.20}$$

If the data were not obtained under isothermal conditions, an equation similar to Eq. 7.20, relating the change of temperature of the reaction δT to the specific heat of the components Cp, the heat of reaction ΔH, and the concentration must be written

$$\delta T = f_T(\Delta H, Cp_p, Cp_Q, \cdots)\delta C_A \tag{7.21}$$

The computer can now be programmed to carry out the calculation of these changes in concentration and temperature for the time interval δt. These calculations are carried out for successive δt steps until the desired time range has been covered. This procedure is called stepwise numerical integration. Because the computer calculates very quickly, a very small time interval δt can be used in the calculation and accurate numerical integration can be achieved. The calculation is repeated with different values of the rate constants, activation energies, and so on, until these variables give results that match the experimental data.

Now that we understand how to obtain and analyze rate data for the purpose of generating a rate expression, let us see how the rate law can be used to study chemical reactions as they occur in both bench scale and pilot plant size equipment under various process conditions. First we consider batch reactors and then extend our study to include two types of flow reactors.

7.1.5 Batch Reactors

In the batch operation shown in Fig. 7.1 there are no streams entering or leaving the reactor. Therefore according to the mass balance equation given in Chapter 3,

$$\left\{ \begin{array}{l} \text{accumulation of} \\ \text{component in the} \\ \text{volume element} \end{array} \right\} = \left\{ \begin{array}{l} \text{quantity of} \\ \text{component formed} \\ \text{or converted in} \\ \text{the volume element} \end{array} \right\} \qquad (7.22)$$

The mass of a key component converted in time dt depends on the rate of reaction applicable to the volume element V. The accumulation term expresses the resultant change in mass of a key component in time dt caused by the reaction. The terms in Eq. 7.2 now take on the added meaning such that

$$r = \frac{1}{V} \cdot \frac{dN}{dt}$$

where N is the moles of the key component formed, V is the volume of the reacting mass, and r is the rate of reaction expressed in terms of the formation of the key component. Since formation of the key component by chemical reaction equals accumulation, Eq. 7.22 can be written in the form

$$rV\,dt = dN \qquad (7.23)$$

Application of the equation to batch reactors is demonstrated in Example 7.1.

Example 7.1

The condensation of the species B to form D occurs via a homogeneous liquid-phase reaction according to the following rate expression:

$$2B \longrightarrow D; \quad r_B = -k(C_B)^2$$

where C_B is the concentration of B in moles per liter and r_B is defined by Eq. 7.2. At 150°C the equilibrium constant for this reaction is 10^8, while the rate constant k equals 0.09 liter/mole · s.

A 400-liter batch reactor is charged with 300 liters of solution containing 100 moles of B. If the temperature is maintained at 150°C, how long does it take for 75 moles of B to react?

Solution

Since the equilibrium constant is large, the reaction will essentially go to completion and we do not have to be concerned with the effects of a reverse reaction in the rate expression.

Recognizing that C_B is equal to (N_B/V), we can write Eq. 7.23 for the above reaction as

$$r_B V dt = -k\left(\frac{N_B}{V}\right)^2 V dt = dN_B$$

Separating variables leads to,

$$dt = -\frac{V}{k} \cdot \frac{dN_B}{(N_B)^2}$$

when $t = 0$, $N_B = 100$, and when $t = t$, $N_B = 25$. Integrating the above equation leads to

$$\int_{t=0}^{t=t} dt = -\frac{V}{k} \int_{N_{B_1} = 100}^{N_{B_2} = 25} \frac{dN_B}{(N_B)^2}$$

$$t = \frac{V}{k}\left(\frac{1}{N_{B_2}} - \frac{1}{N_{B_1}}\right) = \frac{V}{k}\left(\frac{1}{25} - \frac{1}{100}\right)$$

Since V is the volume of solution that is equal to 300 liters, we can solve for t

$$t = \left(\frac{300 \text{ liters}}{0.09 \text{ liter/(mole)(s)}}\right)(\tfrac{1}{25}\text{ mole} - \tfrac{1}{100}\text{ mole}) = 100 \text{ s}$$

7.1.6   Flow Reactors

In many chemical processes the reactors are operated under continuous, steady flow conditions, rather than batchwise. This means that the feed enters the reactor and the product stream leaves the reactor with no accumulation occurring within the reactor. Even though temperature, pressure, and concentration may vary from one point to another in the reactor, they are constant with time at any fixed point. If this steady-state condition does not exist, analysis of the kinetics is impossible.

As discussed in Section 7.1.2, two contrasting types of flow reactors exist, the "tube reactor" and the "continuous stirred tank reactor." In the ideal tube reactor it is assumed that no axial mixing occurs and plug flow exists. In the idealized case of the continuous stirred-tank reactor, perfect mixing is assumed to occur. While these two extremes are idealizations, some reactors do approximate plug flow, whereas others approximate perfect mixing. In the following paragraphs we discuss these two extremes in reactor design.

Plug-Flow Reactors (PFR).   Consider the tube reactor of length L and cross section A, as shown in Fig. 7.5, in which the generalized homogeneous reaction in Eq. 7.1 is being carried out in either the gas or liquid phase.

Using the concepts developed in Chapter 3, the mass balance for a key reactant passing through a flow reactor such as shown in Fig. 7.5 can be written. For a time element Δt and a volume element $A \cdot \Delta L$ this balance takes the form

$$\begin{Bmatrix} \text{mass of key reactant} \\ \text{fed to volume} \\ \text{element} \end{Bmatrix} - \begin{Bmatrix} \text{mass of key reactant} \\ \text{leaving volume} \\ \text{element} \end{Bmatrix} - \begin{Bmatrix} \text{mass of key reactant} \\ \text{converted in the} \\ \text{volume element} \end{Bmatrix}$$

$$= \begin{Bmatrix} \text{accumulation of} \\ \text{key reactant in the} \\ \text{volume element} \end{Bmatrix} \quad (7.24)$$

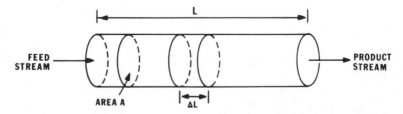

Figure 7.5   A tube flow reactor representing a volume element $A \cdot \Delta L$.

Equation 7.24 applies to a volume element $A \cdot \Delta L$ extending over the entire cross section of the reactor, as shown in Fig. 7.5. This is because there is no variation in properties or velocity in the radial direction. Suppose that the *key reactant* feed enters the PFR reactor at a constant rate of F moles per unit time and the conversion of the reactant at the entrance to the volume element is y. In the absence of axial mixing, reactant can enter the element only by bulk flow of the stream. Hence, the first term in Eq. 7.24 is $F(1 - y) \Delta t$. If the conversion leaving the element is $y + \Delta y$, then the second term is $F(1 - y - \Delta y) \cdot \Delta t$. Since the operation is at a steady state, there is no accumulation and the fourth term is zero. For the third term the amount of key reactant converted in the volume element $A \cdot \Delta L$ in time Δt is $-r(A \cdot \Delta L) \cdot (\Delta t)$, where $-r$ is the rate of disappearance of key reactant by chemical reaction; r is evaluated in terms of the average concentration of the reactant within the volume element. Placing these terms in Eq. 7.24 gives

$$F \cdot (1 - y) \cdot \Delta t - F \cdot (1 - y - \Delta y) \cdot \Delta t + r \cdot (A \cdot \Delta L) \cdot (\Delta t) = 0$$

which reduces to

$$F \cdot \Delta y + r \cdot (A \cdot \Delta L) = 0 \qquad (7.25)$$

Dividing by $A \, \Delta L$ and taking the limit as $\Delta L \to 0$ gives

$$\frac{dy}{A \, dL} = -\frac{r}{F} \qquad (7.26)$$

Equation 7.26 establishes the relationship between the extent of reaction and reactor length for PFR type reactors. It should be noted that the rate r in a tube reactor varies with longitudinal position whereas in the batch reactor expression in Eq. 7.23, the rate varies with time.

Since F is a constant (i.e., the molar flow rate of a single *key reactant* measured under conditions existing at the reactor inlet), Eq. 7.26 can be written in the integrated form as

$$\frac{AL}{F} = \frac{V_R}{F} = \int_{y_1}^{y_2} \frac{dy}{-r} \qquad (7.27)$$

where V_R represents the volume of a reactor with length L and cross section area A. Normally the conversion of the key reactant at the reactor inlet, y_1, is zero. Integration of Eq. 7.27 can be performed when r is known as a graphical or analytical function of y, as illustrated in Ex. 7.2.

Example 7.2

Ethane is decomposed in a PFR reactor according to the reaction

$$C_2H_6 \longrightarrow C_2H_4 + H_2$$

The reaction when carried out at 650°C is found to follow first order kinetics, with a specific rate constant of 1.2 sec^{-1}.

It is planned to build a PFR reactor of sufficient length to convert 25% of the ethane feed to ethylene and hydrogen. The feed rate is to be 0.01 moles/s. The reactor is to be operated at a temperature of 650°C and 1 atm of pressure absolute.

If the reactor is constructed of $\frac{1}{2}$ in. i.d. 316 stainless steel pipe, what length must it be to give the desired conversion?

Solution

The rate equation for this reaction may be written as

$$r = -k[C_2H_6] = -(1.2) \cdot (C_2H_6) = \frac{g \cdot moles}{(liter)(s)}$$

At a point in the reactor where the conversion is y, the mole flow rates are (where F is the feed rate of the reactant in moles/s):

$$C_2H_6 = F(1 - y)$$
$$C_2H_4 = F \cdot y$$
$$H_2 = F \cdot y$$

Choosing C_2H_6 as the key component and a basis of 1 mole of feed we can set up the following table. (Note: in this problem the feed only contains C_2H_6, which is the reactant and key component. Therefore, the reactant feed rate and the total feed rate are equal. If an inert component had been in the feed mixture, then the total feed and reactant feed rates would not be equal.)

Moles

Component	Feed	At Any Conversion	At 25% Conversion
C_2H_6	1.0	$(1 - y)$	0.75
C_2H_4	0	y	0.25
H_2	0	y	0.25
Total, N_t	1.0	$1 + y$	1.25

If we assume ideal gas behavior, and that the volume of one mole of gas at 650°C and 1 atm is represented by $\mathscr{V}$, then for constant pressure operation the rate equation at any conversion can be expressed as:

$$r_{C_2H_6} = -k[C_2H_6] = \frac{-k(1-y)}{(1+y)\mathscr{V}}$$

This expression results from the consideration that the $(1 - y)$ moles of ethane are contained in a volume of $(1 + y) \cdot \mathscr{V}$.

Introducing this into Eq. 7.27 gives

$$\frac{AL}{F} = \int_{y_1}^{y_2} \frac{dy}{-r} = \frac{\mathscr{V}}{k} \int_{y_1}^{y_2} \frac{(1+y)\,dy}{(1-y)}$$

The integration limits are $L = 0$ when $y_1 = 0$ and $L = L$ when $y_2 = 25$. Integration and substitution of these values gives:

$$\frac{AL}{F} = \frac{\mathscr{V}}{k}[-y - 2\ln(1-y)]_0^{25} = \frac{\mathscr{V}}{k}[-0.25 - 2\ln(0.75)]$$

$$= (0.325)\frac{\mathscr{V}}{k}$$

The specific volume $\mathscr{V}$ of an ideal gas at 650°C and 1 atm pressure is 75.7 liters/mole. The cross sectional area of a $\frac{1}{2}$-inch pipe is 1.27 cm². Substituting these values at a feed rate F of 0.01 mole/s gives:

$$L = \frac{(0.325)(\mathscr{V})(F)}{(k)(A)} = \frac{(0.325)(75.7 \times 10^3 \text{ cm}^3/\text{mole})(10^{-2} \text{ mole/s})}{(1.2/\text{s})(1.27 \text{ cm}^2)}$$

$$L = 161.4 \text{ cm}$$

This is the reactor length required to achieve 25% conversion of the ethane feed.

Space Time and Space Velocity. The term V_R/F evaluated from Eq. 7.27 determines the size of the flow reactor necessary to process a given reactant feed rate F. More often, however, the size of the reactor is evaluated by using an expression that involves the total volumetric fluid feed rate to the reactor. For continuous flow reactors the ratio of the reactor volume (V_R) to the total volumetric fluid flow rate at the reactor inlet (Q_0) is called the space time (t_F).

$$t_F = \frac{V_R}{Q_0} = \frac{V_R \cdot F}{C_F} \tag{7.28}$$

The molal feed rate, F, of a key reactant may be written as the product of the volumetric flow rate and the concentration of the key reactant, C_F, (i.e., $F = C_F \cdot Q_0$). The terms F, Q_0, and C_F must be measured at the same conditions. When calculating the space time by Eq. 7.28, the conditions [i.e., the temperature, the physical state (gas or liquid) and pressure] at which the volumetric flow rate to the reactor is measured must be indicated. For most problems the volumetric flow rates are chosen to be those that prevail at the reactor inlet conditions. The reciprocal of the space time is the space velocity v_F

$$v_F = \frac{1}{t_F} = \frac{Q_0}{V_R} = \frac{C_F}{V_R \cdot F} \qquad (7.29)$$

The space velocity gives an indication of relative reactor size required to obtain a given conversion. For example, a high space velocity means that the reaction can be accomplished with a small reactor, on the other hand, a low space velocity indicates that a large reactor is required. (Note: in using Eqs. 7.28 and 7.29 the total key feed rate must be used, not the key reactant feed rate, although in many situations these will be the same.)

The space time is usually not equal to the actual time an element of fluid resides in the reactor. Variations in the temperature, pressure, and moles of reaction mixture can all cause the local density of the reaction mixture to vary throughout the reactor. Also, deviations from ideal tubular flow behavior, such as mixing in the longitudinal direction, result in a distribution of residence times of the fluid elements in the reactor. Therefore, it is necessary to develop a concept of the mean residence time t_θ. The mean residence time is equal to the space time t_F only under the following conditions:

1. The feed rate is measured at the temperature and pressure in the reactor.
2. The temperature and pressure are constant throughout the reactor.
3. The density of the reaction mixture remains constant throughout the reactor.
4. There is no change in the number of moles in a gaseous reaction.

If the above conditions are not met, then the residence time can be calculated only if the effects of these variations are known. We can evaluate the consequence of some of these effects by considering that the actual time dt_θ required for a particle of fluid to pass through a volume element dV is

$$dt_\theta = \frac{\text{volume}}{\text{volumetric flow rate}} = \frac{dV}{F_m / \rho} \qquad (7.30)$$

where ρ is the local mass density at the point dV in the reactor and F_m is the *total* mass feed rate. For gaseous reactions it is more convenient to write Eq. 7.30 in terms of the total molar flow rate, F_t, and the volume per mole of gas, $\mathscr{V}$. (Note: F_t is total moles of fluid entering the volume element dV per unit of time. The term F used in previous equations was the molal feed rate of a key reactant. Of course at the reactor inlet $F_t = F$ if the feed contains only a single reactant and no inert materials.) In these terms the residence time in Eq. 7.30 becomes

$$dt_\theta = \frac{dV}{F_t \mathscr{V}} \tag{7.31}$$

When ρ, F_t, and $\mathscr{V}$ are constant, Eqs. 7.30 and 7.31 show that $t_\theta = t_F$. When these terms are not constant, then the residence time t_θ must be obtained by integration. This can be accomplished by recognizing that $A \cdot dL$ is equal to dV and that dV from Eq. 7.26 is

$$dV = \frac{F dy}{-r}$$

Substituting this value for dV into Eq. 7.31 gives

$$dt_\theta = \frac{F dy}{-rF_t \mathscr{V}} \tag{7.32}$$

Integrating over the length of the reactor yields

$$t_\theta = F \int_{y_1}^{y_2} \frac{dy}{-r\mathscr{V} F_t} \tag{7.33}$$

The time t_θ in Eq. 7.33 is the mean time that is required for a slug of fluid of differential thickness to travel from one end of the tube reactor to the other. It should be noted that the term F_t appears inside the integral sign. The reason for this is that F_t will only be constant if there is no net change in the total number of moles due to the reaction. Otherwise F_t will be a function of the conversion y. Example 7.3 illustrates the application of Eq. 7.33.

Example 7.3

A general homogeneous gas phase decomposition reaction and its corresponding rate expression are as follows:

$$S \longrightarrow U + W; \qquad r_s = -k[C_s]^2$$

r_s is the rate of the reaction and C_s is the concentration of S in moles/liter. The second order rate constant k is 0.25 liter/(s)(mole) at 500°C. At this temperature the equilibrium constant for this reaction is very large. The reactor is 2.5 cm i.d. and 100 cm long and is maintained at a constant temperature of 500°C. The pressure in the reactor is maintained essentially at 1 atm absolute. While feeding only S, one set of data at a space velocity of 45 $(hr)^{-1}$ (calculated at reaction conditions) registered a 20% decomposition of S. Under these conditions, calculate the mean residence time and compare it to the space time.

Solution

Letting y represent the degree of conversion, we can set up the following molal flow rate table:

	Moles	
Component	Feed	At Conversion y
S	F	$F(1 - y)$
U	0	Fy
W	0	Fy
Total	F	$F(1 + y)$

The rate expression can be converted into a useful form by expressing the concentration of S in terms of the conversion y. The molal concentration C_s is equal to the ratio of moles of S at conversion y to the total volumetric flow rate.

$$C_s = \frac{F(1 - y)}{F_t \mathscr{V}}$$

where F_t, as indicated previously, is the total moles of fluid at conversion y and $\mathscr{V}$ is the specific molar volume. From the molar table, $F_t = F(1 + y)$, and if we assume ideal gas behavior $\mathscr{V} = (R_g)(T)/P$. Therefore

$$C_s = \frac{F(1 - y)}{F(1 + y)(R_g T/P)} = \frac{1 - y}{1 + y} \cdot \frac{P}{R_g} \cdot T$$

Then the rate expression becomes

$$r = -k \left(\frac{1 - y}{1 + y}\right)^2 \cdot \left(\frac{P}{R_g T}\right)^2$$

Substituting the above values for r, $\mathcal{V}$ and F_t in Eq. 7.33 gives

$$t_\theta = F \int_{y=0}^{y=0.20} \frac{(1+y)^2 \, dy}{Fk(1+y)(R_g T/P)(1-y)^2(P/R_g T)^2}$$

$$= \frac{R_g T}{Pk} \int_{y=0}^{y=0.20} \frac{(1+y)\, dy}{(1-y)^2}$$

$$t_\theta = \frac{R_g T}{Pk} \left[\frac{2}{(1-y)} + \ln(1-y) \right]_0^{0.20}$$

$$t_\theta = \frac{(0.082)(500+273)}{(1)(0.25)} \left[\frac{2}{(1-0.20)} + \ln(1-0.20) - 2 \right]$$

$$t_\theta = 71 \text{ s}$$

The space time can be calculated from Eq. 7.29 and the space velocity given in the problem

$$t_F = \frac{1}{v_F} = \frac{(1)(3600)}{45} = 80 \text{ s}$$

The difference between the space time and the active residence time t_θ is due to the increase in the number of moles in the reactor during the course of the reaction.

Continuous Stirred-Tank Reactors (CSTR). Our discussion is restricted to CSTR reactors of the type shown in Fig. 7.2 operating under steady-state flow conditions. In this type of reactor mixing may be assumed to be complete, so that the properties of the reaction mixture are uniform in all parts of the vessel and are the same as those in the exit stream. This means that the composition and tempeature at which the reaction takes place are the same as the composition and temperature of the exit stream.

A relation analogous to Eq. 7.27 can be developed for the ideal CSTR reactor by carrying out a mass balance on the primary reactant. Consider the simple case of a single feed stream and a single product stream, as shown in Fig. 7.2. Since the properties of these streams do not change with time, the first two terms in the mass balance expression given by Eq. 7.24 are constant. If the molar feed rate of a key reactant component is again represented by F and its conversion in the feed stream is y_F, then over the time element Δt the first term in Eq. 7.24 is $F(1 - y_F)\Delta t$. The second term represented by $F(1 - y_E)\Delta t$, where y_E is the conversion of reactant in the exit stream. The third term in the mass balance expression can be evaluated by recognizing that since the reaction mixture in the vessel is at the same composition and temperature as the exit stream, the rate of the reaction must be constant and can be evaluated at the temperature and composition of the product stream. If the rate of conversion of a key reactant is r_p, with the subscript p indicating product conditions, the third term in Eq. 7.23 is $(-r_p)(V_{\text{stirred tank}})(\Delta t)$, where $V_{\text{stirred tank}}$ is the volume of the CSTR reactor. Since there can be no accumulation of mass of reactant in the reactor at steady-state conditions, the fourth term is zero. Therefore, Eq. 7.24 can now be written as

$$F(1 - y_F) \cdot \Delta t - F(1 - y_E) \cdot \Delta t + (r_p)(V_{\text{stirred tank}})(\Delta t) = 0 \qquad (7.34)$$

Rearranging gives

$$\frac{V_{\text{stirred tank}}}{F} = \frac{y_E - y_F}{-r_p} \qquad (7.35)$$

This expression can be used to calculate the exit conversion for known conditions F, y_F, $V_{\text{stirred tank}}$ and knowledge about the rate r_p. Likewise, this expression can also be used to calculate the volume of the reactor needed to give a specific conversion of reactant y_E if the reaction rate is known and a feed rate is assumed.

If the volume of the reaction mixture (i.e., the density) is constant, Eq. 7.35 can be written in simpler concentration terms. It can be shown (for a more thorough treatment, see Reference 1) that Eq. 7.35 under these conditions can be expressed as

$$\frac{V_{\text{stirred tank}}}{Q} = \frac{C_0 - C}{-r} \qquad (7.36)$$

where Q is the volume flow of feed, C_0 is the initial concentration of reactant in the feed, and C is the concentration of reactant in the product stream. The rate $-r$ refers to the rate of disappearance of the reactant. Normally only liquids are processed in CSTR reactions where volumetric expansion effects are usually negligible. Consequently Eq. 7.36 is applicable to most CSTR problems.

When the density of the reaction mixture is constant the term $V_{\text{stirred tank}}$ in Eq. 7.36 is the average residence time $\bar{t}_\theta$ for fluid particles flowing through the CSTR reactor. It is easily shown experimentally that there is a wide variation in the residence times of different particles of a liquid stream flowing through a well-stirred reactor. For example, assume that A and B are two inert but miscible liquids with the same density. Initially the CSTR reactor in Fig. 7.2 with a volume of $V_{\text{stirred tank}}$ liters is filled with liquid A. In addition, A is being charged and discharged from the vessel at a steady rate of Q liters per unit time. Without changing flow rates, and assuming perfect mixing, liquid B is suddenly substituted for liquid A in the feed

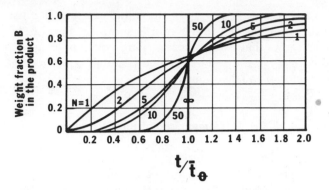

Figure 7.6 Distribution times for a series of stirred-tank reactors.

streams, consequently B is instantly mixed into the contents of the reactor and starts to appear in the product stream. The experimentally determined volume fraction of B in the reactor and in the product stream is plotted against the ratio of elapsed time t to the average residence time $\bar{t}_\theta = V_{\text{stirred tank}}/Q$, as shown by curve 1 in Fig. 7.6. Since $t/\bar{t}_\theta$ is zero at the moment of changeover from A to B, points on curve 1 represent the fraction of the contents of the reactor that have been present for $t/\bar{t}_\theta$ units or less. For example, when $t/\bar{t}_\theta$ is 1.65, the mixture in the reactor is found by analysis to be 80% B, indicating that 80% of the contents have been present in the tank for 1.65 reduced time units, or less. While we have analyzed the experimental method for generating curve 1, it is also possible to reproduce this plot by using a mathematical treatment. For additional discussions on this subject you should consult References 1 to 3.

In a series of CSTR reactors as shown in Fig. 7.7, the single CSTR reactor of volume $V_{\text{stirred tank}}$ discussed in the preceding paragraph is replaced by two or more identical vessels of such size that the total volume of the vessels

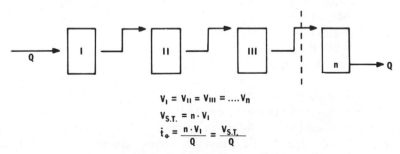

$$V_I = V_{II} = V_{III} = \dots V_n$$
$$V_{S.T.} = n \cdot V_I$$
$$\bar{t}_\theta = \frac{n \cdot V_I}{Q} = \frac{V_{S.T.}}{Q}$$

Figure 7.7 A series of stirred-tank reactors.

is equal to the volume $V_{stirred\,tank}$ of the initial reactor. The distribution of residence times for various values of n is shown in Fig. 7.6. Inspection of these curves shows that as n increases, plug type flow is approached; and indeed when $n = \infty$, the residence time becomes equal to that of a plug flow tube reactor of equal volume.

7.1.7 Comparison of Flow Reactors

In a CSTR reactor the reaction occurs at a rate determined by the composition of the product stream from the reactor. Since the rate generally decreases as conversion increases, the stirred-tank reactor operates at the lowest rate in the range between the high rate corresponding to the reactant concentration in the reactor feed and the low rate corresponding to the concentration in the product stream. In the PFR reactor high rates can be achieved in the first section of the reactor where the conversion is low. This means that for a given feed rate, the CSTR reactor must have a larger volume. Of course, it must be remembered that the total volume required for the CSTR reactor can be reduced by using several small units in series.

As previously indicated, because the contents of a CSTR reactor are mechanically mixed, a uniform temperature, pressure, and composition may be attained. In certain situations this gives the CSTR reactor a definite advantage over the PFR reactor. For example, if it is desirable to carry out a reaction under isothermal conditions, it can be easily accomplished in a CSTR reactor, even when the heat of reaction is high. This could not be accomplished in a long PFR reactor, for it is seldom feasible to supply internal cooling to this type of reactor. Operation under isothermal conditions permits the running of the reaction at an optimum temperature, which is definitely an advantage for the CSTR reactor. However, as discussed in Section 7.1.6, the residence time distribution for these two reactors is different. Thus, when consecutive reactions or side reactions can occur, the small distribution in residence times for the PFR reactor may give a greater selectivity than the isothermal operation of the CSTR reactor. The next section illustrates how selectivity varies with residence time.

Another factor to consider when comparing reactors is that high pressure gas reactions are normally carried out in PFR reactors. This is based primarily on cost factors. A CSTR reactor operated at high pressure would require thick walls of significant area and a complex sealing arrangement for the mixer shaft, both of which lead to high initial and maintenance costs. Because of problems in providing the proper type of heat-transfer surface as discussed in Chapter 5, CSTR reactors are not generally used when high reaction temperatures are required (i.e., $> 500°C$).

7.1.8 Selectivity in Flow Reactors

Both simultaneous and sequential reactions are frequently encountered in reacting systems. This is especially true of industrial organic operations. For example in the chlorination of benzene to produce chlorobenzene, additional chlorination of the product leads to para- and ortho-dichloro-benzenes. In any operation where simultaneous and sequential reactions may exist, whether homogeneous or heterogeneous, the question of selectivity is important.

Selectivity is generally defined as the moles of desired product formed, divided by the moles of a given reagent that is consumed. The selectivity of a reaction may be significantly affected by the residence time as illustrated in the following paragraphs.

Let us consider the simplest case of a sequential reaction, the isothermal conversion in a batch reactor of A to B and of B to C by two first order homogeneous reactions:

$$A \xrightarrow{\ k_1\ } B \xrightarrow{\ k_2\ } C \tag{7.37}$$

Let A, B, and C also represent the mole fractions of these materials at any time t and when $t = 0$, the starting concentration of A is A_0, then

$$A_0 = A + B + C \tag{7.38}$$

The change in A in time dt is

$$dA = -k_1 A \, dt \tag{7.39}$$

which integrates to

$$A = A_0 e^{-k_1 t} \tag{7.40}$$

The net change in B is the result of formation from A, and loss to form C, giving,

$$dB = k_1 A \, dt - k_2 B \, dt \tag{7.41}$$

Introducing Eq. 7.40 and rearranging, gives

$$dB/dt = k_1 A_0 e^{-k_1 t} - k_2 B \tag{7.42}$$

This is a linear equation of the first order in B. Using an integrating factor of $e^{k_2 t}$ gives,

$$B \cdot e^{k_2 t} = \int_0^t (A_0 k_1 e^{-k_1 t} - e^{k_2 t}) \, dt + \text{constant} \tag{7.43}$$

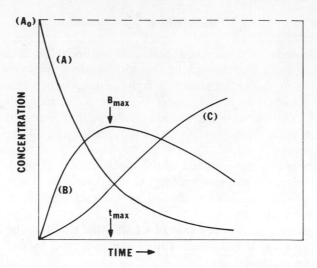

Figure 7.8 Variation of the concentration of A, B, and C with time, for a reaction of the type $A \rightarrow B \rightarrow C$.

This integral can be solved by recognizing that $B = 0$ when $t = 0$, which gives

$$B = \frac{A_0 k_1}{k_2 - k_1} \cdot (e^{-k_1 t} - e^{-k_2 t}) \qquad (7.44)$$

Then, combining Eqs. 7.38 and 7.44, where $t = 0$ and $C_0 = 0$ gives

$$C = A_0 \left[1 + \frac{1}{k_1 - k_2} \cdot (k_2 e^{-k_1 t} - k_1 e^{-k_2 t}) \right] \qquad (7.45)$$

The variation of Eq. 7.40, 7.44, and 7.45 with time is shown in Fig. 7.8. The concentration of A decreases with time in a simple exponential manner. That of B is initially zero, and it passes through a maximum, ending at zero since all of B is finally converted into C. The curve for the concentration of C is S-shaped with an initial induction period during which time very little C is formed. The existence of such induction periods is an indication that a product is not formed directly, but via some intermediate.

Particular notice should be made of the occurrence of a maximum in the intermediate of concentration B_{max} at time t_{max}. At this point $dB/dt = 0$ and from Eq. 7.41.

$$k_1 A = k_2 B \qquad (7.46)$$

Table 7.1 Comparison of Selectivity for Various Types of Reaction Systems in PFR and CSTR Reactors[a]

Reaction System	Reaction Type	Selectivity
1. A → B B → D	Consecutive first order	Selectivity of B with respect to D. Selec. CSTR < PFR
2. A → B A → C	Parallel equal first order	Selectivity of B with respect to C. Selec. CSTR = PFR
3. A + B → C 2A → D	Parallel equal order with respect to A	Selectivity of C with respect to D. Selec. CSTR > PFR

[a]Modified from Reference 6.

Substituting into this equality the values for A and B from Eq. 7.40 and 7.44, it can readily be shown that (see References 1 to 3),

$$t_{max} = \frac{(k_1)\ln(k_1/k_2)}{1 - (k_2/k_1)} \qquad (7.47)$$

It is apparent from Fig. 7.8 that for maximum selectivity in the conversion of A to B via reaction 7.37 the residence time in the reactor should be set at the value t_{max} as calculated by Eq. 7.47. Now as we have seen in Section 7.1.6, while the average residence times in PFR and CSTR reactors may be the same, the distribution of residence times is different. Since in a PFR reactor all reactants could be held for the time t_{max}, and this cannot be done with a single CSTR reactor, we would expect the selectivity towards B to be greater in the PFR reactor.

It has been shown that depending on the kinetics and the nature of the sequential and simultaneous reactions, selectivity obtained in a CSTR reactor may be less, the same as, or greater than for a PFR reactor.[1] Examples of selectivities in several different types of sequential and simultaneous reactions are given in Table 7.1. The order of the rate equation is assumed to follow the stoichiometry for each reaction.

7.2 HETEROGENEOUS REACTIONS

The kinetics and reactor systems we have considered thus far in this chapter are for homogeneous reactions. However many important chemical processes involve heterogeneous reactions. This is because many reactions require a catalyst to achieve practical rates, and the catalyst is commonly in a phase different from that of the reactants and products. Therefore it is important that we learn some of the basic characteristics of heterogeneous catalysis and the kinetics of heterogeneous catalytic systems.

7.2.1   Heterogeneous Catalysis

Most industrial catalytic processes are based upon solid catalyzed vapor phase reactions. These reactions can be visualized as occurring in five steps.

1. Diffusion of the reactants through the bulk fluid around the surface of the solid, and diffusion through the pores of the solid to the catalytically active surface.
2. Adsorption of reactants on the catalytic surface.
3. Reaction of the adsorbed reactants to form products.
4. Desorption of adsorbed products.
5. Diffusion of the products back through the pores and surface film to the bulk fluid around the solid.

At one time it was generally believed that one of the diffusion processes, either 1 or 5, must be the slowest step and therefore determined the overall reaction rate in all heterogeneous catalyst systems. However, more detailed studies of surface reactions have shown that this is primarily true only for very porous catalyst systems. For other less porous heterogeneous catalysts it has been found that appreciable activation energies exist. Since diffusion in the gaseous state involves no activation energy; the diffusion process is much more rapid than the overall process and cannot constitute the slow step.

The processes of adsorption or desorption are much more likely to be the slow steps in heterogeneous reactions, since both may involve appreciable energies of activation. The activation energies for desorption are particularly high, and it is likely that in many reactions the desorption of the products is the rate determining step. However, in practice it is not always convenient to separate steps 3 and 4 because the rate of desorption of the products is usually not known; therefore, it is usual to regard the reaction as taking place on the surface with gaseous products being produced in a single step.

Chemists, in general, believe that catalysts operate by providing new reaction pathways that are different than the noncatalyzed pathways. To illustrate the concept that a catalyst provides an alternate mechanism for accomplishing a reaction, and that this alternate path is a more rapid one, let us examine the stoichiometric reaction $A + B \rightarrow D$. The kinetics of the heterogeneous catalyzed reaction can be represented by the steps:

1. *Adsorption*

$$A + X_1 \longrightarrow AX_1$$

$$B + X_2 \longrightarrow BX_2$$

2. *Surface reaction*

$$AX_1 + BX_2 \longrightarrow D(X_1 + X_2)$$

3. *Desorption*

$$D(X_1 + X_2) \longrightarrow D + X_1 + X_2$$

where X_1 and X_2 are neighboring active sites on the solid catalyst. In the adsorption step A reacts with an active site X_1 on the catalyst to give AX_1, and B reacts with the active site X_2 to give the adsorbed form of B. On the surface AX_1 and BX_2 react to give $D(X_1 + X_2)$, the adsorbed form of product D. D is desorbed and releases the active sites X_1 and X_2. The reaction is truly catalytic if the sequence of steps is such that the sites X_1 and X_2 are regenerated after they have caused the formation of D.

The energy changes during the heterogeneous catalytic reaction of $A + B \to D$ are compared in Fig. 7.9 with the changes for a noncatalytic

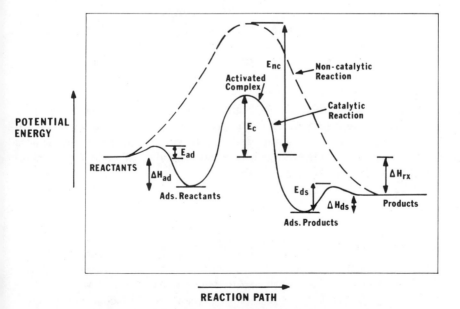

Figure 7.9 Energy profile for catalytic and noncatalytic reactions.

homogeneous mechanism. Consider first the noncatalytic reaction path. The potential energy of the reactant rises to a maximum as the reaction proceeds, and the difference between the initial energy and the activated complex is the activation energy E_{nc} of the noncatalytic reaction. The difference in energy ΔH_{Rx} between reactants and products is of course the heat of the reaction. In this case since ΔH_{Rx} is negative, the reaction is exothermic.

Now, consider the heterogeneous catalyzed reaction path. This path contains the three steps described previously. First, there is adsorption, which has its own activation energy E_{ad} and its own heat of reaction ΔH_{ad}. The second and third steps involving the surface reaction and the desorption also have characteristic activation energies and heats of reaction. If the surface reaction is rate controlling, the overall activation energy, E_c, for the catalytic path is the difference between the maximum energy for the activated complex and the initial reactant energy. As seen in Fig. 7.9, the activation energy for the catalytic reaction is significantly less than the activation energy for the noncatalytic reaction. Consequently, since the reaction rate varies inversely and exponentially with activation energy, the catalytic reaction will be much faster at a given temperature.

7.2.2 Heterogeneous Rate Equation in a Fixed Bed

It has been observed that most surface reactions between two substances, such as the reaction of A and B described in the preceding section, occur by reaction between two molecules that are adsorbed on neighboring surface sites. If the surface reaction is rate controlling (i.e., not diffusion limited) then the rate of a reaction between A and B is proportional to the probability that A and B are adsorbed on neighboring sites. Assuming that all catalytic active sites are alike, then for relatively low surface coverage the probability of A and B being on an active site is proportional to C_A and C_B, the concentrations respectively of A and B above the surface, the total surface area S_C, and the fraction of the total surface that is catalytically active F_S. Therefore, the rate of formation of D in the reaction

$$A + B \longrightarrow D$$

can be written as

$$r_D = -r_A = -r_B = k' S_C F_S (C_A)^{\alpha} (C_B)^{\beta} \tag{7.48}$$

The moles of D formed per unit time and per unit weight of a catalyst bed are determined by the area of the catalytically active surfaces in the bed. For a given catalyst of any stated particle size, the active surface area is proportional to the bed weight. Since catalytic active area is not easily measured, the rate of the catalytic reaction is conveniently expressed as moles of compound produced (or consumed) per unit time and per unit weight of the particular catalyst used. Therefore, the rate of formation of D as given in Eq. 7.48 can be expressed in the form

$$r_D = \frac{dn_D}{W\,dt} = kC_A^{\alpha} C_B^{\beta} \tag{7.49}$$

The term n_D represents the moles of D present, and W is the catalyst weight.

Consider that a fixed catalyst bed is contained in a PFR reactor of the type discussed in Section 7.1.6. The catalyst bed in the reactor has a cross-sectional area A, a volume V, and a packed density ρ_c. The reaction mixture moves through the reactor in plug flow, and the system is under steady-state operation. The feed of a key reactant component to the bed is constant at F moles per unit time. If the conversion of reactants is y at any cross-sectional volume element located at a distance L from the entrance, then, as we have done before by using a mass balance on a key reactant such as B over a differential length dL of the catalyst bed, we obtain

$$F \cdot dy = -r_B \cdot dW = -r_B \cdot \rho_c \cdot A \cdot dL \tag{7.50}$$

If y_1 is the conversion of key reactant that enters the reactor when L is zero and y_2 is the conversion when the key reactant exits the reactor, then

$$L = \frac{F}{A \cdot \rho_c} \int_{y_1}^{y_2} \frac{dy}{-r} \tag{7.51}$$

It should be noted that this expression is very similar to Eq. 7.27 developed in Section 7.1.6.

7.2.3 Reaction Rates over Porous Catalysts

As discussed in the next section, the majority of commercial fixed bed catalysts consist of spherical or cylindrical pellets in the size range of $\frac{1}{16}$ to $\frac{1}{2}$ in. The lower size limit is determined by the pressure drop through the bed and the upper limit is set by diffusional limitations within the pellet pores. As the reactants diffuse into the pores they begin to react, resulting in a decrease in their concentration. If the pores in the catalyst are deep, the reactant concentration may drop to zero before the middle of the pellet is reached. In such a case only the outer layers of the pellet are being effectively used. Any active catalyst sites beyond the region at which the reactant concentration reaches zero are wasted. Since catalyst sites often are made of valuable metals representing a sizable investment, and also because the unused portion of catalyst takes up reactor space, it is desirable to prepare catalysts that have activity throughout the entire pore depth.

The level of activity of a pelletized catalyst in which this type of diffusion problem exists is indicated by an effectiveness factor, η. The effectiveness factor is defined by Eq. 7.52:

$$\eta = \frac{\text{rate of reaction for whole pellet}}{\text{rate of reaction for completely pulverized pellet}} = \frac{r_p}{r_s} \qquad (7.52)$$

The rate equation (per unit mass of catalyst) as developed in the preceding section can be expressed functionally as $r = f(C, T)$, where C represents, symbolically, the concentrations of all the involved components and T the temperature. Applying this concept to Eq. 7.52 gives the reaction rate over a pellet as

$$r_p = \eta r_s = \eta f(C_S, T) \qquad (7.53)$$

In order to utilize Eq. 7.53, we must have some method for determining η and r_s. However, r_s is just the rate of the surface reaction as discussed in Section 7.2.2, because the pulverization step exposes all of the internal area of the catalyst to the same reactant concentration that exists at the pellet surface.

The effectiveness factor is a function of the diffusion coefficient D_e and the rate constants associated with the chemical reaction at the catalyst site. The relationships between η and these parameters have been formulated for a number of reactions of various orders. We will illustrate their application with a simple example. For a more complete discussion on this subject, References 1 to 4 should be consulted.

We can begin by considering a flat catalyst pellet as shown in Fig. 7.10. The pores are cylindrical with radius R, extend through the entire pellet,

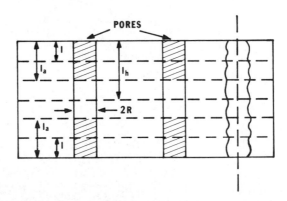

Figure 7.10   Catalyst pellet with cylindrical pores. (The cross-hatched regions represent the active section of the pore.)

and therefore have a length equal to the pellet thickness. The nonreversible reaction of $A \to B$ is occurring at a constant temperature on the catalyst surface. For the purpose of this discussion, let us assume that the reaction is zero order. Consequently, the number of moles of A reacting per unit time per unit pore area is constant at k regardless of the concentration of A until all of A has been consumed. The section of pore wall that is actively catalyzing the reaction is shown as the cross-hatched region in Fig. 7.10 with a length of l_a. No reaction occurs in the pore between the two active sections because the reactant has been completely consumed before diffusing into this region.

The rate of disappearance of A by reaction on the pore walls between l_a and the depth l must equal the amount entering this section of the pore by diffusion at l.

$$\begin{pmatrix} \text{Amount of } A \\ \text{diffusing past } l \end{pmatrix} = \begin{pmatrix} \text{amount of } A \text{ reacting} \\ \text{between } l \text{ and } l_a \end{pmatrix} \tag{7.54}$$

If we let D_e represent the coefficient for diffusion of A we can express this equality as

$$\frac{-D_e \pi R^2 \, dC}{dl} = 2\pi R(l_a - l)k \tag{7.55}$$

where C is the concentration of reactant A. If C_0 is the concentration of A at either entrance to the pore when l is zero, we can write Eq. 7.55 as follows

$$-\int_0^{C_0} dC = \int_{l_a}^0 \frac{2k(l_a - l) \, dl}{D_e R} \tag{7.56}$$

Integration of this equation gives

$$l_a = \sqrt{\frac{C_0 D_e R}{k}} \tag{7.57}$$

The effectiveness factor, by definition, must equal the ratio of the active pore length, l_a, to half the total length of the pore l_h. Therefore,

$$\eta = \frac{l_a}{l_h} \tag{7.58}$$

Substituting Eq. 7.57 into this expression gives

$$\eta = \frac{\sqrt{\dfrac{C_0 D_e R}{k}}}{l_h} \tag{7.59}$$

When $\eta < 1$ the central portion of the catalyst pellet is not being utilized. Inspection of Eq. 7.59 shows that such a situation is caused by large pellets (i.e., when l_h is large), by small values of D_e (i.e., low rate of diffusion), by low reactant concentrations in the zone above the catalyst surface, and by fast reactions (i.e., a large value for the rate constant k). This last factor shows that low effectiveness factors are more likely to exist with a very active catalyst that has large k values.

Expressions very similar to Eq. 7.59 have been obtained for pellets in which first and second order reactions occur.[1] In addition, they have also been obtained for different shaped pellets, such as spherical and cylindrical forms.

7.2.4   Catalyst Preparation and Characterization Methods

The preceding discussion should have provided a basis for understanding the importance of producing and maintaining a large surface area in heterogeneous catalyst systems. Catalysts currently being used in industrial processes generally range in total surface area between 4 and 1400 m²/g. In this section we consider the methods and materials used to produce these high surface area catalysts and some of the more common methods used to measure the surface area.

Catalyst Preparation. Industrial heterogeneous catalysts are generally complex materials. A knowledge of the chemical composition is not sufficient to define the activity of the catalyst. To a great extent activity is also determined by such physical properties as particle size, surface area, pore size, and pore volume. These properties are determined by the detailed preparation procedure; and, unless great care is taken to reproduce these procedures, materials showing wide variations in performance are produced.

Heterogeneous catalysts can basically be grouped into two broad classes: those in which the entire catalyst is composed of one or more of the active components, and those in which the active component is dispersed upon a support or carrier having a large surface area. In industrial processes supported catalysts find the greatest and most diverse use of all catalyst systems.

When the active component makes up the entire catalyst it is usually made by coprecipitation of two or more active components. The high internal area of the catalyst is then developed by the thermal decomposition of the precipitated compound. The most common precipitates used are metal hydroxides, carbonates, nitrates, and acetates. During the heat treatment a network of very fine interconnecting pores is created inside the catalyst particle. The walls of these pores make up the largest fraction of the total surface area of the catalyst. The pore structure of the final catalyst is influenced by such variables as concentrations of precipitating solutions and the time and temperature of the heat treatment step. If each step in a tested catalyst preparation recipe is not followed exactly, a material with a completely different set of activity characteristics than expected may result.

Supported catalysts are most commonly prepared by either the impregnation or ion-exchange technique. The steps involved in preparing a catalyst by the impregnation technique include: (1) contacting the support material with a solution of the appropriate metal salt, (2) removing the excess solution, (3) drying, (4) calcining the deposited material in air so as to form the oxide and/or then reducing the oxide in a hydrogen atmosphere to the metallic state. For example, a ruthenium hydrogenation catalyst can be prepared on alumina by soaking small alumina particles in a ruthenium trichloride solution, draining to remove the excess solution, and then drying the wet alumina pellets in an oven at 110°C. The final step involves reduction of the ruthenium salt to ruthenium metal. This procedure is best carried out by placing the impregnated catalyst in a tube furnace through which hydrogen is passed. The reduction is carried out in the temperature range of 375 to 400°C for four hours. The catalyst is then ready for use in hydrogenation reactions such as the conversion of olefins to saturated hydrocarbons.

In preparing a catalyst by the ion exchange technique, the metal must be present as a cation and the pH of the exchange solution must be greater than 7. The exchange solution is brought into contact with the support material where exchange of metal ions in solution with those on the support occurs. The exchange solution is separated from the support material and then the drying and reduction steps described for the impregnation technique are carried out.

In both the impregnation and exchange methods, the amount of metal on the support can be calculated if the compositions of the solutions are known. Atomic absorption spectroscopy is often used to determine the metal content of the solutions after the impregnation or exchange step. It is most important that every detail of the preparation procedure be adhered to if reproducible catalyst specimens are to be obtained. For example, when using the impregnation technique, catalyst performance has been shown[5] to depend on the concentration of solution used, how long the solution contacts the support, whether the support is soaked in solution once or more often, how the solvent is removed, washing procedures, drying temperatures, reduction times and temperatures, and so on.

The details of the impregnation and ion-exchange procedures (such as metal concentration in solution, number of impregnations, etc.) will determine the final state of dispersion of the metal. The degree of dispersion of the metal can be defined as the ratio of the number of surface metal atoms to the total number of metal atoms. In principle the dispersion of a supported metal can have any value between 0 and 1. Low dispersion values (less than about 0.3) indicate that the metal on the support is composed of relatively large crystallites with most of the metal atoms occupying positions within the crystals. Dispersion values of about 0.4 to 0.7 correspond to very small crystallites whose average diameters might lie between about 20 and 50 Å. Dispersion values approaching unity mean that all metal atoms are on the surface. Determination of the degree of dispersion of a catalyst requires information about the metal surface area. Methods for carrying out such measurements are discussed in the section on catalyst characterization.

It should also be recognized that the nature of the support can affect catalyst activity and selectivity. The effect arises because the support can influence the surface structure of the atoms of the catalytic agent and because support materials are not necessarily inert and therefore may play a direct role in the catalytic reaction itself. An example of such a system consists of the platinum on alumina reforming catalysts used in the petroleum industry. In these catalysts the acidic alumina support functions as the seat of carbonium ion isomerization activity, while the platinum provides a hydrogenation/dehydrogenation function. Because of such possibilities, the support should always be treated as an integral part of a supported metal catalyst. In other words, in order to obtain similar results from different batches of catalyst, it is necessary to keep the physical and chemical characteristics of the support constant.

In the chemical industry a very wide range of conditions exists under which supported metallic catalysts are used; and as a consequence, there is a wide range of possible support materials. Most of these are particulate or granular, although fibrous materials have also been used. Depending on the support, the surface area per unit mass and the pore structure can vary enormously. Table 7.2 gives the internal physical properties for some of the more common support materials used in industry.

Catalyst Characterization.   As discussed in the preceding paragraphs, small differences in materials and preparation procedures can result in catalysts with quite different catalytic activities. The differences in catalytic activity are most often due to differences in either the surface area of the active metal or the support material. Therefore, in comparing the activities of catalysts, it is necessary to know the surface areas of the active component so that comparisons can be made on a common basis.

Methods currently available for surface area measurements can be grouped into two categories: those based on adsorption of gases and those based on physical methods. We will give a brief discussion of these methods, but for a more extensive discussion, References 7 and 8 should be consulted.

Table 7.2   Typical Industrial Catalyst Support Materials and Their Physical Properties

Support	Area (m^2/gm)	Pore Diameter (Å)
(1) Activated Carbon	500-1500	20-10
(2) Silica Gels	200-800	70-20
(3) Kieselguhr (naturally occurring silica)	15-20	7000-2000
(4) Low area silica (powdered silica glass)	0.1-0.6	400,000-20,000
(5) Activated alumina	150-400	--
(6) Low-area alumina	0.1-5.0	20,000-5000
(7) Silica-alumina	200-700	50-30
(8) Zeolites and natural clays	500-700	7.5-2.5
(9) Asbestos (chrysotile)	10-20	--

1. *Adsorption methods.*   Techniques for the measurement of surface area by adsorption fall into one of two types. The first gives the total surface area of the sample by measuring the amount of gas required to form a monolayer of adsorbed gas so that by multiplying the number of adsorbed molecules at this point by the effective area per molecule, the total surface area is obtained. This technique is referred to as physical adsorption of gases. The second technique referred to as chemisorption measures only the metal surface area of the supported catalyst. This is achieved by measuring the uptake of a gas that is chemisorbed on the metal surface but that is not adsorbed on the *non*metallic components.

(*a*)   *Physical adsorption.*   In the measurement of total surface area of a catalyst by physical adsorption, simple nonpolar molecules such as the rare gases or nitrogen are used. In a typical physical adsorption experiment a known weight of catalyst sample is placed in an evacuated glass sample tube. The sample tube is connected to an apparatus that is capable of adding and measuring gas volumes and pressures. The data obtained are gas volumes added to a sample adsorption chamber at a series of measured pressures. The observed volumes are normally corrected to cubic centimeters at 0°C and 1 atm and plotted against the pressure in millimeters of mercury. A typical adsorption isotherm obtained in this manner is shown in Fig. 7.11.

Brunauer, Emmett, and Teller[9] concluded that the lower part of the linear region in curves of the type shown in Fig. 7.11 corresponds to complete monomolecular adsorption. The object then becomes one of locating with some degree of precision, the point at which monomolecular adsorption becomes complete. For this purpose Brunauer and co-workers developed the now famous B.E.T. expression for multilayer adsorption

$$\frac{P}{V(P_o - P)} = \frac{1}{V_m \cdot C} + \frac{(C - 1)P}{C \cdot V_m \cdot P_o} \tag{7.56}$$

where V_m is the volume of one monomolecular layer of gas, P is the pressure in the adsorption cell when V amount of gas has been added, P_o is the vapor pressure of the gas at the operating temperature, and C is a constant for the particular gas-solid system.

According to Eq. 7.56, a plot of $P/V \cdot (P_o - P)$ versus P/P_o should give a straight line. It has been found that data of the type given in Fig. 7.11 when replotted in this form give straight lines that can be extrapolated to $P/P_o = 0$. The intercept obtained from this extrapolation along with the slope of the straight line gives two equations from which V_m can be calculated.

Since V_m is recorded at standard temperature and pressure, it can be readily converted to the number of molecules adsorbed. However, to deter-

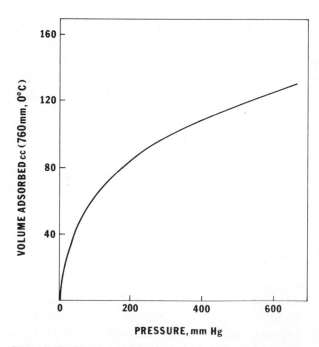

Figure 7.11   The shape of a typical physical adsorption isotherm.

mine the absolute surface area, it is necessary to know the area covered by one adsorbed molecule. If this is a_m then the total surface area per gram of catalyst sample S_p is given by

$$S_p = \left[\frac{V_m N}{\mathscr{V}}\right] \cdot a_m \tag{7.57}$$

where N is Avogadro's number, $\mathscr{V}$ is the volume per mole of gas at conditions of V_m and V_m is the monomolecular layer volume based on a 1-g sample.

Emmett and Brunauer[10] proposed that a_m could be calculated by the following expression

$$a_m = 1.09 \left[\frac{M}{N \cdot \rho}\right]^{2/3} \tag{7.58}$$

where M is molecular weight, ρ the density of the adsorbed molecules, and N is Avogadro's number. The density is normally taken as that of the pure liquid at the temperature of the adsorption experiment. In addition to using Eq. 7.58 to calculate a_m, tables of recommended best general values of a_m have been compiled[7] and are often used in Eq. 7.57.

(b) *Chemisorption.* Briefly, chemisorption is carried out in the same manner as physical adsorption, except that a gas is used that is chemisorbed on the metal surface but not on the support material. In addition to determining the volume of gas required to form the monolayer, knowledge of the average number of surface metal atoms X_m associated with the adsorption of each gas molecule is required. Also, a knowledge of the number of metal atoms per unit of metal surface area n_s is required. Thus for monolayer chemisorption, if V_m is the total volume of adsorbate uptake per gram of catalyst sample, the metal surface area S_c per gram of catalyst is

$$S_c = \frac{[V_m N/\mathscr{V}]X_m}{n_s} \tag{7.59}$$

where N and $\mathscr{V}$ have the same values as called for in Eq. 7.57. Note that the chemisorption stoichiometry X_m is defined with respect to the adsorbate molecule so that if, as with hydrogen, dissociative adsorption occurs and if each chemisorbed hydrogen atom is associated with one surface atom the value of X_m will be two. Values for n_s have been obtained from surface crystallography for a number of metals and are given in Reference 7.

Once the metal surface area has been determined, it can be used to estimate the average crystallite size of the metal if it is assumed that the crystallites are of regular geometrical shape (e.g., spherical or cubic). However, it should be realized that this method yields no information on the size distribution of the crystallites and therefore should ideally be complemented by physical techniques that may give such information.

2. *Physical methods.* The two most commonly used physical techniques for obtaining information about catalyst surface areas are X-ray line broadening and electron microscopy. The X-ray technique is based upon the fact that an X-ray powder diffraction line broadens when the crystallite size falls below about 1000 Å. Below 30 Å the diffraction lines are too broad and diffuse to be of any use, while above 500 Å the change in peak shape is small and the method is therefore insensitive. By suitable calibration and standardization Moss and co-workers[11] have shown that it is possible to calculate the average crystallite size from line broadening data. The surface area of the catalyst can be calculated from the average crystallite size by carrying out a reversal of the procedure outlined for obtaining average crystallite size by surface area measurements.

The second physical method, and one that is finding increasing use, is electron microscopy. The metal particles in a supported catalyst can be observed and measured directly by using an electron microscope operated in the transmission mode.[12] With a sensitive instrument and with adequate contrast between metal and support, it is possible to see and size a crystallite of 10 Å.

Suitable specimens for electron microscopy can be prepared by cutting extremely thin sections (300 to 500Å) with an ultramicrotome from catalyst particles embedded into "Araldite" resin. By using magnifications of the order of 100,000 the resulting micrographs show the metal particles as dark regions against the gray background of the support.

The main problem in the use of electron microscopy for a quantitative estimation of particle size lies in ensuring that the results are representative of the catalyst. Therefore it is imperative that a large and a representative sample of crystallites be observed.

A measurement of the images on the micrographs yields a value for the average crystallite size and also the size distribution. Often the images of crystallites appear as irregular shapes and consequently there is a problem in choosing a measurement to represent the size. Also, since the image of the particle is two dimensional, a false estimate of the mean particle size can be obtained if the particles are not randomly oriented in the specimen being examined. However, in spite of these problems, electron microscopy is a very powerful tool in catalyst characterization studies, for it is the only technique that yields a direct value for the crystallite size distribution.

7.2.5 Catalyst Evaluation Program

Now that we understand how to prepare and characterize catalysts, the next step is to know how to evaluate them for their utility.

During the development of an industrially useful catalyst, several factors must be considered before a decision can be made whether the catalyst is economically useful. The primary factors to be considered involve the catalytic properties of activity, selectivity, and stability and the mechanical properties of strength and abrasion resistance.

Activity and Selectivity. In attempting to develop a catalyst for a chemical process, normally the first place to begin is with a screening program of some of the known catalysts that have been found to be effective for similar processes. For example, if the new process involves a hydrogenation reaction the initial catalysts to be screened might contain the metals Ni, Rh, Pt, or Ru, which are known to be active hydrogenation catalysts.

The initial screening program should be carried out on a bench scale using fresh catalyst samples. For heterogeneous catalyst systems these tests are normally carried out in tubular reactors less than 1 in. in diameter, and for homogeneous catalysts the reactions are normally carried out in stirred-tank reactors with capacities of 1 liter or less. After the bench-scale studies are completed the next step is to evaluate the most promising catalysts, as determined by their activity and selectivity, in a pilot plant reactor. In pilot plant studies it is advisable to simulate the final commercial catalyst as to size of particle, pore size, and metal content since these features are known to affect catalyst activity and selectivity. The diameter of the pilot plant reactor is normally chosen to be 10 to 15 times the diameter of the catalyst pellets being tested. This is done in order to minimize bypassing effects between the catalyst and the wall.

During the pilot plant study the effect of residence time, pressure, temperature, recycle rate, and reactant ratios on the reactant conversion and product yields must be established. Based upon the yields and the process conditions required, the economic potential of the catalyst and the process can be estimated.

Stability Tests. If the economic evaluation based on the activity and selectivity studies looks promising the next, and most difficult, step in the catalyst evaluation program is to obtain accurate information about catalyst life. Catalysts can lose either activity or selectivity, or both, with continued use. Loss in activity occurs primarily because of: (1) loss in surface area of the active component because of agglomeration of metal particles due to thermal sintering of the solid surface; and (2) covering of the catalytic active sites by a foreign substance.

In some catalyst systems the thermal aging process can be simulated in the laboratory by aging for short times at high temperatures. However, the results of accelerated aging tests have to be carefully interpreted to be meaningful. In the catalyst systems that cannot be studied by accelerated stability tests, it is necessary to carry out the tests over periods of time that may extend over many months. In such cases it is particularly expensive to establish the life of the catalyst.

The selectivity of a catalyst often changes with aging because the surface picks up poisons that catalyze new reactions forming undesirable products. For example, during the cracking of heavy petroleum feed stocks to gasoline by the use of silica-alumina catalysts, the catalyst surface picks up nickel, vanadium, and iron from the feed stocks. These metals catalyze dehydrogenation reactions. As their concentration in the cracking catalysts increases, less gasoline is produced and more hydrogen and coke are formed. In processes such as these, where it is established that the catalyst is susceptible to poisoning by certain agents, it is necessary to pretreat the feedstocks in order to obtain satisfactory catalyst life.

Strength Tests. For a fixed bed heterogeneous process, catalysts in pellet form with diameters of $\frac{1}{16}$ to $\frac{1}{2}$ in. are normally used. The most common diameter is about $\frac{1}{4}$ in., which seems to represent a good balance between pressure drop and diffusion limitations for many chemical reactions. Choice of the optimum pellet diameter depends upon the operating conditions, the pore size, and the reaction being catalyzed.

Frequently test catalysts are used in granular rather than pellet form. The restrictions on the size of the catalyst particles in such cases is set by pressure drop limitations and by the problems associated with keeping very small catalyst particles from being carried out of the reaction zone. Normally the smallest particle size used in fixed bed processes is 40 mesh.

Strength requirements for fixed bed catalysts are set in order to reduce breakage and dust formation during loading. For $\frac{1}{4}$-in. pellets, the crushing strength should normally be in excess of 10 psi.

Catalyst Cost. After the activity, selectivity, stability, and strength tests have been completed, the catalyst cost to the chemical process can be estimated. The catalyst cost is equal to the net price of the catalyst per unit weight, divided by the life in product units per unit weight of catalyst. For example, if the net cost of a catalyst is $10/lb and the life of the catalyst is such that 1000 barrels of product can be produced per pound of catalyst, then the catalyst cost per barrel of product produced is 1 cent. Normally if catalyst life is sufficiently long (i.e., greater than one year), catalyst costs constitute only a small fraction of the total operational costs involved in producing a commercial product.

Decision Time—What Catalyst To Use? The final goal of a catalyst evaluation program is the selection of a catalyst that will give the chemical process the greatest profitability. This does not mean that of those catalysts meeting the minimum required activity, selectivity, stability, and strength standards the catalyst giving the greatest yield will necessarily be the most economical. This is because a comparison of yields does not take into account the processing costs. For example, a high yield catalyst may require the expenditure of large amounts of energy in its use, whereas the energy requirements for a medium yield catalyst for the same reaction may be small. In an economic evaluation of these two catalysts it might be determined that the high yield catalyst results in a less profitable process because of the high energy costs. Accordingly, then, the choice of the most profitable catalyst can only be made after a complete economic evaluation of the process is made.

7.3 REFERENCES

1. J. M. Smith, *Chemical Engineering Kinetics*, 2nd ed., McGraw-Hill, New York, 1970.
2. H. P. Meissner, *Processes and Systems in Industrial Chemistry*, Prentice-Hall, Englewood Cliffs, New Jersey, 1971.
3. Gordon M. Harris, *Chemical Kinetics*, Heath, Boston, Massachusetts, 1966.
4. Keith J. Laidler, *Chemical Kinetics*, McGraw-Hill, New York, 1965.
5. E. I. Gil'debrand, *Int. Chem. Eng.* **6**, 449 (1966).
6. T. E. Corrigan, G. A. Lessells, and M. J. Dean, *Ind. Eng. Chem.* **60**, 62 (1968).
7. J. R. Anderson, *Structure of Metallic Catalysts*, Academic, New York, 1975.
8. *Catalyst Handbook*, Springer-Verlag, New York, 1970.
9. S. Brunauer, P. H. Emmett, and E. Tiller, *J. Am. Chem. Soc.* **60**, 309 (1938).
10. P. H. Emmett and S. Brunauer, *J. Am. Chem. Soc.* **59**, 1553 (1937).
11. D. Pope, W. L. Smith, M. J. Eastlake, and R. L. Moss, *J. Catal.* **22**, 72 (1971).
12. R. L. Moss, *Platinum Met. Rev.* **11**, 141 (1967).

7.4 PROBLEMS

1. The saponification of ethyl acetate by sodium hydroxide in aqueous solution is a relatively rapid reaction; and the reverse reaction is negligible. For a second-order reaction when the initial normality of sodium hydroxide is 0.05, the rate constant is given by

$$\log k = \frac{-1780}{T} + 0.00754T + 5.83$$

where k is in units of $1/(g \cdot mole)(min)$ and T is in degrees Kelvin. Calculate the time required to saponify 95 % of the ester at 50°C when the initial concentration of ethyl acetate is 3.0 g/liter, and the initial normality of sodium hydroxide is 0.05.

2. Suppose the following reaction occurs at 800°C and a constant pressure of 1 atm in a plug flow reactor:

$$CH_4 + S \longrightarrow CH_3SH$$

The rate of the reaction at 800°C is given by the expression

$$r_{CH_3SH} = k(C_{CH_3})(C_S)$$

where C_{CH_4} = moles/1 of CH_4 and C_S is moles/1 of S. The value of k at 800°C is 10 liters/mole · s. The gas feed contains 25 mole % CH_4, 25 mole % S, and 50 mole % N_2.

(a) For a 60% conversion of reactants to CH_3SH calculate what the residence time must be.

(b) If the feed flow at the entrance of the reactor is 0.1 mole/s · cm² and the diameter of the reactor is 2.5 cm, how long must the reactor be in order to achieve 60% conversion?

3. The liquid-phase reaction between trimethylamine and n-propyl bromide can be studied by immersing sealed glass tubes containing the reactants in a constant-temperature bath. The results at 150°C are shown in the table below:

Run	t (min)	Conversion (%)
1	12	10.3
2	35	26.8
3	60	38.0
4	120	55.2

The reaction may be written as

$$N(CH_3)_3 + CH_3CH_2CH_2Br \longrightarrow$$
$$(CH_3)_3(CH_2CH_2CH_3)N^+ + Br^-$$

Initial solutions of trimethylamine and n-propyl bromide in benzene, 0.2 molal, are mixed, sealed in glass tubes, and placed in the constant-temperature bath. After various time intervals the tubes are removed and cooled to stop the reaction, and the contents are analyzed. The analysis is accomplished by titrating the completely ionized bromide ions.

Using both the integration and a differential method, calculate the first order and second-order rate constants. Compare your results and determine which rate equation best fits the experimental data.

4. Tetrafluorothylene, the monomer of 'Teflon,' is made by the pyrolysis of monochlorodifluoromethane ($CHClF_2$) in an empty tubular type reactor. The reaction stoichiometry is as follows

$$2\,CHClF_2 \longrightarrow C_2F_4 + 2\,HCl \quad \Delta H° = 18 \text{ kcal}$$

At the reaction temperature of 700°C, the equilibrium constant is very large. The reaction is first order in $CHClF_2$ with a rate constant of 0.97/s at 700°C.

In a laboratory experiment, $CHClF_2$ vapor at 1 atm pressure is fed to a 30 cm long tube reactor. The $CHClF_2$ vapors are preheated in the reactor and reach a temperature of 700°C at the entrance. Reaction products were analyzed by gas-chromatography, where the analysis indicated that the primary product of the pyrolysis reaction was C_2F_4. However, small amounts of other products were observed, primarily members of the homologous series $H(CF_2)_nCl$.

(a) During a run when $CHClF_2$ was fed at a rate of 100 g/hr, with the tube reactor at 700°C and 1 atm of pressure, 30% conversion of $CHClF_2$ was obtained with a residence time of 0.40 seconds. Calculate the approximate inside radius of the tube reactor. Ignoring the side reactions, calculate the amount of heat that must be supplied, in kilocalories per square centimeter of internal tube area per hour.

5. The first-order gas reaction

$$SO_2Cl_2 \longrightarrow SO_2 + Cl_2$$

has $k_1 = 2.20 \times 10^{-5}\,s^{-1}$ at 593°K. In one experiment 10 liters/min of SO_2Cl_2 measured at 100°C and 1 atm pressure is fed to a 2 cm i.d. tube reactor that is 75 cm long. The reactor is maintained at 593°K along the entire 75 cm length. What is the yield of Cl_2 leaving the reactor in units of grams per minute when operating under the above conditions?

6. The gas phase reaction between A and B produces an addition product C,

$$A + B \longrightarrow C$$

where the rate of the reaction is given by

$$r_c = k(C_A)(C_B)$$

The reactants are supplied by two gaseous streams. One stream contains a pure A and is available at a rate of 300 l/min. The second contains 50% B and 50% inert gas and is available at a rate of 150 l/min. These streams are mixed instantaneously and fed to a flow reactor. The reactor contents and feed streams are maintained at a temperature of 150°C and a pressure of 1 atm. Under these conditions, the gases behave ideally.

(a) If the reactor is a tubular-plug flow reactor with a volume of 500 liters and 75% of the B is converted to C, calculate
   (1) The reactor effluent volumetric flow rate.
   (2) The space time in the reactor.
   (3) The rate constant k.
   (4) The average residence time in the reactor.

(b) If the reactor is a continuous stirred tank reactor with a volume of 500 liters and 50 % of the B is converted to C, calculate
   (1) The space time for the reactor.
   (2) The reactor effluent volumetric flow rate.
   (3) The average residence time in the reactor.

7. Fukuoka and Kominami [CHEMTECH, 671 (Nov. 1972)] describe a process for producing dimethylformamide from hydrogen cyanide and methanol. As your first industrial assignment, you are asked to determine whether the yield of dimethylformamide, based upon HCN usage, can be increased above that reported by Fukuoka and Kominami. Propose a plan for carrying out this assignment. Your project plan should include viable catalyst candidates, reactor setup, scale of experiment and equipment, type of equipment needed, flow rates and reaction conditions, and the variables that are to be studied.

8. Kladko [CHEMTECH, 141 (March 1971)] gives a case history of how a real industrial problem was solved. This case history demonstrates the interplay between chemistry and engineering and exemplifies the pitfalls that many new and unsuspecting industrial chemists encounter.

   Using Kladko's data, calculate the size of the reactor, the feed rate, and the heat load on each reactor that is necessary to produce 1,500,000 lb of B in 5000 hours of operation. Your calculations should consider a single-stage stirred tank reactor, a two-stage stirred tank reactor, and a three-stage stirred tank reactor.

9. Skinner and Tieszen [Ind. Eng. Chem. 53, 557 (July 1961)] describe a process for producing maleic acid by the vapor phase oxidation of 2-butene over a mixed phosphomolybdate and phosphovanadate catalyst system. Using data from this article, design a small-scale process for producing 1 kg of maleic anhydride per hour from a 2-butene feed. Your design should indicate all flow rates, temperatures, method of feed introduction, product collection, analyses, and any other details that you consider important.

10. Mocearov and co-workers [CHEMTECH, 182 (March 1973)] describe a process for alkylating benzene with a dilute olefin stream. Using the process flow scheme and data supplied in the paper, calculate the volume of the transalkylation reactor (i.e., stirred tank) that would be required to produce 100 lb of ethylbenzene per day when the reactor is operated at 60°C.

8

SEPARATION PROCESSES

ISOLATING THE PRODUCT

As we have learned from discussions in previous chapters, chemical reactions of industrial importance normally produce a mixture of products, of which only one or two are desired and these in some degree of purity. Consequently the final stage of most chemical processes involves some type of separation. In fact, the difficulties of separating the desired product in sufficiently high purity and at an acceptably low cost can present the major obstacle to commercial exploitation of a new process.

One method for judging the relative importance of the reaction and separation stages in a chemical process is to compare the investment in plant capital for the two activities. At one end of the range we find those processes that involve difficult or complex reaction stages, followed by a relatively simple separation operation to recover a single product. An example of such a process is the manufacture of ammonia by a multistage reactor assembly and product recovery by simple distillation. In a process such as this, 20% of the total capital may be invested in the separation equipment. At the other end of the range are processes in which a simple reaction step precedes a difficult multistage separation for the recovery of several desired products. An example of this process is the manufacture of pure olefin products by cracking a naptha feedstock. In such a process the capital invested in compressors, distillation columns, liquid extraction equipment, and so on can easily exceed 80% of the total investment.

In addition to the imposition of separation requirements on a chemical process to meet product specifications, environmental factors may dictate requirements. Because of federal and state regulations limiting chemicals that can be discharged into the environment, "process wastes" can no longer be simply dumped into the air or water streams leaving the plant site. The removal of regulated chemicals from these discharge streams again may be accomplished by the application of simple separation techniques and equipment, or the separation problem may be difficult, requiring the expenditure of large sums of money. In fact, the inability to economically reduce concentrations of regulated chemicals to acceptable levels can lead to shutting down a production plant.

We hope that the preceding paragraphs have helped to draw your attention to the importance of separation processes in the chemical industry. Many of the solutions to separation problems lie in the province of the chemical engineer, but there is a large area of common interest in which the chemist and chemical engineer must work together in order to achieve practical economic solutions. The cooperation between the chemist and the chemical engineer can be most effectively utilized when both parties recognize and appreciate the significance that each discipline contributes to the field of chemical separations. For the chemical engineer, this requires a basic understanding of the principles of physical chemistry, the foundation upon which most separation processes are built. For the chemist, it requires an understanding and ability to relate laboratory phenomena to the economical operation of large-scale separation processes.

Because of the large number of different separation processes utilized in the chemical industry and because of the vast amount of knowledge that has been generated and published in this area, it would be impossible to discuss adequately, within the confines of this chapter, each separation method. In fact the quantitative treatment of what might appear to be a simple separation technique, such as distillation, would require more pages than we allowed for the entire chapter on separations. Therefore, we present a qualitative discussion of some of the more important characteristics that are common to most separations, primarily for the purpose of illustrating the various factors that must be considered in selecting the most practical and economical separation process. For a more extensive and in-depth discussion of separation processes used in the chemical industry, you should consult one or more of the excellent references listed at the end of this chapter.[1-3]

8.1 CHARACTERISTICS OF SEPARATION PROCESSES

Basically, almost all separation processes can be represented by the simple schematic shown in Fig. 8.1. As illustrated, a stream of matter is fed to a separation unit where it is separated into two product streams that differ in composition from one another. The separation is caused by the addition of a separating agent to the unit. The separating agent may be another stream of matter, or it may be in the form of energy.

The fundamental concept of a separation requires that the product

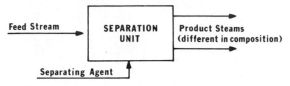

Figure 8.1 Schematic of a separation process.

streams differ in composition. If they have the same composition a separation has not taken place. Consequently, we should define separation processes as those operations that transform a single mixture of substances into two or more products that differ from one another in composition.

Most of the more common separation processes encountered in the chemical industry are based on the introduction or formation of a second phase of matter by the separating agent. For example, in an absorption process the feed stream is a gas mixture, and the separating agent is a liquid stream that preferentially dissolves one or more components of the gas mixture. The product streams from the separations unit are the liquid containing the dissolved gas components and the depleted gas stream. Distillation is an example of a separation process in which the separating agent causes the formation of a second phase. In such a process the feed is a liquid stream and the separating agent consists of energy in the form of heat. The heat causes the liquid to boil thus forming a vapor phase. The products of the separation unit are the vapor stream, which is richer in the more volatile components, and the depleted liquid stream.

In some separation processes the second phase already exists in the feed stream, and the separation unit and separating agent serve simply to separate the phases from one another. For example, a filter serves to separate solid and liquid phases from a feed that may be in the form of a suspension of fine crystals. An electrostatic precipitator removes dust particles from a stack gas stream by means of an imposed electric field. Separations of this type are often referred to as mechanical separation processes.

A listing of some of the more important separation processes utilized in the chemical industry is given in Table 8.1. Also given along with the name of the process is the required separating agent, the product of the separation, an example of the application of the separation technique, and the chemical basis upon which the separation is effected.

In reviewing Table 8.1 it can be seen that all of the nonmechanical separation processes involve a partitioning or transfer of components between two or more distinct phases. These phases may be formed entirely from the components that we wish to separate, as in the distillation of a two-component mixture. Such separations are often referred to as being based on

Table 8.1 Industrial Separation Processes[a]

Name	Feed	Separating Agent	Products	Basis of Separation	Example
1. Distillation	liquid	heat	liquid & vapor	difference in volatilities	crude oil separation
2. Evaporation	liquid	heat	solid or liquid & vapor	same as (1)	drying of crystals
3. Absorption	gas	liquid	liquid & vapor	preferential solubility	CO_2 removed from H_2 by absorption into ethanol-amines
4. Solvent Extraction	liquid	immiscible liquid	two liquids	solubility differences	removal of aromatics from paraffins
5. Adsorption	gas or liquid	solid adsorbent	fluid & solid	difference in adsorption potentials	drying of gases by solid desiccants
6. Crystallization	liquid	cooling	liquids & solids	difference in freezing tendencies	separation of p-xylene from m-xylene
7. Ion Exchange	liquid	solid resin	liquid & solid resin	law of mass action	water softening
8. Drying	liquid-solid	heat	dry solid & vapor	evaporation of water	food dehydration

No.	Operation	Feed	Separating agent	Products	Basis	Application
9.	Leaching	solids	solvent	liquid & solid	preferential solubility	leaching of soluble components from coffee beans
10.	Flotation	mixed powdered solids	rising air bubbles plus surfactants	two solids	preferential absorption of surfactants on one solid species	ore flotation
11.	Gel	liquid	solid gel	gel phase and liquid	difference in molecular size	purification of pharmaceuticals
12.	Filtration	liquid & solid	filter medium	liquid & solid	difference in size of liquid and solid	recovery of crystalline products
13.	Settling	liquid & solid	gravity	liquid & solid	density difference	process water treatment
14.	Centrifuge	liquid & solid	centrifugal force	liquid & solid	density difference	recovery of solid products
15.	Cyclone	gas & solid	flow	gas & solid	density difference	recovery of fluidized catalyst fines
16.	Electrostatic precipitation	gas & fine solids	electric field	gas & fine solids	charge on fine solid particles	dust removal from stack gases

[a]Modified from listing given in Reference 3.

phase equilibrium phenomena. Alternatively one or both phases may consist largely of chemical species that are not part of the mixture to be separated, such as in solvent extraction separations. These separations are referred to as being based on distribution equilibrium phenomena.

Separations based either on phase equilibria or distribution equilibria can be achieved only if there is a difference in the distribution of components between the two phases. A brief review of some of the fundamental principles used to describe separation equilibria of this type is given in the next section.

8.2   SEPARATION EQUILIBRIA

8.2.1   Phase Equilibria Separations

Phase equilibria separations involve phases that are composed of the mixture to be separated. On an industrial scale such separations are favored since no species that require subsequent removal are added. Phase equilibria separations are also best suited for relatively simple mixtures containing only a few components of interest.

An important relationship in understanding phase equilibria is given by the Gibbs phase rule:

$$P + F = C + 2 \tag{8.1}$$

This expression relates the number of components C in a system, to the number of phases P and the number of degrees of freedom F that can coexist. The degrees of freedom correspond to the intensive variables of the system, such as temperature, pressure, or the concentrations of the components in the phases. The phase rule is completely general, being applicable to all types of phase behavior. In the following we will apply it to systems containing one, two, and three components.

One-Component Systems. From the phase rule, a single-component system consisting of a single phase requires that two degrees of freedom (e.g., temperature and pressure) be specified to define the system. The most convenient method for accomplishing this is by the use of a phase diagram. Such a diagram for a single component is illustrated in Fig. 8.2.

The physical state of the component at any given temperature and pressure can readily be ascertained from the phase diagram. For example, point 1 in Fig. 8.2 represents a pressure and temperature at which only the liquid phase exists. Similar domains for the solid and gas phases are also shown. The lines represent pressures and temperatures at which two phases are in equilibrium with each other. In such cases where two phases are specified,

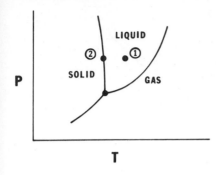

Figure 8.2   A typical phase diagram for a single component.

$F = 1$, so that a value of P or T determines the other variable. Point 2, for example, corresponds to a pressure and temperature at which solid and liquid coexist in equilibrium with one another. Consequently a single-component phase diagram can be very useful in establishing what types of separations are possible under ordinary temperature and pressure conditions. For example, a phase diagram for NaCl would show that its separation from a salt mixture by distillation could only be effected at extremely high temperatures. Because of corrosion and energy costs such a separation would not be desirable, and therefore a search for a lower temperature separation scheme should be made.

Two-Component Systems.   The introduction of a second component into a system leads to a more complex phase diagram. In such systems when only one phase is present, $F = 3$ from the phase rule, and the three independent variables are P, T, and the mole fraction of one component. If two phases are present, only two independent variables are needed to describe the system. When this is the case, it is convenient to take the mole fraction of one component and either temperature or pressure as the independent variable.

An example of a two-component phase diagram is shown in Fig. 8.3 where a diagram for vapor-liquid equilibrium is represented for an ideal mixture. Figure 8.3 illustrates the normal phase equilibrium observed for a lower-boiling component i and a higher-boiling component j. The diagram is divided into three regions, a vapor, a liquid, and a two-phase liquid plus vapor $L + V$ region. Consider a mixture with composition $x = A$; at all temperatures less than T_2, such as T_1, the mixture will exist as a single liquid phase. At a temperature of T_2 boiling of the liquid occurs so that a gas with composition C and a liquid phase coexist. As can be seen, the vapor C is richer in the lower boiling component i than the liquid phase in equilibrium with it. As the temperature is raised above T_2 an increasing fraction of the

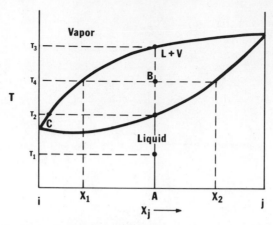

Figure 8.3 Phase diagram for an ideal two-component vapor-liquid system.

mixture will exist in the gaseous phase, until at temperature T_3 all of the original liquid will have been converted to the vapor phase.

At point B in the two-phase region, corresponding to temperature T_4, vapor of composition X_1 is seen to be in equilibrium with liquid of X_2. The horizontal line through point B is called a tie line. Application of the so-called lever rule to the tie line can be used to calculate the relative amounts of liquid and vapor at any point within the two-phase region. At point B application of the lever rule gives:

$$\frac{n_v}{n_l} = \frac{X_2 - X_A}{X_A - X_1} \tag{8.2}$$

where n_v represents the number of moles in the vapor phase and n_l the number of moles of liquid.

In the application of the preceding principles to separation phenomena, we can see that in its broadest terms distillation is a separation process based on the difference in composition between a liquid and the vapor in equilibrium with it. The separation efficiency of a distillation process is thus directly proportional to this difference. It is highly desirable, therefore, to introduce a concept of volatility that expresses the relationship between the composition of the liquid and vapor phases when the two phases are in equilibrium.

The volatility constant, K_v, of a given component is defined by the equation

$$K_v = \frac{y}{x} \tag{8.3}$$

where x and y represent the mole fraction of the component in the liquid and vapor phases, respectively. For a pure substance obviously the volatility constant is unity, and the term has real significance only if applied to mixtures. For a simple binary mixture, such as i and j in Fig. 8.3, the volatility of the two components can be expressed as

$$K_{vi} = \frac{y_i}{x_i} \quad \text{and} \quad K_{vj} = \frac{y_j}{x_j} \tag{8.4}$$

The degree of separation that may be obtained between i and j by distillation is best indicated by the ratio of the individual volatility constants. The resulting constant as shown in Eq. 8.4 is referred to as the relative volatility α

$$\alpha_{ij} = \frac{Kv_i}{Kv_j} = \frac{y_i/x_i}{y_j/x_j} = \frac{y_i x_j}{y_j x_i} \tag{8.5}$$

Upon rearrangement

$$\frac{y_i}{y_j} = \frac{x_i(\alpha_{ij})}{x_j}$$

and since $x_j = 1 - x_1$ and $y_j = 1 - y_i$ Eq. 8.5 reduces to

$$\frac{y_i}{1 - y_i} = \frac{(\alpha_{ij})x_i}{1 - x_1} \tag{8.6}$$

which is a frequently used expression in solving distillation problems. It is customary to let x_i and y_i represent the mole fractions of the more volatile constituent in the liquid and vapor, respectively. In this way α_{ij} is conveniently set greater than unity.

We will see in section 8.2.2 that α_{ij} is also referred to as the separation factor and is used to indicate the degree of separation that may be obtained with any particular separation process. For example, if $\alpha_{ij} = 1$ no separation of components i and j can be accomplished. However, if $\alpha_{ij} > 1$ component i tends to concentrate in one phase and component j in the other.

As you might expect, it would be desirable if α could be calculated from known physical properties of the pure compounds and not have to be determined, as for example, in the case of distillation, by an actual analysis of vapor and liquid phases in a state of equilibrium. For separation by distillation there are several approaches by which α can be approximated from physical constants of the pure components. For example, if Raoult's law holds for the solution and if Dalton's law for partial pressures is applicable in the vapor phase, the relative volatility can be shown to equal

$$\alpha_{ij} = \frac{P_i}{P_j} \tag{8.7}$$

where P_i and P_j are the vapor pressures of the two pure components at the temperature of the system. This equation implies that for a perfect or near-ideal system one can calculate the vapor-liquid-composition diagram such as that given in Fig. 8.3 if the vapor pressures of the pure components are known.

For nonideal mixtures more complex relationships involving activity coefficients are available for predicting the relative volatility of nonideal mixtures.[1]

Three-Component Systems. If a system contains three components the phase rule states that when one phase is present, the number of degrees of freedom is 4 and when two phases are present, the number is 3. The latter situation is of particular interest in several important industrial separations, such as solvent extraction and crystallization. The phase diagrams for three-component two-phase systems are most often drawn with temperature, pressure, and the mole fraction of one component as the independent variables.

An example of a phase diagram for a three-component two-phase liquid extraction system is shown in Fig. 8.4. In this diagram temperature and pressure are held constant and the mole fractions of the components are represented along the axes of the triangular coordinates. Fig. 8.4 shows that j is completely miscible with the other two liquids (i and k), but i and k are only partly miscible. Some typical examples of such three-component systems that give two-phase mixtures are: acetic acid–chloroform–water and ethanol–benzene–water.

The curve shown in Fig. 8.4 is referred to as the binodal curve; any point falling outside this curve (e.g., M) represents a composition of complete

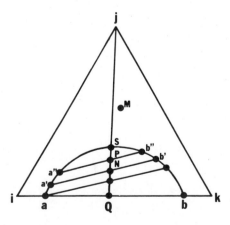

Figure 8.4 Representation of a three-component heterogeneous system on triangular coordinates.

miscibility, while points within the curve (e.g., N) correspond to two phases. The composition and amounts of each phase can be determined from the tie line through that point. Unlike the tie lines in the vapor-liquid phase diagram of Fig. 8.3, the tie lines in phase diagrams of the type shown in Fig. 8.4 are not generally parallel to the base of the diagram and must be determined experimentally. For point N the concentration of the two phases are given by the intersections of the tie line with the binodal curve. As previously indicated, the amount of each phase is given by the lever rule.

For example, in Fig. 8.4 if solvents i and k are added in such proportions as to form the binary mixture Q, two phases with compositions a and b will separate. Suppose a third component j that is miscible with i and k is added such that mixture N is formed. The resulting mixture will separate into two phases with compositions a' and b'. The line $a'b'$ is a tie line. Its upward slope indicates that j is more soluble in the phase rich in component k. The tie line becomes shorter as more j is added. When the ends of the tie lines meet at composition S the system becomes homogeneous.

It is now apparent, just as the phase rule predicted, that in a three-component two-phase system at equilibrium (constant temperature and pressure) only the composition of one phase needs to be stated to define the system. For example, in Fig. 8.4 the mixture P separates into two phases with compositions a'' and b''. If the composition of one phase is established, the composition of the other phase is fixed by the point where the tie line and binodal curve interact. As indicated earlier with the distillation separation process, the phase composition data obtained from diagrams such as Fig. 8.4 can be used to calculate the separation factor α and hence the extent of separation possible for the system under question. Further discussion of α with regard to solvent extraction systems is given in Section 8.2.2.

8.2.2 Distribution Equilibria Separation

In separations based on distribution equilibria the separate phases are composed primarily of compounds other than the components to be separated. Separations based on distribution equilibria have certain advantages over corresponding separations based on phase equilibria. This is because in distribution equilibria systems the composition of one or both phases can be varied over wide limits. This in turn affects the way in which the individual components distribute themselves between the phases and thereby may allow an otherwise difficult separation to be made.

Equilibrium Distribution Curves. The distribution of a solute between two phases at a particular temperature can be described by a plot of solute concentration in one phase y, C_y, versus concentration in the other phase x,

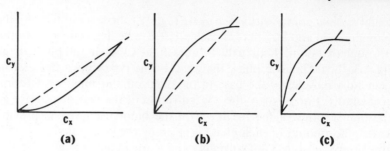

Figure 8.5 Typical shapes of equilibrium distribution curves.

C_x. Such plots, as illustrated in Fig. 8.5, are referred to as equilibrium distribution curves. In addition to showing what phase the solute concentrates in these curves also show that the ratio (C_x/C_y), often called the distribution coefficient K_D, varies with sample concentration. As can be seen from Fig. 8.5, K_D is constant only at very low solute concentrations, and that the distribution curve may be either concave or convex.

Distribution curves such as shown in Fig. 8.5*a* and 8.5*b* are normally observed for liquid extraction systems. In the case of adsorption, often there is competition among solute molecules for a fixed number of adsorption sites. As these adsorption sites become saturated with adsorbed solute molecules, the concentration of solute in the adsorbed phase approaches a limiting value, and the distribution curve flattens out as in Fig. 8.5*c*.

Capacity Factors. In a distribution separation process, the effectiveness of the separation is not only dependent on the value of the distribution coefficient K_D but is also influenced by the relative amount of each phase present. For example, in the case of a solute with a K_D value of 100, with equal volumes of each phase, 99% of the total solute is in one phase. However, if the volume of the preferred phase is reduced by a factor of 100 while the volume of the other phase is held constant, only 50% of the solute will be contained in this phase.

The ratio of the total amount of solute in phase x to that in phase y is referred to as the capacity factor k' and can be written as

$$k' = \frac{\text{total amount of solute in phase } x}{\text{total amount of solute in phase } y} \tag{8.8}$$

$$k' = \frac{C_x V_x}{C_y V_y} \tag{8.9}$$

and since $K_D = (C_x/C_y)$, we can write

$$k' = K_D\left(\frac{V_x}{V_y}\right) \tag{8.10}$$

The fraction of the total solute ϕ^x in a given phase is given by

$$\phi^x = \frac{C_x V_x}{C_x V_x + C_y V_y} \tag{8.11}$$

Written in terms of the capacity factor, Eq. 8.11 reduces to

$$\phi^x = \frac{k'}{1 + k'} \tag{8.12}$$

The interrelationship of K_D, ϕ^x, and the volume ratio V_x/V_y for a particular distribution system must be fully known before the system can be efficiently used in a separation process on the industrial scale. Very often it is the job of the industrial chemist to determine these relationships. Methods for accomplishing this have been established and can be found in several of the references cited at the end of this chapter.

Separation Factor.   In our discussion thus far, we have only considered the distribution of a single solute between two phases. However, in practice many separations involve two or more solutes. In such cases separations can only be achieved if the distribution coefficients of the individual solutes differ. As indicated earlier the maximum degree of separation that can be achieved in any given process is represented by the separation factor α. For a pair of solutes i and j in a distribution equilibrium system, α is given by the ratio of the concentrations of these two components in phase x divided by the ratio in phase y.

$$\alpha_{ij} = \frac{C_{ix}/C_{jx}}{C_{iy}/C_{jy}} \tag{8.13}$$

or

$$\alpha_{ij} = \frac{K_{Di}}{K_{Dj}} \tag{8.14}$$

The concentrations C_{ix}, C_{jx}, C_{iy}, and C_{jy} refer to solute i or j in phase x or y, and K_{Di} and K_{Dj} are the corresponding distribution constants.

We see that an effective separation is accomplished to the extent that the separation factor is significantly different from unity. If $\alpha_{ij} = 1$ no separation of components i and j has been accomplished. If $\alpha_{ij} > 1$ component i tends to concentrate in product x more than does component j, and component j tends to concentrate in product y more than does component i. On the other hand, if $\alpha_{ij} < 1$, component j tends to concentrate in product x preferentially, and component i tends to concentrate preferentially in product y.

The separation factor can also be calculated from the capacity factors, that is:

$$\alpha_{ij} = \frac{k_i'}{k_j'} \tag{8.15}$$

For the purpose of illustration, assume that $k_i' = 100$ and $k_j' = 0.01$. This means that 1 % of i is in phase y and 1 % of j is in phase x, generally considered to be a good separation. This is indicated by the fact that α_{ij} in this case is 10^4. As a general rule of thumb, in choosing a separation process α_{ij} should be as large as possible, and for any given value of α_{ij} optimum separation of two components i and j in a single equilibration occurs when $k_i' k_j' = 1$.

8.3 THE SEPARATION FACTOR AND ITS RELATIONSHIP TO MOLECULAR PROPERTIES

Separation factors for most separation processes reflect the ability of that method to discriminate on the basis of one or more properties of the components at the molecular level. For example, in distillation processes the separation factor is a measure of differences in vapor pressure, which is primarily a function of the molecular weight and dipole moments of the species involved. On the other hand, the separation factor in a molecular sieve adsorption process is primarily a function of molecular size and shape. The molecular properties given below represent the greatest influence on the magnitude of separation factors attainable with the various types of separation processes presented in Table 8.1:

1. *Molecular weight.* Many separation methods exhibit some effect due to differences in this property, but this characteristic is most dominant in distillation and gas-liquid chromatography separation techniques. Molecular weight differences are also important in bringing about separation by such methods as solvent extraction and crystallization, but here other types of molecular properties are of comparable importance, as indicated in Table 8.2.

2. *Molecular shape.* Separation by molecular shape usually is important only when the molecular weights (i.e., also the molecular sizes) of the sample components are similar. Molecular shape, as used in separation phenomena, refers to the qualitative description of whether the molecule is long and thin, short and branched, nearly spherical, and so on. Several separation techniques as indicated in Table 8.2 show a marked dependence on molecular shape. For example, separation by adsorption on molecular sieves of straight

Table 8.2 Dependence of Separation Factor upon Difference in Molecular Properties

Separation Process	Molecular Weight	Molecular Shape	Properties of Pure Substances Dipole Moment and Polarizability	Molecular Charge	Chemical Reaction
Distillation	2	3	1	–	–
Absorption	–	–	1	–	1
Solvent extraction	–	–	1	–	1
Precipitation	–	–	–	–	1
Crystallization	3	1	2	1	–
Ion Exchange	3	2	–	1	1
Adsorption by molecular size	–	1	2	–	–
Ordinary adsorption	–	1	1	–	1
Gas-liquid chromatography	1	2	2	–	1
Dialysis	1	2	2	–	–
Zone melting	1	2	2	–	–
Electrophoresis	1	2	–	1	–

Key: 1 Significant effect
      2 Intermediate effect
      3 Small effect
      – Little or no effect

chain alkanes from branched or cyclic alkanes of similar molecular weight is based on differences in molecular shape; more specifically, differences in minimum cross-sectional area of the molecule. The separation of xylene mixtures by fractional crystallization is possible because of the difference in lattice energy and melting point of the ortho, meta, and para isomers attributable to their different shapes.

3. *Dipole moment and polarizability.* Both of these properties characterize the strength of intermolecular forces between molecules. If a net separation of charge is present in a molecule, then it is said to be polar and the dipole moment is a measure of its polarity. For example, hydrocarbons are relatively "nonpolar," esters and ketones have intermediate "polarity," and alcohols and amines are highly "polar." Polar molecules interact more strongly with other molecules than do nonpolar molecules of the same size. As a result, a polar substance has a lower vapor pressure than a nonpolar species of about the same molecular weight, and polar substances are dissolved more readily into polar solvents. Thus the separation factor in such procedures as distillation, solvent extraction, absorption, and chromotography is significantly affected by the polarity difference between the sample components.

The polarizability of a substance reflects the tendency for a dipole to be induced in a molecule due to the presence of a nearby dipolar molecule. Polarizability depends upon the size of a molecule and upon the mobility of electrons within the molecule. A more polarizable molecule will tend to have a lower vapor pressure and will have a greater solubility in a polar solvent. Consequently, polarizability influences the same separation techniques as listed for the dipole effect.

4. *Molecular charge.* Some molecules because of their ionization in solution carry a net charge. Several separation techniques, such as electrophoresis, are based on phenomena involving these charged species. The separation factor in such cases is strongly influenced by both the magnitude of the charge on the sample component and the sign of the charge.

5. *Chemical reaction.* Several separation techniques as indicated in Table 8.2 are based upon the difference between molecules in their ability to take part in chemical reactions. Acid-base equilibria, complex formation, and precipitation are the most commonly employed phenomena for utilizing chemical reaction separation techniques. For example, in separation methods involving ion exchange or extraction with acids or bases, components are separated according to their dissociation constants.

The relative importance of the different molecular properties in determining the value of the separation factor for various separation processes is summarized in Table 8.2. However, because of the numerous exceptions and because the dividing line between strong and weak influences of a given property is not definite, in many cases the categorizations given cannot be expected to be religiously exact. Nevertheless, basic differences in the importance of different molecular properties in determining the separation factors for various separation processes can be ascertained from Table 8.2. For example, the separation factor in crystallization reflects primarily the ability of molecules of different kinds to fit together, and therefore factors such as size and shape are important. On the other hand, the separation factor in distillation is due to differences in vapor pressure, which in turn can be related primarily to the strength of intermolecular forces.

8.4 SELECTION OF A SEPARATION PROCESS

In this section we will briefly consider the logic leading to the selection of suitable candidate processes for carrying out a particular separation. However, before beginning it should be understood that because of the variety of separation problems encountered in the chemistry industry no single selection procedure exists that can be reliably applied to the solution of all problems. However, it is possible to identify several rules of thumb that, if followed, can be helpful in identifying a few separation processes that should be particularly strong candidates for evaluation in any given situation. The factors to be considered when looking for candidate separation processes are:

Feasibility
Selectivity
Separation capacity
Product quality
Product value

8.4.1 Feasibility

The first question to be answered in choosing a candidate separation process is whether it has the potential of affecting a separation on the mixture in question. Accordingly, a large number of separation schemes can normally be ruled out by first applying a feasibility requirement. A simple example will serve to illustrate the method. Suppose we wish to separate a mixture of ortho-, meta-, and paraxylenes. A comparison of the properties of these compounds to the bases of the separation techniques listed in Table 8.1 immediately allows us to rule out several different separation schemes. For example, the xylenes are nonionic compounds; therefore it is immediately apparent that such separation processes as ion exchange, electrostatic precipitation, and flotation, would not be viable candidates since the phenomena on which these separation techniques are based cannot be applied to this mixture.

Feasibility criteria can also be used to rule out separation schemes that require extreme conditions. Definite guidelines for establishing what is extreme and what is not extreme are difficult to draw, but the general idea is that a separation process that requires very high or very low pressures or temperatures, very high voltages or currents, or other such conditions will be less desirable than one which does not require extreme conditions. For example, separation of a propane-butane mixture by any process that requires a solid feed (for example, crystallization, filtration, or leaching) would require that the feed be frozen, which in turn would require exceedingly low-temperature refrigeration. Since refrigeration is expensive and difficult to carry out on an industrial scale, its avoidance is desirable if possible. Likewise, the earlier cited example of the separation of NaCl and KCl by distillation would not be considered a feasible industrial separation, because of the extreme conditions of low pressure and high temperature.

8.4.2 Selectivity—The Separation Factor

This is a major consideration in the choice of candidate methods for a separation process. The selectivity required is in turn determined by the nature of the sample mixture. For illustration purposes, suppose we know that a mixture contains compounds of widely different molecular weights, then a separation process that exhibits pronounced molecular weight selectivity will probably be required. In reviewing Table 8.2 we see that such methods as electrophoresis, distillation, and zone melting might be chosen as likely candidates. On the other hand, if the mixture is composed of different compound types, one or more separation techniques that exhibit polar and/or polarizability selectivity might be required, for example, absorption or solvent extraction.

In choosing candidate separation processes from a comparison of molecular properties one should always give primary consideration to those processes that emphasize molecular properties in which the components differ to the greatest extent. For example, consider the separation of a mixture containing the three xylene isomers shown below:

Ortho   Meta

Para

Table 8.3   Some Physical Properties of the Xylenes

Property	o-xylene	m-xylene	p-xylene
Molecular wt.	106.2	106.2	106.2
Boiling point °C	144	139	138
Freezing point °C	-25	-47.4	+13
Dipole moment, 10^{-18} esu	0.62	0.36	0.0

Since the species shown are isomers of one another, they have identical molecular weights, and hence any separation process dependent upon molecular weight differences will not give a separation. However, it is apparent that the shapes of the molecules are different. In addition, we might expect that the polarity of the xylene isomers might differ because of the polar nature of the bond between the methyl group and the benzene ring. In p-xylene these two polar bonds oppose one another and the net dipole moment should be zero. In o-xylene the polar bonds are aligned in nearly the same direction and a net dipole moment should result. The m-xylene isomer should have a polarity somewhere between the ortho- and para-isomer. Consequently, in applying the foregoing discussion to the selection of candidate separation processes from Table 8.2, we conclude that distillation and crystallization look like good candidates.

Some of the physical properties of the xylene isomers pertinent to this problem are shown in Table 8.3. As can be seen from the data, fractional crystallization would be a feasible process for separating the three isomers. Another possible method for accomplishing the separation would be to combine a distillation with a crystallization. For example, the boiling point of o-xylene is sufficiently different from the other isomers so that separation by distillation is feasible. Therefore, p-xylene with its high melting point could be separated from the other two by crystallization, with the separation of the resulting ortho-meta mixture by distillation. This process would eliminate the need for low-temperature refrigeration, which the total separation by fractional crystallization would require. The combined crystallization-distillation scheme is the separation process that is actually used in the commercial production of the pure xylene isomers.

8.4.3   Separation Capacity

The quantity of material to be separated will have significant influence on the choice of a separation process. Some separation processes (for example, gas chromatography) are ideally suited for milligram quantity separations, but are difficult to scale up for large quantity separations. On the other hand, some separation processes (for example, distillation) can be used on the milligram scale or on the thousands of gallons a day scale.

Clearly then one of the requirements of an industrial separation process is that its capacity be compatible with the intended feed rate.

8.4.4 Product Quality

All chemical products must meet certain specifications. For some products (such as pharmaceuticals, fine chemicals, monomers, or food products) these specifications may be very rigid. On the other hand, some products (such as cleaning solvents) may have specifications that allow for a slight variation in properties and/or the presence of small amounts of foreign substances that are not detrimental to the use for which the product is intended.

Such differences in product specifications can play a major role in the selection of a separation process. For example, separation processes that require the addition of a mass-separating agent are not often used in food processing because of the possibility of contamination from any residual separating agent in the product. Contamination, of course, means that the product does not meet specification. However the production of an aromatic solvent by the separation of its components from a catalytic reformer stream by solvent extraction with ethylene glycol is perfectly acceptable in terms of meeting product specifications, for trace quantities of ethylene glycol in the final product would not damage the product.

Many chemical products are susceptible to decomposition when exposed to heat. Therefore, it is often necessary to protect a product from thermal damage during its separation stages. Thermal damage may be manifested through the formation of an unwanted color, reduced storage stability, offensive odor, polymerization, and so on. For such heat sensitive materials, those separation processes requiring high temperatures would not likely be ranked at the top of potential separation schemes. However it is often possible to reduce the thermal effect in these separation processes by designing the equipment in such a way that the holdup time at high temperature for material passing through it is very short. Alternatively, when thermal damage is a problem in separation by distillation, a common solution is to carry out the distillation under vacuum so as to keep the boiling material at the lowest possible temperature.

8.4.5 Product Value

As discussed in earlier chapters, the economic value of a product will have a significant influence on the design and operation of the chemical process used in its production. As you might expect, the selection of the separation stage will also be strongly influenced by the economic value of the products being isolated. For example, n-butane is worth about 10 cents/lb, while many pharmaceuticals are valued at $200 to $300/g. Clearly there will be many separation processes that would be suitable for substances with a high value that cannot be considered for a substance with a low value. The general rule is that the lower the economic value of the product, the more important it will be to select a process with the lowest possible energy consumption and to select a process where the cost per pound of any added mass separating agent is relatively low.

8.5 SUMMARY

In concluding this chapter we would like to reemphasize our earlier statement that the field of industrial chemical separations is a very broad and complex area of study. Consequently, it is impossible to do the subject justice within the confines of one chapter. However we hope that the information presented here will help to orient your throughts to the information that is required to solve problems in industrial separations. Once oriented to the goals of a separation process, specific solutions can be made by the combined use of the vast amount of information available in the literature, your own laboratory data, and the appropriate laws and relationships obtained from physical chemistry theory.

8.6 REFERENCES

1. B. L. Karger, L. R. Snyder, and C. Horvath, *An Introduction to Separation Science*, Wiley, New York, 1973.
2. E. W. Berg, *Physical and Chemical Methods of Separation*, McGraw-Hill, New York, 1963.
3. C. J. King, *Separation Processes*, McGraw-Hill, New York, 1971.

9

INSTRUMENTATION

CONTROLLING THE PROCESS

In the early days of the chemical industry, prior to the 1940s, chemical processes were rather unsophisticated, and therefore they could be maintained and controlled by means of simple devices and procedures. For example, indicators and recorders displaying the basic measurements of temperature, pressure, level, and flow were used in many cases, with the operator observing the measurements and adjusting valves as necessary to achieve desired results.

Modern chemical processes have become very sophisticated and simple control procedures are no longer practical. Today's chemical plants utilize the latest electronic hardware, automatic controllers, computer controls, advanced analytical monitoring, and advanced control concepts. In order to appreciate the need for this type of control and instrumentation, we must first understand the motive for developing highly automated chemical processes.

9.1 REASONS FOR AUTOMATING A CHEMICAL PROCESS

Some or possibly all of the following basic benefits are realized when automatic control is introduced into a chemical process:

1. The process, whether it is being carried out in bench-scale equipment, a pilot plant, or a full production scale plant, can be run continuously without attention from operators or chemical technicians. This reduces manpower requirements and therefore lowers labor costs.
2. The reduction in need for operating personnel results in an elimination or decrease in human errors.

230

3. Chosen adherence to optimum conditions results in an improvement in overall process quality.

4. Necessary adjustments to operation can be made from one central location often resulting in a reduction in the space requirements for the process unit.

5. Safety in operation is increased by providing warning of abnormal conditions and automatically taking corrective action. In addition, automatic control eliminates the need for personnel to be in the immediate vicinity of hazardous equipment or conditions.

9.2   WHAT IS AUTOMATIC CONTROL?

A greater appreciation and understanding of an automatic control system can be obtained if we first consider a simple manual control procedure. As a way of illustration, suppose we wish to control the temperature of a solution contained in a beaker on a hotplate at a value of 50°C ± 2°C. This might be accomplished by placing a mercury-filled thermometer in the solution, observing the temperature, and then manually adjusting the voltage to the hotplate heating element by means of a rheostat so that the temperature is maintained within the desired range. This manual control system contains four basic elements, (1) a detecting device, the mercury bulb; (2) a measuring device, the mercury column and matching calibration scale; (3) a controlling device, the observer; and (4) the final control element, the rheostat.

Basically, the function of this control system is to measure the value of a variable, the temperature, and then produce a counter response to limit its deviation from a reference point, in this case 50°C. This is, for all practical purposes, the definition of automatic control. However, in an automatic control system, the observer, or more specifically the operator, is replaced in the control loop by a device known as the automatic controller.

The basic layout of an automatic control loop is shown in Fig. 9.1. As can be seen, this system contains the same four basic elements of the manual control system described in the preceding paragraph. In addition to these four elements, a transmitter element is often added. However, depending upon the particular variable to be controlled, more than one element may be designed into the same instrument, so that a loop does not always contain four or five separate units.

Most of the remaining material in this chapter deals with the construction and operation of the basic elements in the control loop, but before considering these topics, let us take a look at what kinds of things can be controlled in a chemical process.

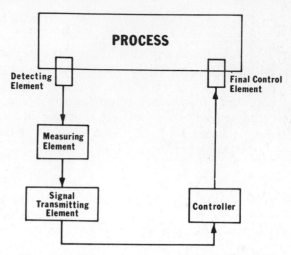

Figure 9.1 Basic layout of an automatic control loop.

9.3 WHAT IS CONTROLLED IN A CHEMICAL PROCESS?

Some of the more important variables that are controlled in chemical processes follow.

Flow. As might be expected, it is important that the material balance requirements of a process be maintained at all times. In a continuous chemical process this requires the control of material flows. Since most chemical reactions are sensitive to the ratio of reactants, it is very often a requirement that accurate flow control be maintained so that product quality and yield standards can be met.

Temperature. The control of reaction temperature is very important because, as discussed in previous chapters, conversion, yields, and product qualities are functions of temperature. Also, proper temperature control is often essential to the successful operation of many separation processes, such as distillation and crystallization.

Pressure. Since many chemical reactions are sensitive to pressure conditions, pressure control is a requirement in most chemical reactors. Most chemical separations also require that the pressure be controlled. For example, distillations are often performed under a reduced pressure, whereas absorptions and adsorptions are carried out at elevated pressures.

Liquid Level.   Level control is often associated with flow control; however, there are instances where it is associated with the proper operation of a piece of equipment, such as the level of solvent in a solvent extraction tower or the liquid level in a reactor.

The four control parameters cited above comprise the majority of all control applications in the chemical industry. However, the control of variables relating to composition of matter are also frequently encountered.

Composition Control.   The control of the composition of matter has been accomplished, depending on the properties of the substances, by a number of different techniques. The majority of these techniques are based upon one of three different types of analysis: component analysis, physical property analysis, or chemical property analysis. Most of the important control loops based upon component analysis utilize the analytical technique of chromatography to determine the mixture composition. Although infrared and other forms of spectroscopy also find use in certain processes. A mixture of definite composition often possesses a unique set of physical properties. These properties may actually form part of the product specifications or they may be used as a measure of product composition. Some of the more common physical properties that are measured and used for control purposes include density, initial and final boiling points, color, freezing point, and viscosity. The composition of a mixture is often related to a unique set of chemical properties. Chemical properties that are most often monitored for control purposes include pH, redox potential, and conductivity.

9.4   INSTRUMENTATION USED TO ACHIEVE CONTROL

Now that we have some idea as to what can be controlled, let us consider how it is accomplished. To achieve control we must first detect and measure the variable we wish to maintain at the desired value. As discussed in the preceding paragraphs, the variable may be flow, temperature, pressure, liquid level, viscosity, mole per cent, pH, or some other parameter. Let us begin our study of control loops by taking a closer look at some of the basic characteristics of the different measuring systems. Measuring system, as used here, is taken to mean both the detecting and measuring elements.

The overall performance of a control system can never exceed that of its associated measuring system. Good control depends not only on the static accuracy of the measuring system but also on its speed of response. If the measuring system is not capable of responding as fast as the measured variable can change, then the correcting action dictated by the control system will be too slow to achieve the most effective control.

9.5 RESPONSE OF MEASUREMENT SYSTEMS

When a change is made in a chemical process, an immediate and complete indication of the change as observed by a process operator or automatic controller is an ideal that cannot be achieved in practice. This is due to the fact that time lags exist in all real systems. The time lag is made up of three contributing factors: the lag inherent in the process, the response lag associated with the detecting element, and the lag associated with signal transmission. The contribution that each makes depends on the particular process and control loop. However, in most cases one of the three sources has a much greater effect than the other two. In this section we only consider time lags associated with detection and transmission. Time lags associated with the chemical process itself are discussed in a later section.

9.5.1 Temperature Measurement

Because temperature is one of the most important controlled variables in chemical processes and because its method of measurement should be familiar to all chemists, we will consider its response factors first and in greater detail than for other measuring devices. We will also use temperature measurements to illustrate the effect of measurement lags on control loop performance.

Most temperature measurements made in the chemical industry are in the range of $-150°F$ to $+3000°F$. To meet the requirements of such a broad spectrum of temperature measurements, several physical principles have become well established as the basis of temperature measuring instruments. The normal range of these instruments is shown in Fig. 9.2.

Thermocouples.   The thermocouple is one of the simplest and most commonly used methods for determining temperatures in chemical processes. When remote temperature indication is required, it is also the most inexpensive method available.

A thermocouple as shown in Fig. 9.3 consists basically of two dissimilar metals such as iron and constantan wires joined to produce a thermal electromotive force when the junctions are at different temperatures. One junction (the measuring junction) is immersed in the medium whose temperature is to be measured, while the other junction (the reference junction) is connected to the terminals of a measuring instrument.

Sensitive measuring instruments are required since the emf levels developed in most temperature measurements are generally in the range of -10 to $+50$ millivolts. Since the emf of a thermocouple increases as the difference in junction temperatures increases, the measuring instrument can

TEMPERATURE MEASUREMENT

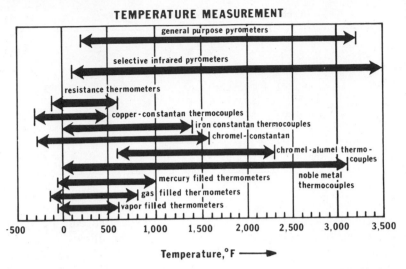

Figure 9.2   Useful range of temperature measuring devices used in the chemical industry.

be calibrated to read temperature directly. The newest measuring instruments basically consist of self-balancing potentiometers employing Zener diode regulated power supplies and temperature sensitive copper or nickel compensating resistors. With these devices, accurate temperature measurements can be made without using an ice bath reference junction and without the need for manual balancing of the bridge circuit.

Several different types of thermocouples are used in the chemical industry; however, the iron constantan and copper constantan types find the most use. Table 9.1 lists several of the more common types of thermocouples and their useful ranges.

Very often the measuring junction of the thermocouple is located a considerable length from the reference junction. In such cases the use of thermocouples made of noble metals, for example, like types R and S, could become very expensive if it were not for the fact that when a third metal is added to a

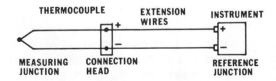

Figure 9.3   A typical thermocouple setup.

Table 9.1 Thermocouple Types and Ranges

Type	Metal	Range °F
E	Chromel - Constantan	-300 to 1600
J	Iron - Constantan	0 to 1400
K	Chromel - Alumel	+600 to 2300
R	Platinum and 10% Rhodium - Pt	0 to 2700
S	Platinum and 13% Rhodium - Pt	0 to 2700
T	Copper - Constantan	-300 to 650

thermocouple circuit it has no effect on the generated emf, provided the temperatures of its two junctions are the same. This allows the use of less expensive lead wire when expensive metals are used to make thermocouples. Normally copper lead wire in sizes varying from #14 to #20 gauge are used. However, with iron constantan and copper constantan thermocouples lead wire of the same material is often used.

A thermocouple junction can be made simply by twisting the bare ends of the two dissimilar metals together and insulating them from other conductors. However, the most common method of fabrication is to butt weld the ends of the two wires together. The junction must be insulated from other conductors in order to prevent the accidental formation of new couples and thereby introducing erroneous emf's into the circuit. Sometimes it is possible to immerse the bare thermocouple junction directly into the medium to be measured. However, because of the corrosive nature of most process streams, it is good practice to equip the thermocouple with a protective well. The wells are normally made of 304 or 316 stainless steel, but are also available in special metals when required. The junction and leads are electrically insulated from the well by the use of various ceramic materials.

Now that we understand the basis of at least one measuring device, we are in a position to analyze some of the factors that affect the speed and accuracy of response of a detecting element such as a thermocouple. For example, experience has shown that the thermal capacity of the thermocouple at the measuring junction has a significant effect on its speed of response. The thermal capacity of a detecting element is a function of its size, shape, and material. An illustration of mass effect is shown in Fig. 9.4 where the response curve for a thermocouple that has a butt-welded junction is compared to one that has a twisted junction. As can be seen, the speed of response of the butt-welded thermocouple is much faster than the twisted thermocouple. The reason for this observed behavior is that the mass of

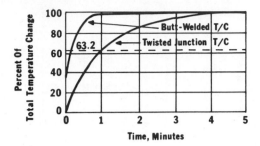

Figure 9.4   Speed of response of a thermocouple with and without a welded junction.

the measuring junction in the butt-welded thermocouple is less than in the twisted junction.

In comparing thermocouples such as those in Fig. 9.4, the instrumentation industry has generally agreed to express the speed of response in terms of the instrument's response to an arbitrary 63.2% of the total change. The time for this 63.2% change is called the lag coefficient. In Fig. 9.4 the lag coefficient for the welded thermocouple is 0.2 min and for the twisted thermocouple 1.0 min.

The effect of measuring junction thermal capacity on the response of a thermocouple becomes even more apparent when we compare the response of a bare butt-welded thermocouple to one that is immersed in a stainless steel protective well (Fig. 9.5). The delayed temperature rise of the element in the protecting well is caused by the increased thermal capacity, primarily because of the added mass of the protecting well. However, the increased mass is not the only factor affecting the speed of response, for the walls of the protecting well offer a resistance to the flow of heat to the detecting element. As we learned in Chapter 6, this resistance is even greater if an air space exists between the well and the thermocouple element. These resistances slow down the transmission of heat to the element and thereby slow down its response.

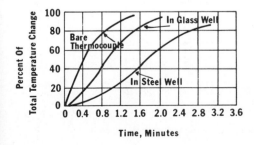

Figure 9.5   Speed of response of a thermocouple with and without a protective well.

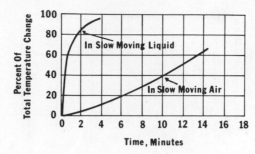

Figure 9.6 Comparison of the speed of response of a thermocouple in flowing air to one in flowing water.

In addition to the well wall affecting the rate at which heat is exchanged between the element and its surroundings, the nature of the surrounding and its velocity also has an effect. For example, as discussed in Chapter 6 heat transfer through a surface by a combined conduction-convection process is a function of the thermal conductivity, the specific heat, and the velocity of the fluid. The effect of these parameters in a well-protected thermocouple is shown in Figs. 9.6 and 9.7.

In Fig. 9.6 we can see the effect of thermal conductivity and specific heat on the response factor. Small thermal conductivity and specific heat values for the surrounding fluid lead to long response times. Thus, it is difficult to obtain rapid temperature responses in gaseous atmospheres.

The speed of response, as shown in Fig. 9.7, increases as the velocity of the surrounding fluid increases. For example, the lag coefficient for the thermocouple in Fig. 9.7 decreases from 0.1 min at a flow of 5.0 ft/min to 0.04 min at a flow of 80 ft/min.

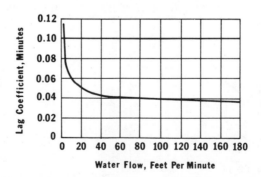

Figure 9.7 Speed of response of a thermocouple versus fluid velocity.

In summary, for temperature measurements with thermocouples, the fastest response and therefore the minimum amount of lag can be achieved by:

1. Using a thin-walled protective well.
2. Using a close fit between the element and its protecting well.
3. Using a welded measuring junction for the thermocouple element.
4. Maintaining a high velocity of the fluid to be measured.

Filled System Thermometers. One of the early and still used methods of measuring temperature is based on the principle that fluids expand when heated. For example as the gas in the bulb shown in Fig. 9.8 is heated, an internal pressure that is proportional to the temperature of the bulb builds up in the system. This pressure causes the spiral to uncoil and move the pointer along the scale. In practice the pointer is most often replaced by a pen that moves across the chart of a recorder, or a transmitter coil, flapper, or switch mechanism of a controller.

A filled thermometer system may contain a liquid, vapor, or gas as the expansion fluid. However, as shown in Fig. 9.8, each system consists basically of the same elements:

1. A temperature-sensing bulb immersed in the medium to be measured.
2. A capillary tube connecting the bulb to a readout or transmitting device.
3. A readout or transmitting device that responds to pressure changes caused by temperature.

Filled systems find their major application today as temperature transmitters for pneumatic control systems. In such applications, liquid-filled systems are useful in the temperature range -300 to $600°F$, vapor filled from -425 to $600°F$, gas filled from -450 to $+1400°F$, and mercury filled from -38 to $1200°F$.

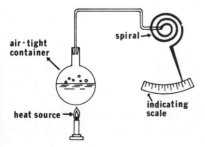

Figure 9.8 Principle of gas-filled thermometer system.

The positive features of the filled systems are that they are simple, inexpensive, and generally have short response times. The disadvantage of these systems is that they are subject to error caused by variations in ambient temperature.

Resistance Temperature Detectors.   Resistance thermometers are based on the fact that the electrical resistance of a conductor changes as its temperature varies. For most metals the change is constant over a specific temperature range. Therefore a coil of wire can act as a temperature sensor with a direct relationship established between resistance and temperature.

Commonly used metals include nickel, copper, and platinum. Platinum has a broad temperature range ($-100°F$ to $1650°F$) over which it gives a linear response. Nickel exhibits a nonlinear response but its large temperature coefficient allows for accuracies within $0.1°F$.

The detecting element of a resistance thermometer consists of a length of fine wire wound around a mica support. This coil is then placed in a protective sheath, and the entire assembly is referred to as a resistance thermometer bulb. The bulb is connected to a measuring instrument containing a Wheatstone bridge.

The primary advantage of resistance thermometers is their high accuracy, and accordingly they find their greatest application in services requiring this type of accuracy. A disadvantage in using resistance thermometers is that resistance bulbs are not interchangeable and therefore new calibration curves are required when they are changed. The speed of response of resistance thermometers is also slower than for thermocouples. This is due to the greater mass of the resistance bulb. As in the case of thermocouples, transmission lags are negligible because the electrical signal travels through the wire at the speed of light.

Radiation Pyrometers.   Radiation pyrometers are used to measure temperatures that are above the average range capabilities of other measuring devices. Their operation is based on the principle that the intensity of radiant energy emitted from the surface of a body increases proportionately to the fourth power of the absolute temperature of the body. Consequently, a radiation pyrometer does not require direct contact with the material whose temperature is to be measured.

A schematic illustrating the principle of operation of a radiation pyrometer is shown in Fig. 9.9. Sensors other than thermopiles are sometimes used. Some of these include photocells, transistor-like photovoltaic cells, and bolometer sensors.

While radiation pyrometers are ideally suited for high temperature measurements and for measuring temperatures of objects that cannot be

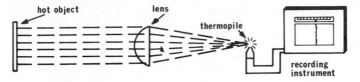

Figure 9.9   Principle of radiation pyrometer operation.

contacted with other measuring devices, it is subject to error in sensor output. The most important one is radiation attenuation due to absorption of certain frequencies by intervening substances such as gases, vapors, and particulate matter in fumes.

Since electromagnetic radiation is being detected, the response lag time is negligible.

9.5.2   Flow Measurement

As indicated in earlier discussions, flow measurements account for a very high percentage of the process variables measured in the chemical industry. This is because the entire material balance in a chemical process is based upon flow measurements.

In our discussion of fluid transfer concepts in Chapter 5, several methods for measuring flow were presented. As judged from this discussion, the most common method for measuring flow is the insertion of a primary element in the stream thereby changing the fluid velocity, which is sensed as a differential pressure. For this method, flow rate is proportional to the square root of the differential pressure generated by flow through the restriction. Other flow rate measurements discussed in Chapter 5 include turbine meters, magnetic meters, positive displacement meters, and rotometers.

Once a differential pressure is created by a primary device, it must be converted to some usable force or signal if flow indication or control is to be established. Several different types of devices are used for this purpose. Basically all of these devices fall into one of two classes, those that use a liquid in a typical manometer setup to measure differential pressure and those that do not.

While the typical "U" tube mercury manometer works very well in measuring differential pressures in simple laboratory experiments, it is not very practical or convenient to use in control loops or in plant operations. Consequently it has been replaced almost exclusively by other types of measuring devices. The most common is a force-balance system utilizing either a diaphragm or bellows as the differential pressure sensing element.

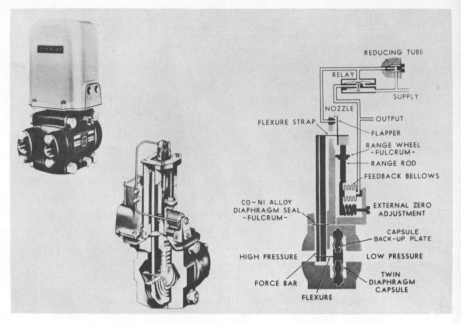

Figure 9.10 A pneumatic d/p cell transmitter. Courtesy of the Foxboro Co.

Figure 9.10 shows a typical force-balance device, which detects differential pressures and transmits a pneumatic signal proportional to the measurement to a receiving device such as a recorder, controller, or indicator.

The transmitter in Fig. 9.10 is referred to as a d/p cell. In operation, the differential pressure to be measured is applied across a diaphragm. The resulting force is transmitted to an attached force bar to which is applied an opposing force from a pneumatic feedback bellows (see Fig. 9.10). The motion of the force bar repositions the flapper which in turn causes a change in the output pressure. The output pressure is used in the feedback bellows to establish equilibrium of the force bar. The output pressure is therefore proportional to the differential pressure and is transmitted to a pneumatic receiving device.

An electronic version of the d/p cell is similar to the pneumatic type shown in Fig. 9.10. The difference is that the output is an electric current rather than a pneumatic signal.

D/p cell transmitters are available in differential ranges from 1 to 1000 in. of water. They are also economical, reliable, and easy to calibrate.

The response of differential type flow elements is almost instantaneous. So is the response of other types of flow measuring elements such as turbine meters, rotometers, and magnetic meters. If the element produces an electronic signal, the transmission lag is essentially nil. However if the element produces a pneumatic signal, the lag may be as much as a few seconds, depending on how far apart the detecting element, measuring element, controller, and control element are. If the transmission lag time in such systems is too long, normal corrective action is to locate the controller closer to the measurement and control point or to increase the diameter of the transmission tubing thereby reducing the resistance to signal flow. If this does not result in satisfactory response times, an electronic system should be used.

9.5.3 Pressure Measurements

A very broad pressure range, from vacuum where the pressure may be as low as 1 micron to pressures of 50,000 psi, is utilized in the chemical industry. Because of this large range, we find that numerous pressure-measuring elements are utilized. Figure 9.11 lists some of the more common elements and their useful ranges.

Manometers. These pressure-sensing devices are most often found in the laboratory, but they also find application in chemical plants; primarily as differential pressure devices. The primary advantage in using a manometer is that it is simple in construction, gives high accuracy and good repeatability, and can be used with a wide range of filling fluids. The most common fluid used is mercury, but fluids of lower specific gravity have also found use.

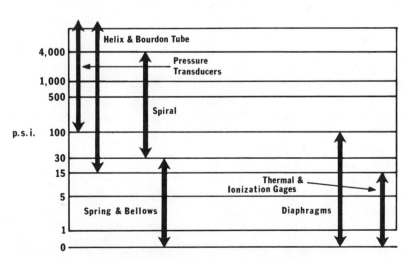

Figure 9.11 Range of pressure-measuring devices.

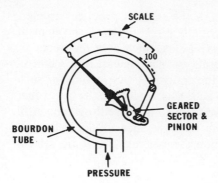

Figure 9.12 C-type bourdon tube pressure element.

The major disadvantages in using manometers are due to problems in leveling, lack of portability, and chemical attack of the filling fluids.

Signal transmission from manometers for remote reading and control can be achieved by methods similar to those described for flow measurements.

Bourdon Tubes.   Bourdon pressure-sensing elements are of three types: the C, the spiral, and the helical. The basic principle of operation is the same in each, when pressure is applied to the thin-walled tube, a deformation takes place that is translated into a pressure measure. Figure 9.12 illustrates the measurement technique for a C-type bourdon gauge.

When internal pressure is applied to the C-bourdon element, the tube tends to straighten out. The movement is balanced against the stiffness of the tube. The motion of the tip of the tube is nonlinear, but linear response is obtained by use of a geared sector and pinion movement.

Bourdon elements possess several desirable features; for example, they are simple in construction and low cost, they have long life in service and are easily adapted to transducer designs for obtaining electrical outputs. However, some of their disadvantages are that their precision is limited when measuring pressures less than 50 psi, they are susceptible to shock and vibration, and they are subject to failure of the element to return to its zero position after every flex.

Transmission signals can be obtained by use of force-balance systems such as described for flow measurement transmission. The response of bourdon elements is almost instantaneous and normally presents no problems in pressure control loops.

Other Pressure-Measuring Elements.   Diaphragms and bellows elements are used primarily to measure pressure in the intermediate or low range from 0 to

50 psi. They find their greatest application in force-balance systems of pneumatic receivers and controllers.

Ionization and thermocouple sensors are used to measure pressures in the vacuum range. McLeod gauges also find application in vacuum measurements, and are quite often used to standardize ionization and thermocouple sensors.

More information on these devices can be found in the references listed at the end of this chapter.[1–5]

9.5.4 Liquid Level Measurement

Many different devices are used to measure liquid levels, but most of those used for control purposes rely upon the measurement of the pressure of the liquid head.

In general, the instruments employed for hydrostatic head level measurements are similar either to those employed as the measuring element of differential pressure flowmeters or to pressure elements discussed in the preceding section. Figure 9.13 shows how a conventional differential pressure transmitter can be used to measure liquid level in a reactor.

The lag time for most liquid level measurement elements is very short and presents no problem in control applications.

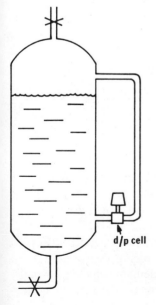

Figure 9.13 Measurement of liquid level with a d/p cell.

9.5.5   Analytical Measurements

Most analytical process analyzers operate on the same basic principles as their laboratory counterparts except that modifications are made to allow for unattended operation. Since most chemistry students are quite familiar with the more common analytical instruments (e.g., chromotograpy, pH, redox potential, and conductivity) used in process control applications, we will not go into a detailed discussion on their makeup and operation. For a complete coverage of analytical process instrumentation, you should consult one of the references given at the end of this chapter.[1-5]

One major difference between a laboratory analyzer and a process analyzer is that the process analyzer must be enclosed in a housing that will give it adequate protection from the environment. In addition, most process analyzers must be made explosion proof for they are often used in hazardous areas.

The method by which the detecting element is brought into contact with the material to be analyzed depends upon the analytical measurement being made. For pH, redox potential, and conductivity measurements, the detecting element (e.g., electrode or cell) can be immersed directly in the material to be analyzed, thereby providing continuous analysis. On the other hand, a measurement like a chromatographic analysis requires that a representative aliquot of the sample be introduced into the analyzing section of the process chromatograph. This type of equipment cannot accept another sample until the previous analysis is complete, which may be several minutes.

The difference in sampling mode for the various analytical measurements leads to differences in the speed of response of the detecting elements. The response time for pH, conductivity, and redox potential elements is almost instantaneous, whereas the response time for a chromatographic response may be several minutes. The signal transmission lag for analytical measurements is nil, for almost universally electronic signal transmission is used.

9.6   THE CONTROLLER

Now that we know how and what types of measurements are made in a control loop and basically how they are transmitted, the next step is to see what the controller does with these measurements and how it accomplishes it.

The controller is the most complicated element in the control loop. It receives information continuously from the measuring device, and must send a continuous signal back to the final control element, which is normally a control valve. The job of the controller is to compare the value of a mea-

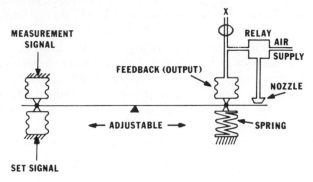

Figure 9.14 Schematic of a force-balance control unit.

sured variable (signal input) to its desired value (set point) to produce an output signal that maintains the desired value. The signals produced by the controller may be either pneumatic or electronic.

A schematic of a typical pneumatic controller of the force-balance design is shown in Fig. 9.14. A picture of the actual controller is shown in Fig. 9.15. The measurement signal is a 3 to 15 psi transmitted signal representing the

Figure 9.15 An electronic controller. Courtesy of the Foxboro Co.

span of true measurement and is opposed by a 3 to 15 psi set signal. Measurement and set bellows are selected so that equal pressures produce equal forces on the flapper. Any difference in force between the measurement and set bellows produces a torque that is the activating signal. Suppose, for example, that an increase in the measurement signal occurs. This will cause the flapper to rotate counterclockwise around the fulcrum until the increased output from the relay in the feedback bellows produces a torque exactly balancing the activating signal torque. The output signal change will be in proportion to the activating signal change. Since flapper position is related to the balance of forces acting on the flapper bar, this type of controller is referred to as being of the force-balance design.

From our discussion up to this point, we might be inclined to believe that we can control a variable by simply turning the knob of the controller to the set point (desired value). The controller now knows what to aim for and responds to any changes in the controlled variable by immediately readjusting the position of the final control element, the control valve. If this were possible in a real system, the controller's job would be easy. In practice, none of the instruments making up the control loop has an instantaneous response; consequently, allowance must be made for the inherent time lag.

As indicated earlier in this chapter, time lags in a control loop can be broken down into three classifications: measurement lags, process lags, and transmission lags. The effect of these lags on a control system can be evaluated with the help of Fig. 9.16. In Fig. 9.16a, temperature is being measured at point E and controlled by increasing or decreasing the flow of a small control stream through the valve at A. When the valve at A is opened slightly, the flow of hot fluid entering the main stream at point B increases at once, but no change will be detected at E until time equal to $B \rightarrow E$ flow time has elapsed. This type of time lag is referred to as process lag, because its magnitude depends upon the physical design and dimensions of the process equipment. Additional time lags, as discussed earlier, are associated with the temperature measurement (e.g., thermal capacity of the temperature bulb) referred to as measurement lag and transmission of the measurement, transmission lag. The net effect of these time lags is shown in Fig. 9.16b. When the flow rate of the hot stream is changed, as shown by curve F temperature at the tank exit does not respond as a step function nor does its change follow a straight line relationship. Because the mass of the fluid in the tank serves as a heat sink, its temperature changes slowly and causes the exit temperature to respond as indicated by curve T.

Because of the indicated time lags, it is not sufficient for the controller to give a simple proportional response to the deviation of the measured value from the set point. The controller must be able to compensate for time lags

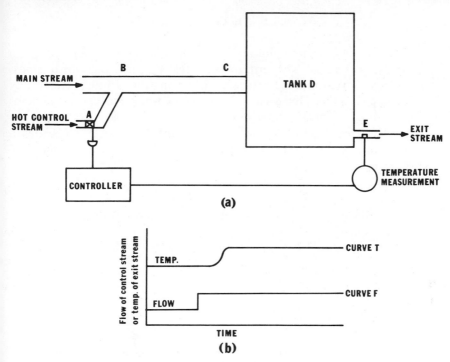

Figure 9.16 (a) Temperature control—indicating time lags; (b) curve T shows how the temperature of the exit stream in (a) changes when a change in the flow of the hot control stream is made.

in the control loop. Depending on how control loop time lags are compensated for leads to the following classification of controllers.

1. On-off control action: a control mechanism having only two discrete values of output, fully on or fully off.
2. Proportional control action: the controller output signal is proportional to the deviation.
3. Proportional-plus-reset control action: the controller output signal is proportional to the time integral of the deviation.
4. Rate control action: the controller output signal is proportional to the rate of deviation change.

Let us take a very brief look at each of these modes of control action; for a more extensive treatment, one of the references at the end of this chapter should be consulted.[1-9]

9.6.1 On-Off Control

This is the simplest form of automatic control and is sometimes referred to as two-position control. The term on-off is really a misnomer, for it is possible to have two-position control without having the positions either fully on or completely off. However, since most two-position control systems are on or off, the term is normally applicable.

In two-position control, whenever the controlled variable deviates a predetermined amount from the set point, the controller moves the final control element to either of two extreme positions. The controller cannot move the final control element to any intermediate position between the two extremes. The amount of deviation of the controlled variable necessary for control action varies with the controller design, but in general is less than 1 % of the instrument range.

Figure 9.17 illustrates the application of an on-off controller in the opera-

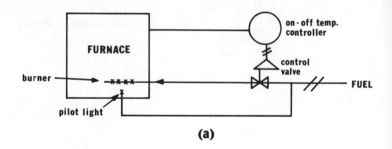

(a)

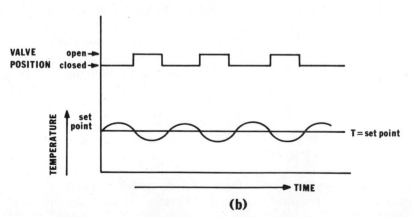

(b)

Figure 9.17 (*a*) Schematic of an on-off control system in the operation of a furnace. (*b*) Curves showing the effect of valve position on furnace temperature.

tion of a gas-fired furnace. When the measured temperature is below the set point, the deviation signal (i.e., difference between set point and measured temperature) is positive and the fuel valve should be open. The fuel valve remains open until the deviation signal becomes zero at which time the valve closes, but because of the large capacitance in the system, the temperature continues to rise and the measured variable rises above the set point, creating a negative deviation signal. After peaking, the temperature falls back below the set point and the valve opens again. The temperature does not respond immediately but continues to drop, reaching a low point, then rises to the set point starting another cycle.

Because of its simplicity on-off control is very popular, and often adequate for many types of process control problems. In general it functions best when the lag time in the control loop is small and the capacitance (e.g., ability to hold a quantity of energy or material) is large and the two extreme positions can be adjusted to permit an input just slightly above and slightly below requirements for normal operation. When these conditions are not met, on-off control will lead to excessive cycling of the controlled variable because the two extreme valve positions will supply either too much or too little of the control agent.

9.6.2 Proportional Control

Some chemical processes cannot tolerate the variation and continuous cycling produced by on-off control. A smoother control action can be obtained with proportional control. In proportional control a fixed linear relationship exists between the value of the contolled variable and the position of the final control element. The proportional controller moves the final control element to a definite position for each value of the controlled variable. This mode of action is illustrated in Fig. 9.18.

Suppose the controller in the temperature control loop of Fig. 9.14 is a proportional controller. Then the relationship between the position of the control valve and the temperature of the tank exit stream can be represented by Fig. 9.18. The set point of this control loop is 200°C, with a range of action extending from 150°C to 250°C. When the controlled variable is at 150°C or less, the valve is wide open. But when the temperature is between 150 and 200°C, the proportional controller moves the valve to a position that is proportional to the value of the controlled variable (i.e., the temperature) as given by the proportional action line shown in the figure. For example, at 175°C the valve is only 75% open; at 200°C only 50% and so on. Finally, when the variable reaches 250°C or more the valve is completely closed.

The range of values of the controlled variable through which proportional control action occurs is called the proportional band of the controller. In

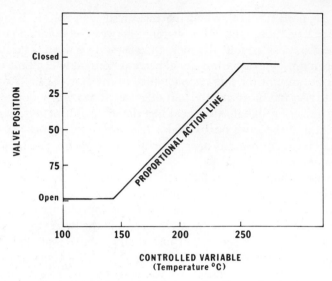

Figure 9.18 Illustration of proportional control action.

Fig. 9.18, for instance, the proportional band extends from 150°C to 250°C. The proportional band is usually expressed as a percentage of the full-scale range of the controller. For example, the full-scale range of the instrument shown in Fig. 9.18 is 100 to 300°C. Since there is proportional control action only between 150°C and 250°C, the proportional band is (250–150)/(300–100), or 50%. If, however, this controller had a full-scale range of 1000°C, then the proportional band would be 100°C/1000°C, or 10%.

The proportional band is adjustable on most controllers and is field tuned to give optimum response to the process changes expected. In tuning controllers, it should be remembered that the width of the proportional band determines the amount of valve motion for any given change in the controlled variable. The rule to follow is: "The larger the proportional band, the smaller the change in valve position for any given change in the controlled variable."

While the proportional controller does provide smoother control than the on-off type, it does have one important limitation. For example, assume that the temperature of the fluid in the tank of Fig. 9.16 is being controlled at a temperature of 200°C by the proportional controller shown in Fig. 9.18. Under these conditions, the temperature is at the set point when the control valve is 50% open. Now if the flow of the main inlet stream is increased, creating an added load on the control system, the temperature of the tank will drop and the opening of the control valve will increase. Eventually, the control valve will find a new position and the process will be back in balance with an exit stream temperature of 200°C. However, when equilibrium is reestablished the control valve position will be at some new position that is offset from the 50% open position where we started to control the process.

This offset is characteristic of all proportional type control systems because the mechanism is unable to cope with process load changes without it. Most proportional controllers have a manual reset adjustment that eliminates offset by shifting the proportional band about the set point. When the reset is done automatically, it is referred to as proportional-plus-reset control action.

9.6.3 Proportional-Plus-Reset Control

In the proportional-plus-reset mode of control, reset is automatic. In this mode of control as soon as the controlled variable deviates above or below the set point (i.e., as soon as some offset develops), there is a gradual and automatic shift of the proportional band to bring the variable back to the set point. Thus while proportional control is limited to a single valve position for each value of the controlled variable, proportional-plus-reset control changes valve position to accommodate load changes.

Figure 9.19 illustrates the primary purpose of the reset function by showing a typical measured variable controlled with and without reset. Since the controller mechanism automatically resets the control point by integrating the deviation in the system, proportional-plus-reset control is often referred to as "integral" control action.

Proportional-plus-reset control finds tremendous application in chemical processes. It is particularly applicable when process load changes are large.

9.6.4 Rate Control

Many chemical processes contain multiple resistances and capacitances resulting in a sizable lag between the time that a change occurs and the time that the measured change is applied to the control mechanism. This problem can partially be overcome by using a control action that is proportional to the *rate* of change of the deviation rather than to the magnitude of the deviation. For example, a rapid deviation indicates that drastic action is occurring, and the controller with rate action will respond by providing an initially large corrective action even though the deviation is small. Because of this characteristic, it sometimes appears that rate action actually anticipates changes; hence, it is sometimes called "anticipatory" action. This mode of control is also often referred to as "derivative" action.

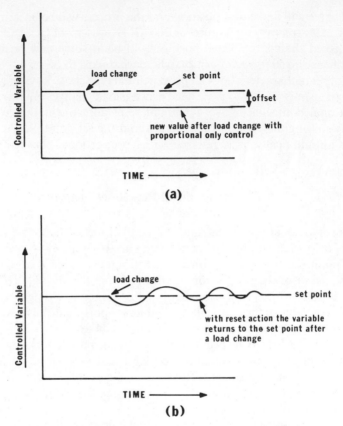

Figure 9.19 (*a*) Control of a variable with proportional-only control action. (*b*) Control of a variable with proportional-plus-reset action.

Rate action is never used alone, but in combination with proportional or proportional-plus-reset action. Its addition to these controllers enhances both the speed and stability of control responses, particularly on slow-responding systems. Rate action is particularly useful during start-up of processes when its initially large correction brings the variable up to the set point very quickly.

9.6.5 Operation of the Controller

During normal operation of a piece of process equipment, the operator adjusts temperatures, flow rates, pressures, and so on, by changing the set point of the appropriate controller to the desired value and allowing the controller to adjust the valve position until the new set point is reached. However, there may be occasions (e.g., during start-up or shutdown) when the operator wishes to be able to change the valve directly. This can be accomplished with most all controllers by simply changing a lever from "automatic" to "manual" control. In this position, manual movement of the set point knob allows the operator to send a direct signal to the valve actuator to set the valve in any desired position.

9.7 FINAL CONTROL ELEMENT—THE CONTROL VALVE

The final control element is the term that has been applied to the item that finally responds to make a change in the measured variable. In the majority of control loops this is a control valve; although it could be a pump, heater, and so on.

A control valve regulates the supply of material or energy to a process by adjusting an opening through which the material flows. Many different types of control valves are available. They vary in body design, flow characteristics, size, and activator type. However, basically they all consist of two major components: the diaphragm actuator and the valve subassembly. Both of these can be seen in the cutaway diagram of a needle style control valve shown in Fig. 9.20.

The diaphragm activator positions the inner valve positively and quickly for any change in controller output. In proportional control applications, the diaphragm area and spring are designed to stroke the valve fully for a specified controller output range. Most initial spring settings are made so that the valve operates on a 3 to 15 psi pneumatic signal change.

When the pressure of the controller signal increases, the diaphragm is moved downward with a force equal to the air pressure multiplied by the area of the diaphragm. As the diaphragm is moved downward, the spring compresses until it creates an equal, opposing, upward force. At this point, the motion stops and the valve stem and plug are in a balanced position.

The "action" of a valve is generally defined as either "air-to-close" or "air-to-open." These terms signify that an increase in air pressure over the effective area of the diaphragm will either close or open the valve—depending on the type of activator used and the plug to seat-ring relationship. The choice of action depends on the "fail-safe" position one wishes the valve to take upon failure of the air supply or pneumatic signal. For example, suppose a control loop is controlling the pressure in a reactor by means of a control valve on the exit line. If overpressurization of the vessel could result from the control valve being in the closed position, then an air-to-close valve should be used. An air-to-close valve will go to the "open" position when there is

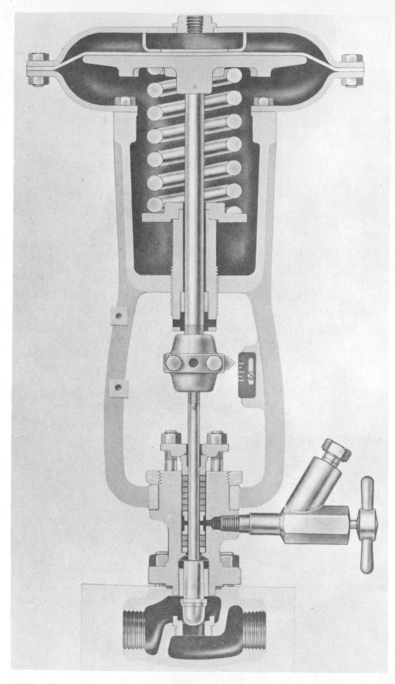

Figure 9.20 Cross-sectional view of an "air-to-close" needle style control valve. Courtesy of Honeywell Co.

no air on its diaphragm and for this application this would be the "fail-safe" position.

Other aspects to consider in selecting a control valve are: (1) environmental factors, such as temperature, pressure, corrosion, and abrasion; (2) size; and (3) percentage travel versus flow that is compatible with the control loop and process characteristics. In general the smallest control valve that will pass the necessary maximum flow provides optimum control and maximum economy.

9.8 OTHER ELEMENTS

In addition to the four basic elements of all control loops, other elements may also be present. These include such devices as indicators, recorders, transducers, and alarms.

Indicators include pointers, floats, liquid columns, and so on. Most instruments in the past utilized fixed scales with moving pens or pointers to show the measured variable. However, with newer instruments, digital indicators that present measured variables in numerical form are finding greater application.

Recorders are used to provide continuous records of measured variables with respect to time. Recorder charts use essentially the same scales that are used with indicators but with an added coordinate to designate time.

Transducers are used to convert signals from one energy form to another or from one signal level to another. For example, a transducer is often used to convert an electronic current signal to a pneumatic signal required by a control valve.

Alarm and/or shutdown units are used to represent abnormal situations in the process. These units are activated whenever the measured variable in the control loop exceeds a certain preselected boundary.

9.9 COMPUTER CONTROL

Because chemical processes have become more sophisticated, a typical pilot plant may have between 20 and 60 automatic controllers and a full-scale plant 60 to 300. The process operators must monitor these controllers in order to assure that operation of the plant falls within prescribed guidelines. However, because the volume of information coming from the controllers is great, process operators can only perform simple calculations with respect to plant performance. Thus the outstanding characteristics of a computer

control system—the ability to acquire, assimilate, analyze, and dissemi-
nate large amounts of information with great speed and accuracy—make the
computer a logical tool for the automation of chemical processes.

A computer can be integrated into the control loop by: (1) feeding the
electrical signal imputs from the measuring units into the computer where
they are stored, and (2) using the computer output to adjust the set points
of the automatic controllers. The computer can then carry out the tasks of
continuous surveying of the operating conditions, frequent calculating
of the optimum set of operating conditions, and adjusting of controller
set points to achieve these conditions. It should be understood, however,
that the computer can carry out optimum operational calculations only
if a mathematic model of the process is available. The model enables the
computer to be programmed to calculate the behavior of the plant in
response to changes in such conditions as temperature, pressure, and feed
rates. The computer can therefore similate plant performance and determine
optimum operating conditions.

9.10 SUMMARY

In this chapter we have tried to give you a representative picture of some of
the basic principles, equipment, and applications of instrumentation and
control used in the chemical industry. The material presented should serve
only to stimulate your interest and we hope it will encourage you to explore
more deeply some of the most important areas by consulting one of the
many references that contain supplementary information on process control
and instrumentation cited at the end of this chapter.

9.11 REFERENCES

1. N. A. Anderson, *Instrumentation for Process Measurement and Control*, 2nd ed., Chilton
   Book Co., Radnor, Pennsylvania, 1972.
2. W. G. Andrew, *Applied Instrumentation in the Process Industries*, Vols. 1 and 2, Gulf Publishing Co., Houston, Texas, 1974.
3. P. S. Buckley, *Techniques of Process Control*, Wiley, New York, 1964.
4. D. R. Coughanour and L. B. Koppel, *Process Systems Analysis and Control*, McGraw-Hill,
   New York, 1965.
5. P. Harriatt, *Process Control*, McGraw-Hill, New York, 1964.
6. E. F. Johnson, *Automatic Process Control*, McGraw-Hill, New York, 1967.
7. D. Permutter, *Introduction to Chemical Process Control*, Wiley, New York, 1965.

8. A. Pollard, *Process Control for the Chemical and Allied Fluid-Processing Industries*, American Elsevier, New York, 1971.

9. T. J. Williams and U. A. Lauher, *Automatic Control of Chemical and Petroleum Processes*, Gulf Publishing Co., Houston, Texas, 1961.

9.12   PROBLEMS

1. Draw a block diagram of your home heating system, indicating the components of the system and their interrelationships. Identify the basic elements of the control system. Identify all factors that contribute to a time lag in the response of the heating system.

2. Draw a block diagram of a continuous distillation assembly that is capable of separating a 60% (weight basis) acetone–40% benzene feed at a rate of 200 liters per hour. Design a control system that would allow the distillation to be run on a continuous unattended basis. Make specific recommendations with regard to the type of elements that should be used in the control loop.

3. Suppose you were assigned the task of evaluating the selectivity of a new heterogeneous olefin oxidation catalyst over a broad range of temperatures. Design a small-scale tube reactor that would be suitable for obtaining the desired information and could be operated with a minimum of attention.

4. Can a thermocouple temperature indicating device be used in place of a thermometer in all cases? Explain.

5. How could you modify a mercury manometer so that it could be used to give pressure measurements at a remote location?

6. Design a control system for conducting a small-scale liquid–liquid extraction on a continuous basis at a temperature of 50°C. Identify the type, specifications, and function of each of the individual elements in the control system.

10

FURTHER CONSIDERATIONS

DEVELOPING THE PROCESS

In Chapter 2 we considered the first stage in the development of a process. The focus was on selecting the most promising synthetic route to the desired product. After this choice has been made it is usually necessary to demonstrate experimentally that the reaction does indeed take place. The next stages involve translating this laboratory synthesis into a form that will be the experimental basis for the engineering design of the production plant. The general sequence of events in the development of a process are:

1. Envisioning the process.
2. Verifying the reaction.
3. Studying the reaction.
4. Defining the process.

The chemical reaction is the key to the process. The greater the understanding of the chemistry of the reaction, the greater the confidence with which the process can be scaled up. But a reaction is not a process. A process is defined as a systematic series of actions directed to some end. In addition to the reaction step, the process development effort must consider the product itself and all of the operations required to isolate and purify the product.

10.1 LABORATORY PREPARATION

The chemical reaction of interest might have been discovered in the course of an exploratory research program. Initially the "product" might be a peak on a gas chromatography plot; or the reaction might be a rather slight variation of a preparation described in the literature for a similar compound.

Regardless of the source, the laboratory preparation to be used as the starting point for the process development effort is very frequently just what the terms imply—a procedure for synthesizing small quantities of the product in the laboratory. It is seldom suitable for direct scale-up to the manufacture of commercial quantities. The objective at this stage has been to verify that the reaction occurs and the desired product is formed. In line with this objective the laboratory work has usually been carried out in a manner that stresses speed, simplicity, and convenience. Anything that expedites the project is practical at this point. The use of reagent grade chemicals, expensive or hazardous solvents, dry ice bath temperatures, and refluxing overnight are common. After the reaction has been demonstrated to have some potential, then the question of large scale practicality and economy can be considered.

A process development program represents a significant commitment of research facilities and talent. Research management cannot initiate such a program unless it is convinced that the proposed project fits with the overall corporate goals and the reaction has the potential of being developed into a practical and economical process. The research chemist plays an important role in this decision. His first obligation is to be objective in his evaluation. His second obligation is to communicate clearly the results and the conclusions of his experimental work. All too commonly the chemist's research report states the facts but fails to clearly point out the significance of the facts.

10.2 THE PROCESS DEVELOPMENT PROGRAM

Process development from the viewpoint of the chemist is an experimental program that investigates the chemistry and the utility of the reaction and provides the information required for the engineering calculations and decisions in the process design effort. For a typical process this information would include:

1. A simple flow diagram of the process.
2. The types of reactors that could be used.
3. The separation and purification operations required.
4. Data for a material and energy balance.
5. The effects of process variables upon conversion, yield, and product quality.
6. The effect of raw material quality upon the process and the product.
7. Schemes for process and product control.
8. Suggested materials of construction.

9. An idea of those points in the process that are most critical from the standpoint of major uncertainties, potential scale-up problems, or major contributions to costs.

Planning a process development program is little different than the planning of any other piece of work. One compares the present situation with the desired status at the completion of the project and then systematically considers the individual tasks that are required.

In process development this can be done conveniently in terms of flow diagrams. At the beginning of the program the process is described in a simple block flow diagram as in Figure 10.1. At the completion of the process design engineering, the flow diagram will depict individual items of equipment as illustrated in Figure 10.2. Temperatures, pressures, residence times, and compositions of each stream will have been established. Detailed discussions between the development chemist and the process design engineer will result in a listing of the information required. The time that the information is needed and the accuracy required should also be considered. Table 10.1 lists types of data that may be required. References 1 and 2 include a more complete listing of information and data requirements. It should be noted that some of the required data may be available in the literature.

With respect to planning the experimental program the special importance of analytical methods should be mentioned. Much of the data described above is based upon the analysis of experimental samples. The development chemist must understand the analytical methods and must assure himself that they are of sufficient accuracy, and indeed that they actually measure what they are intended to measure.

The objective of the process development effort is a process that is practical, safe, and economical. To be economical it must produce a product that a customer is willing to buy at a price that makes the venture attractive to the producer. To be practical and safe the process itself must first of all be stable and controllable. It should be able to withstand a certain amount of variation or error in operation without producing catastrophic results.

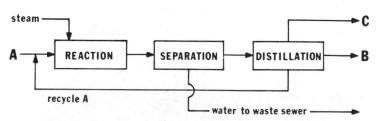

Figure 10.1 Block flow diagram for the reaction $A \rightarrow B + C$.

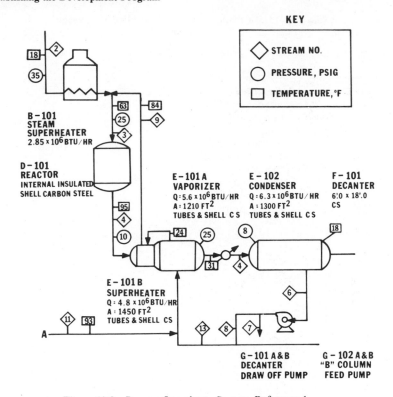

Figure 10.2 Process flow sheet. *Source*: Reference 1.

In short, it should be as "idiot-proof" as possible. Automatic controls *do* fail; operating personnel *can* be unattentive and sloppy.

10.3 ESTABLISHING THE DEVELOPMENT PROGRAM

The development of a new process can involve many different functional groups within a company. The research and development department and engineering department bear the primary responsibility for the program. Production people must be involved because they will have the responsibility of operating the process, and they must accept it and must feel comfortable with it. Since the ultimate object of the whole effort is the sale of the product, the marketing group must be involved to assure that the product is saleable and the projected volumes are realistic. The economic evaluation group not only conducts the financial analysis of the completed process but also provides guidance and interim evaluations during the development program.

Table 10.1   Data Typically Required for Process Design

1. Stoichiometry of the process reaction and side reactions

2. Thermal and thermodynamic data
    a) Heats of reaction
    b) Equilibrium constants, effect of temperature
    c) Specific heat of reactants, products
    d) Heats of vaporization and fusion
    e) Physical properties of reactants and products--density, viscosity, thermal conductivity, surface tension, flash points
    f) Flammability limits
    g) Solubilities and heats of absorption
    h) Adsorption and heat effects
    i) Distillation-vapor pressure activity coefficients
    j) Extraction-solubility, phase diagrams

3. Kinetic data
    a) Rate expression
    b) Activation energies

4. Effect of process variables on yield, conversion and product quality
    a) Concentrations, pressure, temperature, catalysts, agitation, impurities and recycle
    b) For heterogeneous catalysts, the effect of activity, aging, particle size, attrition, poisoning effects, regeneration

5. Other
    a) Safety hazards
    b) Thermal stability of reactants, intermediates, products
    c) Problems with foaming, emulsions, plugging, etc.
    d) Corrosion data

The legal department may provide services in matters pertaining to patents, licenses, labeling, and government registration. In a similar manner purchasing agents, safety engineers, industrial hygienists, waste disposal experts, and corrosion specialists each provide their special input to the development program. Before the process development is completed just about everyone except the manager of the cafeteria has gotten involved in one way or another. With the involvement of so many people with apparently diverse interests, it might seem to be a minor miracle whenever a new plant does start up. Actually, things go more smoothly than might be imagined. Very likely this is because the interests of the people are not diverse at all. For the most part, everyone realizes that new products and new processes are necessary to the well-being of the company. Each is ready to contribute his particular talent to the project. The foregoing is simply to point out that, although basic research or exploratory research might well be conducted in relative seclusion, development work by its very nature involves close and frequent contact with many other groups. This can make the work very stimulating for some chemists and very frustrating for others.

The process development chemist then must maintain close and frequent communications with others during the course of the development program. The primary contact is with the process development engineer. The primary function of the chemist is to develop the general flow scheme of the process and to provide the experimental data required by the engineer for the more detailed process design work. To do this job well he should have some familiarity with *how* the engineer does his job and what he needs from the chemist.

The process development program should be planned carefully. These plans should be reviewed with the engineer so that the data gathered is as relevant as possible and as accurate as needed. It would be wasteful of time and money to study a reaction run at dry ice/acetone temperatures if it were known that there is no possibility that the plant would operate at these temperatures. Similarly, it would be wasteful to determine the solubility of a reaction product to 0.1 g/liter if the design work can be accomplished using data to the nearest 10 g/liter. Product quality must be considered in the same light. It does not make any sense to develop a process that produces a product that does not meet the purity requirements of the potential customer. Nor is it reasonable to expend effort on a purification scheme to produce a reagent grade product if this purity is not needed.

It is the responsibility of the process development chemist to design and carry out the experimental program. It cannot and should not be directed by an engineering or a marketing group. However, the chemist must communicate freely and easily with these groups to assure himself that his program is headed in the right direction. The mark of a good process development effort is a smooth scale-up and a product that meets sales specifications.

10.4 SCHEDULE

A process development program is not an open-ended opportunity to dabble in the laboratory until a major scientific breakthrough happens spontaneously. If the project is worthwhile, there will be considerable pressure for completion within a given time.

Experimental work is difficult to schedule because it involves uncertainties. If there were no uncertainties, there would be no need of an experimental program. In spite of these difficulties, the chemist should review the projected tasks and realistically estimate the time and effort that will be required. It is generally less hassle to argue for, and get agreement on, a reasonable schedule at the start of a program. It is nerve-racking and bothersome to plead for additional time during the course of the program.

The overall program involves a number of objectives and a number of independent tasks. Some of these tasks must be performed in proper sequence, while others can be performed concurrently. Program Evaluation and Review Technique (PERT) is a useful method of organizing and scheduling such development projects.[3]

10.5   THE PRODUCT

Before the completion of the development program it will be necessary to define the product that is to be offered for sale. This must involve the marketing department and must consider the potential customers and the intended end use for the product. It is seldom adequate to describe the product in vague terms such as "very pure" or even ill-defined terms such as "minimum 95 % pure." If the sales specifications will be based upon an assay, the method of analysis should be agreed upon. In certain types of products standardized tests such as those published by ASTM are applicable. If this is the case, these quality criteria should be recognized and worked into the development program. If certain types of impurities, such as trace metals, or certain properties, such as color, are particularly undesirable this should be noted. The earlier these questions can be settled, the easier the development program will be.

10.6   REACTION TYPE

Very early in the laboratory development it should be possible to consider the relative merits of operating the process batchwise, semibatchwise, or continuously. It may be necessary to reconsider this decision as additional data are gathered. Most frequently, however, a sufficient number of factors are known well enough to allow a reliable evaluation at an early stage. The experimental program can then be designed to provide the most relevant data for the reaction type chosen.

In a *batch* process all of the reactants are charged into the reactor and then the reaction is initiated, usually by increasing the temperature or by the addition of a catalyst or initiator. The reaction is allowed to proceed until the desired degree of conversion is achieved. The reaction is then stopped and the reaction mixture treated to isolate and purify the product. An example would be an esterification reaction. Methyl salicylate (oil of wintergreen) is produced by charging a glass-lined, steam-jacketed reactor with salicylic acid, excess methanol, and sulfuric acid catalyst. The mixture is then heated for a number of hours at reflux temperature. The reaction mixture is then cooled, washed to remove catalyst and unreacted reactants, dried, and then pure product obtained by distillation.

In a *semibatch* process one or more of the reactants are charged to the reactor initially and then one reactant is added or metered in over the reaction period. Many halogenation, nitration, and hydrogenation reactions are conducted in a semibatch manner. For example, monochloroacetic acid is produced by charging glacial acetic acid and red phosphorous to a jacketed reactor; chlorine is then sparged into the reactor over an 8-hr period while maintaining the temperature at about 100°C by circulating cooling water through the jacket. The reaction mixture is then discharged to a crystallizer for isolation of the solid product.[2]

In a continuous process the reactants are pumped continuously through the reactor system. As discussed in Chapter 7, the two types of reactors used are the tubular and the stirred tank. Vapor-phase reactions frequently utilize tubular reactors. Unchanged reactants can be recycled to the feed stream. Liquid-phase reactions may be run in a stirred-tank reactor system. In the case of slow reactions where high conversions are desired, a series of stirred tank reactors may be used.

Methanol is synthesized in a tubular reactor by the continuous reaction of hydrogen and carbon monoxide in approximately the theoretical ratio. The mixed gas feed is compressed, heated, and passed through a copper-lined reactor packed with a heterogeneous catalyst such as zinc oxide pellets containing chromium oxide. The temperature of the exothermic reaction is maintained by controlling the space velocity and by external cooling. The reactor effluent is cooled by heat exchange with the feed stream. The product is isolated and purified by distillation. The conversion per pass is about 60%. Unreacted hydrogen and carbon monoxide are recycled.

The formation of tricresyl phosphate, by the reaction of cresol and phosphorous oxychloride, may be carried out continuously in a cascade of stirred-tank reactors. In a three-reactor system the first reactor is maintained at 150°C, the second at 225°C, the third at 300°C. The reactants flow slowly through the system. Hydrogen chloride is evolved and passes through the reflux condensers to a recovery system. The product stream is cooled, washed to remove acids and phenols, and distilled.

The type of process best suited for a particular product depends upon the chemistry of the reaction and upon the scale of the production. The kinetics of the reaction and the degree of conversion required are particularly important. It should be noted also that the product characteristics frequently depend upon the mode of operation. Generally a continuous process will yield a better quality and more consistent product. In regard to scale, production rates below 5 million lb/year tend to favor batch operations; above 20 million lb/year continuous operations are clearly favored.[4] It appears that 10 million lb is the approximate break-even point. Generally a continuous process will give better yields, better quality product, and lower operating costs, but to gain these advantages a certain minimum production volume is required. It is frequently a difficult decision because it involves foreseeing future sales volumes. It is quite easy to be conservative. However, conservatism has a price. If the decision is made to go with a batch process at a volume of 10 million lb/year, and a continuous process would have saved 5 cents/lb in manufacturing costs—that choice will cost $500,000 per year.

10.7 SCREENING VARIABLES

One of the primary purposes of the experimental program is to determine the effect of process variables on the reaction and to define optimum reaction conditions. For many reactions there may well be more than one optimum. That is, one set of reaction conditions might result in the highest yield; another set of conditions might give the highest conversion per unit time; another set of conditions might produce product of highest purity. These various responses are then considered, and possibly compromised, in establishing the design operating conditions for the process. The understanding of these effects is very important, not only in establishing the design of the process but also to the efficient operation of the plant as well. Economic and marketing situations change. It might be most desirable to operate the plant at highest yield conditions to minimize raw material costs. On the other hand, in situations of high sales demand it might be more economical to sacrifice yield and produce as much product per day as possible. Or it might be that an increase in product purity is required to increase or maintain the sales volume. Maximizing the economic return on the production plant requires a full understanding of the reaction and the effects of process variables. The first job, then, is to identify the key variables in the reaction. Practical and statistically sound methods are available for the design of experiments to screen independent variables. The beauty of these methods is that they can be used effectively by chemists who do not have good backgrounds in statistical theory.

The first step is to list all of the possible independent variables. These variables might be continuous, such as temperature, pressure, concentration, or discontinuous, such as catalyst type or additive. This is based upon a knowledge of similar reactions and upon a consideration of the physical processes—such as heat transfer or gaseous diffusion—that might be involved. This listing of candidate variables should be done realistically but diligently. It is particularly bad strategy to *assume*, at this stage, that a variable is not important. A little time and money might be saved at the beginning of the project and then considerably more time and money spent investigating factors of secondary importance. Be bold.

Depending on the complexity of the problem, there might typically be 6 to 25 candidate variables in the list. The next step is selecting the screening experiment.

Experimental designs that have been found particularly useful in screening variables are the Plackett–Burman designs.[5–7] These are termed fractional factorial designs. They represent a great reduction in the number of runs that would be required for a full 2^P factorial experiment (Table 10.2). This reduction in number of runs is accomplished by sacrificing estimates of interactions between the variables. These interactions might be very important. However, the purpose of the screening expriment is to identify the most important variables so that they can then be studied in more detail. Possible interactions will be picked up in these later experiments.

After deciding upon the candidate variables the levels of each are chosen. In the case of discrete variables, these might be two types of additives or it might be the presence or absence of one additive. In the case of continuous

Table 10.2 Comparison of Experimental Designs

Number of Factors in Variables	Number of Runs	
	2^P Factorial	Plackett–Burman
6	64	12
7	128	20
8	256	20
9	512	20
10	1024	20
11	2048	20
12	4096	20

variables, such as temperature, it is good strategy to choose levels that are as far apart as possible and reasonable. This assures that any effect will be clearly observed above the inherent experimental error.

All of the measured responses are subject to experimental error due to errors in setting the reaction conditions as well as errors in measurement of the response. These errors are of two types: bias errors, which tend to remain constant or show a constant change throughout a number of trials, and random errors, which fluctuate from one run to the next, averaging zero over a large number of runs. Bias errors are handled by conducting the experimental runs in random order. Random errors are dealt with by replication or repeating experiments.

Example 10.1   Screening of Variables in a Dehydrogenation Reaction

To illustrate how a screening experiment, using a Plackett–Burman design might be set up and the computations accomplished, let us consider a dehydrogenation reaction conducted in a laboratory scale tubular reactor.

The chemist has identified six potential variables that he wants to screen for effect upon yield. These factors and the levels to be used are listed in Table 10.3. Since there are six factors, a 12-trial Placket–Burman design would be suitable. This design has a nominal capacity of 11 variables or factors. The five unassigned factors (X_7 through X_{11}) will be used in the computation to achieve an effective replication so as to get some measure of the experimental error.

Step 1.   The experiment design and the calculations are illustrated in Table 10.4. Each of the 12 trials of the design are listed in horizontal lines. The experimental runs, however, would be conducted in a random order. The random

Table 10.3   Candidate Variables—Dehydrogenation

Variable		Type	+Level	-Level
X_1	Pre-heat	Discrete	Yes	No
X_2	Catalyst	Discrete	Type 1	Type 2
X_3	Flowrate	Continuous	25 ml/min	2.0 ml/min
X_4	Steam	Discrete	Yes	No
X_5	Temperature	Continuous	375°C	250°C
X_6	Catalyst support	Discrete	Type A	Type B
X_7-X_{11}	Unassigned factors used to calculate standard deviation			
Y	Response--%Yield			

Table 10.4 Screening Experiment—Dehydrogenation

								Unassigned Factors					
Trial	Mean	Pre-heat X_1	Cata-lyst X_2	Flow-rate X_3	Steam X_4	Temp. X_5	Cat. Supt. X_6	UFE X_7	UFE X_8	UFE X_9	UFE X_{10}	UFE X_{11}	Yield Y
1	+	+	+	−	+	+	+	−	−	−	+	−	49
2	+	+	−	+	+	+	−	−	−	+	−	+	70
3	+	−	+	+	+	−	−	−	+	−	+	+	40
4	+	+	+	+	−	−	−	+	−	+	+	−	39
5	+	+	+	−	−	−	+	−	+	+	−	+	33
6	+	+	−	−	−	+	−	+	+	−	+	+	40
7	+	−	−	−	+	−	+	+	−	+	+	+	8
8	+	−	−	+	−	+	+	−	+	+	+	−	65
9	+	−	+	−	+	+	−	+	+	+	−	−	53
10	+	+	−	+	+	−	+	+	+	−	−	−	13
11	+	−	+	+	−	+	+	+	−	−	−	+	78
12	+	−	−	−	−	−	−	−	−	−	−	−	12
Sum +'s	500	244	292	305	233	355	246	231	244	268	241	269	
Sum −'s	0	256	208	195	267	145	254	269	256	232	259	231	
Sum +'s & −'s	500	500	500	500	500	500	500	500	500	500	500	500	
Difference	500	−12	+84	+110	−34	+210	−8	−38	−12	+34	−18	+38	
Effect	41.7	−2.0	+14.0*	+18.3*	−5.7	+35.0*	−1.3	−6.3	−2.0	+5.7	−3.0	+6.3	

$(UFE)^2$: $X_7 = 39.7$, $X_8 = 4.0$, $X_9 = 32.5$, $X_{10} = 9.0$, $X_{11} = 39.7$

$$\Sigma (UFE)^2 = 124.9$$

$$\tfrac{1}{5}\,\Sigma (UFE)^2 = 24.98$$

$$\sqrt{\tfrac{1}{5}\,\Sigma (UFE)^2} = S_{FE} = 5.0$$

$$[\text{MIN}]_{95} = 2.57 \times 5.0 = 12.9$$

order might come from a random number table or by picking numbers out of a hat. (If this does not seem scientific enough, you might pick numbers out of a beaker.) At any rate, by randomization of the trials, we have reduced the chances of bias error affecting the results. The vertical columns labelled X_1 through X_{11} indicate the level of the factor in each trial. In regard to the design, note that in the 12 trials each factor was at a high $+$ level for 6 trials and at a low $-$ level for 6 trials. The yield for each trial is indicated in the Y column on the right.

Step 2. The Sum $+$'s line is then computed by adding the yield values for all lines where the factor was at a $+$ level. (Example: X_1 factor $49 + 70 + 39 + 33 + 40 + 13 = 244$.) This operation is continued across the table for all factors, including the five unassigned factors. In a similar way, the Sum $-$'s line is computed. The next line simply totals the Sum $+$'s and the Sum $-$'s to check the arithmetic.

Step 3. The next line is the difference between the Sum $+$'s and the Sum $-$'s for each factor. This represents the total difference in yield for the six trials where the factor was at the plus level, from the six trials where the factor was at a minus level.

Step 4. The last line represents the average effects of the factor at the plus level and is computed by dividing the difference by the number of plus signs in the column, in this case 6. The absolute values of the calculated factor effects relates to their relative importance. In this example X_5 temperature is clearly the most important variable.

Step 5. In order to determine whether a factor effect is significant, experimental error must be considered. The minimum value for factor effect to be significant is computed using the five unassigned factor effects X_7 through X_{11}. Each unassigned factor effect is squared, totaled, divided by the number of unassigned factors, in this case 5. The square root of this number multiplied by a magic number gives the minimum significant factor effect [MIN]. The magic number used in this computation (2.57) comes from a table of probability points of the t-distribution corresponding to five degrees of freedom (five unassigned factors) and the 95% confidence level. What this means is that if we use 12.9 as the cut-off point, we have a 95 out of 100 chance of being correct in our selection of the significant factor effects.

Using this criteria then, three variables—temperature, catalyst, and flow rate—are important and should be investigated further. Preheating, steam, and catalyst support either have no effect or an effect so small that it is obscured by the experimental error and interaction effects.

10.8 OPTIMIZING CONDITIONS

After determining which of the candidate variables are really significant, the next objective is to develop predictive models or equations that describe the process.

The responses or results (which might be yield, conversion, or product properties such as tensile strength, color, or melting point) are considered to be functions of the process variables. The responses are then thought of in terms of contour plots or response surfaces. Consideration of these response surfaces then allows optimization of the variables to either maximize or minimize the particular response. It is seldom that reliable *theoretical* models are available for the responses of interest. An important objective of the experimental program then is the development of appropriate *empirical* models for each of the important responses. These models can then be used to predict future behavior of the process or system.

These predictive models can be developed in two stages. The first stage results in a limited response surface model and is typically based upon two-level factorial experiments. This model can be considered a planar surface contour plot that represents linear and interaction effects. The mathematical expression contains a constant, linear terms, and cross-product terms. The second stage model is more refined and is based upon three-level factorial designs. This model gives higher quality predications and can be considered to be a curved surface contour plot. The mathematical representations typically are second-order polynomials with quadratic terms.

An approach to the development of these empirical models can be based upon the following generalizations:

1.   Over the experimental range of interest the response function usually is smooth; it may slope and curve upward or downward, but it seldom has complicated bumps or sharp kinks.

2.   Interactions between process variables are common; the slope of the variable/response relation may be significantly changed when the level of a second variable is changed.

3.   Experimental error, which may be due to both bias error and random error, is always present and is always significant; a rational experimental program must include some measurement of this error and evaluate experimental results with respect to this experimental error.

Example 10.2   Optimizing Variables

To illustrate the approach and the strategies that might be used in developing these empirical models, we shall consider an example with three key process variables and two critical responses. Table 10.5 lists the experimental ranges of the variables temperature, pH, and reaction time. The two critical responses of interest are yield and purity.

Step 1.   The experimental design used will be Yates pattern, 3 factor two-level factorial; there are 2^3 or eight trials. A design with 2 factors would require 2^2 or 4 trials; 4 factors would require 2^4 or 16 trials; 5 factors would require 2^5 or 32 trials. The basic Yates design is shown in Table 10.6.

The important strategy considerations in setting up this experiment are:

1.   Setting the high and low levels of each process variable at the practical extremes so that any measured effect will be large compared to the experimental error.
2.   Performing trials in a random order to minimize bias error.
3.   The number of replicates or trials is chosen carefully. In this example, the parameters are set as follows:
   (*a*)   The chance that an apparent effect is not real is set at 5%.
   (*b*)   The chance that a real variable will not be detected is set at 10%.
   (*c*)   The size of the effects to be detected is set at twice the experimental error.

These specifications, in a balanced two-level factorial experiment, lead to the determination that 12 to 16 trials are required. Since the basic 2^3 factorial design involves eight trials, each will be run in duplicate yielding

Table 10.5   Process Variables and Responses

Variable		Low(-)	Mid(o)	High(+)
			Range	
X_1	Temperature (°C)	50	100	150
X_2	pH	4.0	7.0	10.0
X_3	Time (hours)	1.5	3.0	4.5
Responses: Y_1 - Yield,%; Y_2 - Purity,%				

Table 10.6 Yates Pattern Experiment Design

Trial		X_1	X_2	X_3	X_4
1		−	−	−	−
2		+	−	−	−
3		−	+	−	−
4	2^2	+	+	−	−
5		−	−	+	−
6		+	−	+	−
7		−	+	+	−
8	2^3	+	+	+	−
9		−	−	−	+
10		+	−	−	+
11		−	+	−	+
12		+	+	−	+
13		−	−	+	+
14		+	−	+	+
15		−	+	+	+
16	2^4	+	+	+	+

16 trials. The basic two-level factorial design yields a planar response surface. In order to check the lack of fit due to curvature, additional trials will be made at the midpoint level of each factor. The difference between the average centerpoint value and the overall average of the design points will indicate the severity of curvature. Because the test for curvature requires more trials for the same precision, this center point trial will be run four times. These four runs, listed as trial 9 will not be randomized but scheduled to be spread throughout the experiment in positions 1, 7, 14, and 20. This will detect any trends in changing bias errors.

Step 2. Table 10.7 illustrates the two-level 3 factor design with the factors in coded form. The experimental runs for trials 1 through 8 are run in duplicate and in random order; trial 9, the center point trial is run four times, interspersed throughout the experimental runs.

Step 3. The results of this experiment are listed in Table 10.8. Each measured value of yield and purity are listed. In the case of yield Y, the average

Table 10.7 Experiment Design

Trial	Replicates	Design: Temp. X_1	pH X_2	Time X_3
1	2	-	-	-
2	2	+	-	-
3	2	-	+	-
4	2	+	+	-
5	2	-	-	+
6	2	+	-	+
7	2	-	+	+
8	2	+	+	+
9	4	0	0	0

Table 10.8 Results of Three Factor Experiment

	Results					Purity	
	Yield						
Trial	Y_1	Y_2	$\bar{Y}$	Range	Var.	Y'_1	Y'_2
1	85.0	80.0	82.5	5	13	93.1	98.1
2	86.3	92.1	89.2	6	17	93.6	87.8
3	81.0	78.6	79.8	2	3	100.0	99.0
4	92.1	98.3	95.2	6	18	94.5	91.9
5	94.0	91.0	92.5	3	5	94.0	90.2
6	96.8	94.8	95.8	2	2	95.4	93.2
7	79.8	88.8	84.3	9	41	97.1	95.7
8	94.3	89.3	91.8	5	13	98.0	96.6
9	82.7	84.3				95.9	96.8
	83.5	84.9	83.4	2	1	97.0	97.4

$\overline{Y}$, the range, and the variance have been calculated for each trial. The variance, which is an estimate of dispersion of data, is calculated by the following formula:

$$\text{variance} = S^2 = \frac{(Y_1 - \overline{Y})^2 + (Y_2 - \overline{Y})^2 + \cdots + (Y_n - \overline{Y})^2}{n - 1}$$

where Y = response value
$\overline{Y}$ = average or mean of response values
n = number of observations

For example, for Trial 1

$$\text{variance} = \frac{(80.0 - 82.5)^2 + (85.0 - 82.5)^2}{2 - 1}$$

$$S_1^2 = \frac{6.25 + 6.25}{1} = 13$$

and for Trial 9

$$\text{variance} = \frac{(82.7 - 83.4)^2 + (84.3 - 83.4)^2 + (83.5 - 83.4)^2 + (84.9 - 83.4)^2}{4 - 1}$$

$$S_9^2 = \frac{3.5}{3} = 1$$

Step 4. The variances calculated for each trial are then used in the calculation of a weighted average of the individual variances for each trial.

$$\text{pooled variance} = S_{\text{pooled}}^2 = \frac{(n_1 - 1)(S_1^2) + (n_2 - 1)(S_2^2) + \cdots + (n_k - 1)S_k^2}{(n_1 - 1) + (n_2 - 1) + \cdots + (n_k - 1)}$$

$$= \frac{(1)(13) + (1)(17) + \cdots + (3)(1)}{1 + 1 + \cdots + (3)}$$

$$= \frac{115}{11} = 10.45$$

The pooled standard deviation is the square root of the pooled variance:

$$\text{Standard deviation}_{\text{pooled}} = \sqrt{S_{\text{pooled}}^2}$$

$$= \sqrt{10.45} = 3.2$$

The pooled standard deviation is used to calculate the minimum observed effect that is statistically significant.

Step 5. The computational analysis for this experiment is shown in Table 10.9. The design matrix has been supplemented with a computation matrix, which will be used to detect any interaction effects. This computation matrix is generated by simple algebraic multiplication of the coded factor levels. In Trial 1, X_1 is minus, X_2 is minus, therefore X_1X_2 is plus; in Trial 2, X_1 is plus, X_2 is minus, therefore, X_1X_2 is minus. The column at the far right of the table is the average yield for each trial. The Sum +'s row is generated by totaling the response values on each row with a plus for each column.

For X_1 factor, $89.2 + 95.2 + 95.8 + 91.8 = 372$. In a similar manner the Sum −'s row is generated. The sum of these two rows should equal the sum of all the average responses and is included as a check on the calculations. The difference row represents the difference between the responses in the four trials when the factor was at a high level and the responses in the four trials when the factor was at a low level. The effect is then calculated by dividing the difference by the number of plus signs in the column. In the first column, labeled mean, the effect row value is the mean or average of all data points. The average of the center point runs, Trial 9, is then subtracted from the mean effect to give a measure of curvature.

Step 6. The minimum significant factor effect [MIN] and the minimum significant curvature effect [MIN C] are again derived from t-test significance criteria. The relationships are:

$$[\text{MIN}] = t \cdot s \cdot \sqrt{\left(\frac{2}{m \cdot k}\right)}$$

$$[\text{MIN C}] = t \cdot s \cdot \sqrt{\left(\frac{1}{m \cdot k} + \frac{1}{c}\right)}$$

where t = appropriate value from "t table"
s = pooled standard deviation
m = number of plus signs in column
k = number of replicates in each trial
c = number of center points

In our example the t value of 2.20 is from the Students' "t" table for the 95% confidence level and 11 degrees of freedom. The degrees of freedom result from eight trials with two replicates and one trial with four replicates.

$$\text{Degrees of freedom} = 8(2 - 1) + 1(4 - 1) = 11$$

Table 10.9 Computation Matrix for Three Factor Experiment

Trial	Mean	Design			Computation				Response
		X_1	X_2	X_3	X_1X_2	X_1X_3	X_2X_3	$X_1X_2X_3$	$\bar{Y}_1$
1	+	-	-	-	+	+	+	-	82.5
2	+	+	-	-	-	-	+	+	89.2
3	+	-	+	-	-	+	-	+	79.8
4	+	+	+	-	+	-	-	-	95.2
5	+	-	-	+	+	-	-	+	92.5
6	+	+	-	+	-	+	-	-	95.8
7	+	-	+	+	-	-	+	-	84.3
8	+	+	+	+	+	+	+	+	91.8
Sum +'s	711	372	351	364	362	350	348	353	
Sum -'s	0	339	360	347	349	361	363	358	
Sum	711	711	711	711	711	711	711	711	
Difference		+33	-9	+17	+13	-11	-15	-5	
Effect	88.9	+8.3*	-2.3	+4.3*	3.3	-2.8	-3.8*	-1.2	

Curvature = 88.9 - 83.4 = 5.5*

$$[\text{MIN}] = 2.20 \times 3.2 \sqrt{\frac{2}{4 \times 2}}$$
$$= 2.2 \times 3.2 \sqrt{.25}$$
$$= 3.5$$

$$[\text{MIN C}] = 2.20 \times 3.2 \sqrt{\frac{1}{8 \times 2} + 1/4}$$
$$= 2.20 \times 3.2 \times 0.56$$
$$= 3.9$$

The calculations for the minimum significant effects in the example are as follows:

$$[\text{MIN}] = 2.20 \times 3.2 \sqrt{\frac{2}{4 \times 2}} = 3.5$$

$$[\text{MIN C}] = 2.20 \times 3.2 \sqrt{\frac{1}{8 \times 2} + \frac{1}{4}} = 3.9$$

Step 7. Applying these criteria to the calculated effects we see that the effects of temperature (X_1) and time (X_3) are significant; there is a small but significant interaction between pH and time $(X_2 X_3)$; and there is a significant curvature effect. Ignoring the curvature for the present, these results can be expressed as a mathematical model using a first-order polynomial. The values for the coefficients are one-half the factor effects listed in the table since these are based upon coded levels $+1$ and -1 that differ by two units.

$$Y = 88.9 + 4.15X_1 + 2.15X_3 - 1.9X_2 X_3$$

Step 8. In this equation the factors are still expressed in coded units. These may be converted into real world units by substituting:

for temperature

$$X_1 = \frac{T(^\circ\text{C}) - (150 + 50)/2}{(150 - 50)/2} = \frac{T(^\circ\text{C}) - 100}{50}$$

for pH

$$X_2 = \frac{\text{pH} - 7.0}{30}$$

for time

$$X_3 = \frac{t(\text{hr}) - 3.0}{1.5}$$

For the example this substitution yields the following expression:

$$Y = 67.5 + 0.083T + 4.3t - 0.42 \, \text{pH}t + 1.3 \, \text{pH}$$

As noted above, this mathematical model does not consider the curvature of the response surface which was detected in the 2^P factorial experiment. The next step in the development program would be to develop a more accurate model upon a 3^P factorial experiment, such as a Box–Wilson design.[8]

10.9   REGIME CONCEPT

Many chemical reactions involve a series of processes, both physical and chemical. The overall observed reaction rate may be determined by the intrinsic rate of the chemical process or by the rate of some physical process such as mass or heat transfer. The rate controlling process is referred to as a "regime."[9]

In a study of the kinetics of the catalytic water gas reaction, Laupichler expressed the regime concept in terms of "reaction resistance."[10] The measured reaction rate was considered to be inversely proportional to this reaction resistance R, which was defined by the equation:

$$R = \frac{1}{kC_m} + \frac{\delta}{D}$$

where k = velocity constant of the reaction
C_m = mean concentration of water vapor
δ = thickness of gas film on catalyst surface
D = diffusion coefficient of carbon monoxide through gas film

The first term, which is the inverse of the chemical reaction rate, was called the "conversion resistance." In the reaction studied the chemical reaction rate was found to be slow compared to the diffusion rate and therefore controlled the overall reaction rate. This system is said to be subject to a *chemical regime*. Reactions where diffusion resistance is the rate controlling factor are said to be subject to a *dynamic regime*. Knowledge of the controlling regime is important to reliable scale-up calculations.

In a given system the controlling regime may vary with conditions of operation. Consider a liquid-phase oxidation of p-xylene with a homogeneous catalyst. The reaction is conducted in a semibatch manner by bubbling oxygen into a propionic acid solution of xylene and the catalyst. The system may be in a dynamic regime with the rate of diffusion of oxygen into solution controlling the observed reaction rate. With an increase in agitation, however, the diffusion resistance may be diminished to a point at which conversion resistance is controlling and the system is in a chemical regime. At intermediate levels of agitation, there may be a mixed regime where both conversion resistance and diffusion resistance are significant. The scale-up of mixed regimes is very difficult.

When the process development concerns a heterogeneous reaction, the rate controlling processes should be investigated. In many reactions that occur near atmospheric temperature, this is done conveniently by measuring the effect of changes in temperature upon the observed reaction rate.

An increase in temperature tends to increase the rate of both chemical and physical processes. However, chemical processes characteristically show a larger temperature effect than do physical processes. As discussed in Chapter 7, the effect of temperature upon the rate of a reaction is expressed by the Arrhenius equation. The effects of temperature upon the rates of mass and heat transfer are related to changes in diffusion rates and viscosity and can be described by similar equations:

Chemical $\quad k = Ae^{-E_a/RT}$

Viscosity $\quad \dfrac{1}{\mu} = Be^{-E_v/RT}$

Diffusion $\quad D = Ce^{-E_D/RT}$

where k = chemical reaction rate constant
$\quad \mu$ = viscosity
$\quad D$ = diffusivity
$\quad E$ = activation energies
$A, B, C,$ = "constants" over small temperature ranges

The temperature effect on physical processes is less, in part because the "energies of activation" for viscous flow and diffusion are smaller than the energy of activation for a chemical reaction.

In the laboratory the reaction is conducted at two or three temperatures and the relative rate measured. The temperature coefficient of the rate is then calculated and expressed as the increase in rate caused by a 10° increase in temperature. Most chemical reactions have 10° temperature coefficients in the range of 2 to 4. Liquid-phase mass transfer processes have temperature coefficients close to 1.3; liquid-phase heat transfer processes have temperature coefficients that are usually below 1.1; the temperature coefficients of both mass and heat transfer processes in the gas phase approach 1.0.

The effects of changing the degree of agitation can be more complex due to the differences between laminar and turbulent flow and fixed and free interfaces.[9] In a chemical regime, however, changes in agitation will not change the observed reaction rate.

10.10   THE RATE EXPRESSION

The process development project as presented to the engineer is the design of a process to produce a certain number of pounds of product per month or per year. Now, given the basic stoichiometry of the reaction and yield data, he can size storage tanks for the reactants and product. Because the rates of many purely physical processes, such as heat transfer, can be estimated quite accurately, he can proceed to the design of such items of equipment as heat exchangers. However, there is no way he can begin to calculate the size of the reactor that will be required, until he has some experimental data on the rate of the reaction. The accuracy and sophistication of the data that are required may vary depending upon the nature of the reaction and the type of reactor. But, even for the most simple process, some rate expression is needed.

Reactor design involves determining the size and type of reactor, and the operating conditions necessary to yield a specified production level. Typically there might be many combinations of reactor size and operating conditions that will satisfy the specified production rate. The objective of the process engineer then is to establish the optimum design—optimum from the standpoint of the overall economics of the process. The process design thus must consider not only the chemical and physical factors but also economic factors such as cost and availability of raw materials, capital and operating costs, utility and fuel costs, and the value of the product.

It is frequently difficult to specify a precise kinetic order to the reaction, particularly under operating conditions approaching those of interest. The rate data used in process design are typically empirical expressions that relate the reaction rate to reactant concentrations and important variables such as temperature and amount of catalyst. In addition to data on the rate of product formation, it is often necessary to obtain data on the rates of various side reactions leading to by-products.

Rate data from homogeneous, liquid-phase reactions conducted in a batch manner are perhaps the easiest to determine experimentally and to scale up to larger volumes. Complications, however, may arise due to the variations of temperature with time. The times required to heat up and cool down the reaction mixture will increase upon scale-up as the ratio of volume to heat transfer surface increases. Therefore the times required to achieve a particular degree of conversion or yield will not be the same in the laboratory as in the plant. One solution then is to operate the laboratory scale reactor in a manner that approximates the anticipated temperature profile of the plant equipment.

Liquid-phase exothermic reactions are frequently conducted semibatch-wise. Here again the heat transfer surface of the reactor can be a limiting factor. The conversion rates in the plant are often controlled not by the chemical reaction rate but by the rate at which the heat of reaction can be removed. With heterogeneous semibatch operations, it is important to determine whether the reaction is subject to a chemical or physical regime. In the addition of a gaseous reactant to a liquid-phase reaction system, for example, the controlling factor might be the rate of diffusion of the gas rather than the chemical reaction rate. The scale-up of such a reaction would be based upon dynamic similarity and would be mainly concerned with reactor geometry and agitation. Liquid-phase reactions may be run continuously in a stirred-tank reactor. With efficient agitation there are no concentration gradients and isothermal conditions prevail. The kinetics are therefore comparatively straightforward, and the rate expressions obtained from laboratory batch experiments are applicable. For slow reactions a single continuous stirred reactor tends to be quite large. The required degree of conversion can frequently be obtained more economically with the smaller total reactor volume resulting from a series of reactors.

The determination of specific reaction rates in a tubular-flow reactor is complicated by a number of factors. The isothermal conditions that might be approximated in stirred-tank reactors seldom prevail in tubular reactors. Reactions are either endothermic or exothermic. Heat must be either supplied to or removed from the system. The reaction rate varies along the reactor resulting in a varying heat requirement and setting up a temperature gradient along the reactor. Because of incomplete radial mixing, there are also temperature gradients across the diameter of the reactor. Similarly there may be significant pressure and density gradients in the reactor. Since these variables affect the reaction rate, the rate varies from point to point. The measured formation of product in the effluent is an integrated average of these point rates.

In continuous-flow reactors the principle independent variables are temperature, reactor volume, and flow rate of reactants. As discussed in Chapter 7 the flow rate may be related to the reactor volume as "residence" or "contact" time or by "space velocity."

Reactions that utilize heterogeneous catalysts necessitate additional considerations because of physical processes that may affect the observed reaction rate. The processes involve transport of the reactant to the catalyst surface, diffusion through the pores, absorption at the active site; desorption of the products, diffusion through the pores and transport from the surface into the bulk gas stream. These processes will cause the observed reaction rate to differ from the chemical reaction rate at the active site unless the reaction is very slow. To differentiate between these two rates, the observed overall rate that includes the physical processes is termed the *global* reaction rate; the rate of chemical transformation is termed the *intrinsic* reaction rate.

All experimentally determined reaction rates are global rates. One of the first steps in the process engineering calculations is to correct the experimental data for transport and diffusion effects so as to obtain intrinsic rate data. These calculations are discussed and illustrated in References 11 and 12. A number of types of bench scale reactors such as differential, recycle, and rotating basket reactors have been devised to simplify these calculations. The objective is to operate at very low conversions so that changes in temperature, pressure, and composition are very small. The observed reaction rate thus approaches a point rate rather than an average rate. The applicability of this approach is dependent on the precision of the analytical methods used to measure the small changes in composition.

10.11 SUMMARY

A process development program, from the standpoint of the chemist, is a project with specific objectives. It involves painstaking, careful work which must be accomplished within a specified time frame. It requires of the chemist an unusual degree of objectivity toward his own work. There is little room for unfounded optimism. Pesky little problems do not just go away upon scale-up. On the other hand, unwarranted pessimism and doubts can delay a project until it is no longer economically relevant. Process development work demands of the chemist a faith in the value of his own laboratory work. The results of the development will be tested by the cold unemotional criteria of economic reality. There is little room for mistakes, misjudgments, or sloppy work. Process development work is challenging. It is satisfying. Many chemists think it is fun.

10.12 REFERENCES

1. Ralph Landau, (Ed.), *The Chemical Plant from Process Selection to Commercial Operation*, Reinhold Publishers Company, New York, 1966.
2. D. G. Jones, (Ed.), *Chemistry and Industry*, Clarendon Press, Oxford, 1967.
3. Roger Walck, *Chem. Tech.*, June 1972, pp. 336–339.
4. George Dappert, *Chem. Tech.*, December 1975, pp. 734–735.
5. R. L. Plackett and J. P. Burman, *Biometrika* **33**: pp. 305–325 (1946).
6. R. A. Stowe and R. P. Mayer, *Ind. Eng. Chem.* **58**(2) pp. 36–40 (1966).
7. W. B. Isaacson, *Chem. Eng.* **77**(14) pp. 69–75 (1970).
8. G. E. P. Box and K. B. Wilson, *J. R. Stat. Soc.*, Series B, pp. 1–45 (1951).
9. R. E. Johnstone and M. W. Thring, *Pilot Plant, Models and Scale-Up Methods in Chemical Engineering*, McGraw-Hill, New York, 1957.
10. F. G. Laupichler, *Ind. Eng. Chem.* **30**, 578 (1938).
11. J. M. Smith, *Chemical Engineering Kinetics*, 2nd ed., McGraw-Hill, New York, 1970.
12. S. W. Bodman, *The Industrial Practice of Chemical Process Engineering*, The MIT Press, Cambridge, Massachusetts, 1968.

Supplementary References

Bacon, D. W., *Ind. Eng. Chem.*, **62**(7), (1970), p. 27–34.

Murphy, T. E., *Chem. Eng.*, June 6, 1977, pp. 168–182.

Pavelic, V., and U. Saxena, *Chem. Eng.*, October 6, 1969, pp. 175–180.

Stowe, R. A., and R. P. Mayer, *Ind. Eng. Chem.*, **58**(2), (1966), pp. 36–40.

Szonyi, G., *Chem. Tech.*, January 1973, pp. 36–44.

10.13 PROBLEMS

1. Propose plausible schemes for the commercial synthesis of β-diethyl-aminoethanol. Using the approach described in Chapter 2 and current market prices published in the *Chemical Marketing Reporter*, calculate the market value difference for each scheme.

2. There are two basic synthetic routes for the production of vinyl chloride: the direct route based on acetylene and an indirect route based on ethylene.

$$H-C\equiv C-H + HCl \longrightarrow CH_2{=}CHCl \tag{1}$$

$$CH_2{=}CH_2 + Cl_2 \longrightarrow CH_2Cl-CH_2Cl \tag{2a}$$

$$CH_2Cl-CH_2Cl \longrightarrow CH_2{=}CHCl + HCl \tag{2b}$$

Generally, ethylene is a more desirable starting material because it is less expensive and less hazardous. One disadvantage of the ethylene route is the by-product hydrogen chloride which must be disposed of or utilized. An additional possibility is available based upon oxychlorination technology. This is a method of chlorination which utilizes hydrogen chloride and oxygen rather than chlorine:

$$2HCl + \tfrac{1}{2}O_2 + CH_2{=}CH_2 \longrightarrow CH_2Cl-CH_2Cl + H_2O \tag{3}$$

We now have four reactions. Using the values listed in the table below, calculate the market value difference for:

(a) The direct route.

(b) The indirect route.

(c) Various reaction paths based upon combinations of the above four reactions.

Material	Value $/lb.
acetylene	0.20
chlorine	0.04
ethylene	0.03
hydrogen chloride	0.02
vinyl chloride	0.05

3. A new pharmaceutical is ready for small-scale clinical trial. There is an urgent need for 20 lb of a key intermediate for the synthesis of this compound. The intermediate has the following structure:

This is not a new compound. Consult the literature. Considering your time, raw material costs, and the ease of reproducing literature results, propose a scheme for making 20 lb of this compound. Submit a memo describing the procedure you propose to use. Back up your decision with literature references and the prices of the chemicals you will purchase. you may assume that your time is worth $30 per hour.

4. There is frequently considerable difference between a laboratory preparation and a commercial scale process. Consider the laboratory work which would be required to design a process to produce 350,000 pounds per year of 3-aminopyridine. The process will be based upon the Hofmann rearrangement of commercially available nicotinamide. A laboratory procedure for this reaction is described in Organic Synthesis, Coll. Vol. IV. Outline an experimental program which would provide the information required for the process design. This program should include consideration of (a) number and types of individual steps or operations, (b) material handling and measuring, (c) heating or cooling requirements, (d) raw material and solvent costs, (e) hazards, and (f) waste disposal.

5. Listed below are literature references to journal articles describing the development of new processes. Review these articles and write a summary of the process development project. This summary should include:
   (a) The incentive for the project.
   (b) The technological basis or rationale for the approach.
   (c) The key points involved in the program.
   (d) The economic advantages of the new process.

References

1. Dimethyl formamide—Y. Fukuoka and N. Kominami, *Chem. Tech.*, November 1972, pp. 670–674.
2. Acrylamide—F. Matsuda, *Chem. Tech.*, May 1977, pp. 306–308.
3. Isopropyl alcohol—W. Neier and J. Woellner, *Chem. Tech.*, February 1973, pp. 95–99.
4. Methyl isobutyl ketone—Y. Onoue, *et al.*, *Chem. Tech.*, January 1977, pp. 36–39.
5. Acetic acid—J. Roth, *et al.*, *Chem. Tech.*, October 1971, pp. 600–605.
6. Oxamide—W. Riemenschneider, *Chem. Tech.*, October 1976. pp. 658–661.

6. The Cellulose Products Division uses large quantities of ethylene glycol diacetate as a solvent in the production of special grades of cellulose acetate. Current usage is 350,000 lb/mo; this is projected to rise to 900,000 lb/mo over the next four years. This requirement is currently being supplied by the Ethylene Products Department in Texas and by purchase from outside suppliers. The Texas plant has a capacity of 250,000 lb/mo and is based upon an obsolete and high labor usage process. The Production Planning Group has recommended that development and process design work be initiated for a 1,000,000 lb/mo plant to be built in Texas. Sufficient lead time is available to evaluate several alternative processes so as to optimize the economics.

   There have been a number of reports in the literature that strong acid ion exchange resins catalyze this esterification as effectively as the strong mineral acids. If the reaction rates were comparable and the catalyst life were sufficient, there would be significant advantages to the use of an ion exchange resin catalyst particularly with regard to the elimination of the neutralization step. Such a process could operate on a continuous basis, using cascaded stirred tank reactors or a trickle bed column reactor.

   It is requested that you submit a plan for preliminary laboratory studies to investigate the use of ion exchange catalysts in the reaction of ethylene glycol and acetic acid.

7. In section 10.8, an example was given for an experiment design used in optimizing variables. This example included three variables and two responses, yield and purity. Using the data included in Table 10.8, compute the effect of the variables on purity and derive a mathematical expression for this relationship.

8. We have been approached by a reputable pharmaceutical concern regarding our interest in entering into a long-term contract to supply α-bromoacetophenone in quantities between 5 and 6 million pounds per year. It is anticipated that we must declare our interest and enter detailed negotiations within 60 days. Therefore, we request that you consider diverting the efforts of your four-man group to this project on a short-term, high-priority basis. You can anticipate, and will receive, the full support of our analytical, process design, and economic evaluation groups.

   Our interest in the manufacture of α-bromoacetophenone is, of course, based upon the fact that we are basic in the manufacture of both bromine and acetophenone. In assessing the preliminary economics of this product, it would seem reasonable to consider the value of bromine and acetophenone to be the current, lowest-quoted bulk market prices less 15% for sales and packaging expenses. We have learned that the potential customer is now purchasing α-bromoacetophenone for $1.75 per pound.

It is requested that you prepare a plan for a process development study. The results of this study will be used for preliminary process design calculations, economic evaluation, and as a basis for negotiations with the customer.

The qualitative aspects of the reaction are quite well known from the literature. The reaction is rapid and exothermic, which suggests that semibatch and continuous processes should be considered. The by-product hydrogen bromide must be considered from the standpoint of efficient removal since it appears to have a deleterious effect upon product quality, and efficient recovery of this by-product value can have a significant effect upon the economics.

It is suggested that your study should start with the laboratory preparation reported in Organic Synthesis Coll. Vol. I. Your plan should include the following:

(*a*)   Preliminary economic potential.
(*b*)   Consideration of reactor type.
(*c*)   Suitable solvents.
(*d*)   Heat of reaction.
(*e*)   Rate of reaction.
(*f*)   HBr removal and recovery.
(*g*)   Product isolation.
(*h*)   Product stability and analysis.
(*i*)   Consideration of catalyst.
(*j*)   Determination of significant processing variables.
(*k*)   Optimization of variables.

11
CHEMICAL PATENTS

WHAT, HOW, AND WHY

An inventor does not have a natural God-given right to the *exclusive* use of his invention. The only absolute and permanent way of preventing others from making use of his invention is to keep it secret. There are obvious disadvantages to this. First, the inventor usually cannot derive any commercial benefits from his invention. Second, there is the risk that another person may independently make the same invention. However, certain limited exclusive privileges are available to him through specific federal laws governing patents. Exclusivity is granted by patent for a limited period of time and may be considered to be a reward to the inventor for disclosing his invention to the public.

11.1 HISTORICAL BACKGROUND[1]

The foundation for U.S. patent laws lies in Clause 8, Section 8 of Article I of the Constitution which grants to Congress the power "To promote the progress of science and useful arts by securing for a limited period to . . . inventors the exclusive right to their . . . discoveries." Congress has the complete authority to enact laws relating to patents or to modify or abandon present patent laws. Philosophically, the primary purpose of the patent system is "to promote the progress of the science and useful arts" rather than to protect the assests of the inventor. The reward to the inventor is of secondary consideration and incidental to the public interest. The second philosophical point that should be noted is the limited time period. A patent is an asset; it is property; it can have economic value. It can be sold, traded, mortgaged, inherited, or licensed; it can generate income. But it is for only a limited time period. In this respect ownership of a patent is fundamentally

290

different from the ownership of other property such as a plot of land. A third point is that a patent does not grant the owner the right to practice his invention; it grants the owner the legal means to prevent others from making, using, or selling the subject of his invention. It is the responsibility of the patent owner to police his patent and to seek injunction and damages for infringement by others.

The concept of awarding patents for inventions originated in England. A brief description of this historical background is perhaps in order because it illustrates the fundamental difference between granting a monopoly privilege and granting a limited exclusive right.

In England, prior to about 1623, the Crown by virtue of its royal prerogative granted "Litterae Patentes" (i.e., "open letters") addressed to the general public, "to all to whom these presents shall come." These letters indicated that the Crown thereby granted a favored status to someone, commonly a relative or loyal vassal who had done the Crown a service or paid a sum to the Crown's coffers, perhaps a parcel of land, a title, or an exclusive privilege. Practically, it was a means for raising revenue. By way of example, the sole and exclusive privilege of making or selling copper pans in a designated area of the realm would be sold to the highest bidder. If copper pans had previously been made or sold by anyone desiring and having the ability to do so, this amounted to a grant of a monopoly to the Crown's favored subject, *taking* from the public at large a privilege it had previously enjoyed.

In about 1623 Parliament enacted the Statute of Monopolies which stripped this prerogative from James I and subsequent monarchs. It outlawed all existing monopolies and future monopolies *except* for letters patent granted, for not more than 14 years, to the first and true inventor for the privilege of working or making his *new* "manufacture" so long as such a grant was not contrary to law or "mischievous to the state, by raising prices of commodities at home, or hurt of trade, or generally inconvenient." From this it is clear that a patent grant, where a new invention was concerned, created no monopoly in the traditional sense since it took nothing from the public. Instead, it *gave* the public the benefit of the innovation by encouraging the inventor or his licensees to take the risk of commercializing the invention, protected from the destructive effects of competition on an infant trade or industry. It also disclosed the invention to the public, which could then build on that invention to further develop that art or science. Ultimately, the public also acquired the right to make and use a previously unknown invention when the patent grant expired.

The first patent act in the United States was enacted in 1790 and authorized the granting of a patent for any new and useful art, machine, or manufacture. Thomas Jefferson was the first Patent Examiner. In 1793 the act was amended to include compositions of matter. The field of classifications in which patents may be granted has been enlarged, and minor changes and modifications have occurred through the years. However, the basic fundamental legal requirements for patentability have remained unchanged.

The fundamental concept, which ensures that the Constitutional objective will be fulfilled, is

> In return for a full and complete disclosure to the public of a new and useful invention, the government will grant to the inventor the right to exclude others from making, using, and selling the invention for a limited period—namely 17 years.

The application of this fundamental principle to specific cases has presented innumerable problems. The meaning of the terms "new and useful," "full and complete," and even "invention" have been the subject of much thought, discussion, and debate.

The two classifications of most interest to chemists and to the chemical industry are "process" and "compositions of matter." The first patent granted by the United States in 1790 was a chemical patent. It was signed by George Washington and was granted to Samuel Hopkins for a method of making potash and pearl ash.

11.2   LEGAL REQUIREMENTS FOR PATENTABILITY[2]

There are three basic legal requirements for patentability: novelty, utility, and nonobvious subject matter. All three requirements are essential and an applicant must address each of these criteria to secure a valid patent.

Novelty, originality or newness of subject matter has always been a requirement of the patent laws. The present act requires that for a patent to be granted the invention was not—

1.   Before the invention by the patent applicant

   (a)   publicly known or used by others in this country, or
   (b)   patented or described in a printed publication anywhere in the world; or

2.   More than one year before the effective filing date of an application for a patent in the United States

   (a)   in public use or offered for sale in this country, or
   (b)   patented or described in a printed publication anywhere in the world.

This standard of novelty described in the patent laws differs from the ordinary definition. It is in fact possible that a patent could be granted on subject matter that is not truly novel. Prior knowledge or use in a foreign country, for example, does not negate novelty unless it has been previously patented or described in a printed publication. Prior public knowledge or use in the United States by even *one* person, however, will effectively cancel a claim of novelty. It must also be demonstrated that the information contained in the prior knowledge, use, patent, or publication is specific enough to enable another person to understand and reproduce the invention in order to negate the novelty claim. This is not to suggest "importing" ideas from another country. To obtain a patent it is necessary for one to take an oath that he is the true inventor.

Utility has always been a legal requirement for the granting of a U.S. patent. This follows the intent of the Constitutional provision that provides for the promotion of "useful" arts. However, the patent laws enacted by Congress have not established any standards or definition of utility or usefulness. Through judicial interpretations it has been established that the invention is useful if it is capable of performing some beneficial function claimed for it. Beneficial is liberally interpreted to mean any function not considered frivolous or injurious. It is not required that the invention perform perfectly or even well. But for an invention to be considered patentable, the law has required an assertion of utility and an indication of the use or uses intended. Adequate proof or support should be available for each use included in the patent application.

The requirement that the invention be unobvious was first included in the patent laws in 1952 and is based upon prior court decisions that held patents invalid on grounds of lack of invention.

> A patent may not be obtained ... if the difference between the subject matter sought to be patented and the prior art are such that the subject matter as a whole would have been obvious at the time the invention was made to a person having ordinary skill in the art ....

This requires that an invention be truly an invention—that it should involve the conceiving of an original concept or unique relationship rather than be a simple, logical extension of past relationships. Philosophically and morally this requirement seems reasonable. Without it we might anticipate the awarding of a large number of future patents to various computers. Although the requirement is reasonable and desirable, in practice it gives rise to many problems. The interpretation of the requirement and the application in a specific case is always subjective. Such determinations are made by nontechnical courts. Some of the resultant frustrations have been expressed by Judge Learned Hand:

> The test laid down is misty enough. It directs us to surmise what was the range of ingenuity of a person "having ordinary skill" in an "art" with which we are totally unfamiliar. . . .[2]

In the case of chemical patents, many of the problems involve cases of homologs and near homologs of known compounds. Another type of dispute involves the question of obviousness when based upon not one but a combination of references (i.e., disclosures of prior art). In both instances it appears that the strength of the applicant's case is determined by the degree to which the final result is unexpected in light of the prior art.

11.3   THE INVENTOR

The question of who is the *inventor*—or who are the *inventors*—is frequently troublesome. The team approach used on many industrial research projects today tends to cause this question to arise more frequently than in the past, and to make a clear-cut answer more difficult. The eloquent Judge Clarence Newcomer noted that

> The exact parameters of what constitutes joint inventorship are quite difficult to define. It is one of the muddiest concepts in the muddy metaphysics of the patent law. On the one hand, it is reasonably clear that a person who has merely followed instructions of another in performing experiments is not a co-inventor of the object to which those experiments are directed. To claim inventorship is to claim at least some role in the final conception of that which is sought to be patented. Perhaps one need not be able to point to a specific component as one's sole idea, but one must be able to say that without his contribution to the final conception, it would have been less efficient, less simple, less economical, less something of benefit.[5]

It is important that the question of inventorship be answered correctly and honestly; this requires an objective, unemotional review of the contributions of each party to the invention. Courts take the subject of inventorship seriously. A number of patents have been completely invalidated when it was shown that the inventors had been named improperly.

Before an invention can be considered to be an invention in the legal sense, it must be conceived, it must be reduced to practice, and it must be demonstrated that it has utility. Frequently, more than one individual is involved in the course of these three activities. If a single person, in conceiving the invention, conceives the complete and total invention in its operative form and possessing a utility, then he is the sole inventor. Reduction to practice, and demonstration of utility by others does not change this if, in performing these functions, they followed the general plan and concepts of the inventor. The co-workers are considered to be "extended technical arms of the inventor," and not joint inventors.

There are a number of situations which can give rise to joint inventorship. When two or more persons jointly conceive the same idea and concur upon the physical acts and experiments necessary to complete the inventive act, the result is a joint invention. When, in the course of a development process that results in an invention, there have been two important and critical contributions by two persons, they may be named as joint inventors. Legally, an invention is not patentable until utility has been demonstrated. One scientist may synthesize a novel compound and a second scientist discover a use for it. Whether this is a joint invention or not depends upon the circumstances in which the utility was discovered. If it results from a screening program involving standardized test procedures, the first scientist is the sole inventor. If the second scientist conceived and demonstrated the utility using nonroutine procedures, there is strong evidence for joint invention.

The determination of inventorship is often difficult and always important. It can have a serious effect upon the validity of a patent. It can naturally have serious effects upon the morale and working relationships of the individuals who have worked on the invention. Fortunately, the scientists involved generally have a clear picture of the true inventive situation, and, with guidance from a knowledgeable patent attorney, the right decision usually results.

11.4 OWNERSHIP

Inventions are the property of the inventor and the right to this property is protected by law. What is protected is not the idea or concept behind the invention—this belongs to the public; the use of the invention, the reduction to practice, is what is protected and this is for a limited time period—17 years. The inventor, as sole owner, may choose to assign his right to the invention to another party for monetary or other considerations.

There are several circumstances in which the inventor is not entitled to exclusive sole ownership of the invention.[2]

The first case involves what is known as "shop right." If an employee, in the course of other duties, creates an invention on company time and at company expense, the company is entitled to make use of the invention.

The invention is owned solely by the inventor, but the company has the right to use the invention on a nonroyalty basis in the course of its business. The company is not a part owner of the invention and cannot sell or transfer this right to a third party.

In the second case an employee is hired for the particular purpose of inventing or improving the products, processes, or services of the company. All of the costs of creating and developing the invention are borne by the company. Ownership of the invention is awarded to the employee, but he is bound to assign all rights to the company. His obligation to the company has been recognized in a number of court decisions. In order to avoid any possible uncertainty as to the respective rights of the company and the employee, invention assignment agreements came into use.

A well-written and valid employment agreement clearly notifies the employee of his obligation to assign any pertinent inventions made during his period of employment to the company. It also may contain a clause obligating the employee to keep secret any company trade secrets or confidential information. It should be emphasized that both of these obligations are imposed on the employee by common law, and the agreement merely serves to point out these obligations.

Beyond this, some companies have used employment agreements that have broadly obligated the employee to assign rights after he is no longer employed by the company and for inventions outside the normal business areas of the company. These agreements, which have no limit as to time and scope of the obligation of the employee, have been held to be unenforceable.

The matter of the obligation of an employee in maintaining trade secrets is most delicate. Chemists, in the course of their work, are exposed to information that is considered to be secret or confidential concerning the company's activities, processes, or methods of operation. There is a clear, legal obligation to maintain the secrecy of such information even after the employment is terminated. On the other hand, it is recognized that during the course of employment the employees will increase their general knowledge of particular areas of science and technology. This part of their experience and background they are entitled to make use of in any future employment.

11.5 PRIORITY OF INVENTION

In the United States, when several inventors claim the same or overlapping subject matter in patent applications, the inventor entitled to the patent is that one with the earliest *proven* date of invention. The priority in filing the patent application does not necessarily determine the matter. This is a very important point.

Disputes concerning priority of invention are termed interference procedures and are most complex. It is the position of the government that there can be only one valid patent for an invention, and the party filing the earliest application is presumed to be the first inventor. Those filing later must present proof of earlier invention.

In this regard it should be understood that invention for patent purposes involves conception, followed by reduction to practice. It is advantageous to be able to prove an earlier conception date. It must also be proven, however, that in addition to the earlier conception that this was followed by "diligent" efforts to reduce to practice. The proof required in these disputes is considerably more than the testimony of the claimed inventor. Complete and systematic written records of the development effort are important. In addition, corrobative testimony by someone having actual knowledge of the events is required.

Interference proceedings are highly structured, complex procedures that are best understood by legal specialists. For the chemist the best advice is to file a patent application at the earliest possible date; keep complete up-to-date written records of the development work; develop corroborators who can testify as to conception and reduction to practice activities. Objective reporting of results in quantitative terms rather than subjective terms is preferable (e.g., "80 % yield based upon component A converted" instead of "poorer yield than previously obtained"). As a final word—if possible, choose standard well-recognized test procedures in determining the utility of the invention, and try to refrain from excessive modesty when recording the results. A phrase such as "warrants further investigation" has been sufficient cause to lose an interference proceeding.

11.6 PATENT DOCUMENTS[2,4]

There are a number of documents that are associated with the process of obtaining a patent.

Patent Office rules prescribe that the patent application must be accompanied by an oath, a petition for the granting of the patent, and the filing fee. The *application* consists of the specification and the claims. The specification presents the prior art and pertinent background information, a statement of the problem, and a description of the solution to the problem. The claims section delineates the specific claims of the applicant. The *oath* is a formal document in which the applicant identifies himself and swears that in accordance with the patent laws and to the best of his knowledge, he is entitled to the patent. The *petition* is another formal document in which the inventor formally requests that a patent may be granted.

The *power of attorney* is a document that simply empowers the person or persons designated the right to represent the inventor at the Patent Office. The person named must be an agent or attorney registered with the Patent Office.

In those cases involving an interference procedure to determine the priority of invention, each party is required to submit a *preliminary statement*. This document states the dates on which various steps were carried out in reducing the invention to practice. It is of utmost importance that these dates be carefully established and supportable by evidence and testimony.

When further information is required by the Patent Office in processing an application, an *affidavit* may be submitted. There are two types. A Rule 131 affidavit is concerned with establishing conception and reduction to practice dates in order to overcome a cited publication having a date subsequent to the invention. It should be accompanied by supportive documents, drawings, or records. A Rule 132 affidavit is submitted to classify the value of the claimed invention in light of the prior art. In those cases where the application itself does not establish value or utility of the invention to the satisfaction of the Patent Office, this document represents an addendum to application to provide additional information.

The *assignment* is the document assigning the rights to the invention. Other internal or company documents concerned with the patent process are *notebooks* and *conception records* or *invention disclosure forms*. Both can be documents of great legal value if executed properly, dated, witnessed, and capable of corroboration.

11.7   PATENT STRUCTURE AND LANGUAGE

Patents tend to follow a rather stylized format and to use some rather peculiar terms.

The general format used is a response to the Patent Office rules relating to the content of the patent specification:

> The specification must include a written description of the invention or discovery and of the manner and process of making and using the same, and is required to be in such full, clear, concise and exact terms as to enable any person skilled in the art or science to which the invention or discovery appertains, or with which it is most nearly connected, to make and use the same.

> The specification must set forth the precise invention for which a patent is solicited, in such a manner as to distinguish it from other inventions and from what is old.

It must describe completely, a specific embodiment of the process, machine, manufacture, composition of matter or improvement invented and must explain the mode of operation or principle whenever applicable. The best mode contemplated by the inventor of carrying out his invention must be set forth.

The specification must conclude with a claim particularly pointing out and distinctly claiming the subject matter which the applicant regards as his invention or discovery.

The general approach is to build the patent specification towards the ultimate objective of the application, the claim, or series of claims that are listed at the end. After clearly stating the subject matter of the invention, the invention is described in a manner that contrasts it with the old or prior art. This part of the specification attempts to build a foundation for the claims to follow. Since it is desirable to make these claims as broad as is reasonable, language is used that, while descriptive, is not needlessly limiting. Terms such as "preferably," "plurality," "multiplicity" are used frequently. Thus, instead of "The compound is reacted with sodium hydroxide," it would be better to use "the compound is reacted with an alkaline material, preferably an alkali metal hydroxide, such as sodium hydroxide." Preferably has the connotation of desirability but includes the element of reservation. It implies that other materials might be used and this would not alter the invention. Thus, it tends to provide a foundation for broad claims. In a similar way "substantial" is used in lieu of a more specific term. "Substantial amounts" could imply any quantity more than a trace amount.

In order to emphasize that the invention is indeed unobvious to a person skilled in the art, the terms "surprisingly" or "unexpectedly" are sometimes used. Thus, "it was unexpectedly found that the solubility was increased" establishes that the results could not be predicted on the basis of known laws and concepts.

After the description of the invention there follows a number of specific examples or embodiments. These may be prefaced with the admonition that "the following examples will serve to further illustrate the invention but are not meant to limit it thereto." Generally there are a large number of examples, limited only by the laboratory data available. The reason is again the objective of building as broad a base as possible to support broad claims.

The last part of the specification is a listing of the claims. It is highly desirable to have a patent with broad coverage. Claims that are so broad that, in the opinion of the patent examiner, they cannot be supported by the experimental data, will be rejected. Or they may be held invalid during an interference procedure or an infringement suit. Instead of attempting to second guess the examiner, the strategy is to make as many claims as possible, ranging from the very broad to the specific. If the broadest of these claims are rejected, the remaining claims still represent the broadest coverage possible for that particular application.

Chemical patents are a unique art form; they differ in purpose, format, and style from the scientific literature. The laws and regulations governing patents and the patenting process are complex and require the attention of attorneys with training and experience in this highly specialized field.

An unusually brief process patent has been included as Fig. 11.1 to illustrate the general structure of a patent application. Note that at the start of the subject matter the invention is stated and contrasted with the prior art; this is followed by a more complete description of the process that establishes the novelty, unobviousness, and utility of the invention. The largest portion of the application is devoted to a discussion of the possible variables in the process—reagents, solvents, temperature, and catalysts. The objective is to illustrate the broad general nature of the invention. This is followed by a brief but detailed example of the process. The last section is a listing of the claims. In this case there are six claims, starting with the

---

United States Patent [19]

Mitchell et al.

[11] **3,900,451**

[45] **Aug. 19, 1975**

[54] **METHOD OF MAKING SULFHYDRYL-CONTAINING POLYMERS**

[75] Inventors: **Albertha B. Mitchell**, Framingham; **Suzanne V. McKinley**, Wellesley, both of Mass.; **Joseph W. Rakshys, Jr.**, Midland, Mich.

[73] Assignee: **The Dow Chemical Company**, Midland, Mich.

[22] Filed: **Aug. 24, 1973**

[21] Appl. No.: **391,429**

[52] U.S. Cl. .... **260/79.5 NV; 260/778**
[51] Int. Cl. **C08f 27/06**
[58] Field of Search .... 260/79.5 NV, 778

[56] **References Cited**

UNITED STATES PATENTS

2,137,584	11/1938	Ott	260/778
2,563,640	8/1951	Brown	260/778
2,563,662	8/1951	Rothrock	260/778
3,696,083	10/1972	Hwa	260/79.5 C
3,696,084	10/1972	Gordon	260/79.3 R

FOREIGN PATENTS OR APPLICATIONS

1,227,144	4/1971	United Kingdom	252/438

OTHER PUBLICATIONS

Chemical Abstracts, Vol. 53, No. 6, Mar. 25, 1959, page 5730d.

Reid, Organic Chemistry of Bivalent Sulfur, Vol. 1, 1958, Chemical Publishing Co. Inc., N.Y., N.Y., pages 25, 26 & 381.

Primary Examiner—Ronald W. Griffin
Attorney, Agent, or Firm—L. Wayne White

[57]   **ABSTRACT**

A novel process is disclosed for making a functional polymer whose backbone comprises units of the formula

$$+CH_2-CH+$$
$$| $$
$$C_6H_4-CH_2SH$$

In this process, a chloromethylated vinylaromatic polymer (e.g. chloromethylated polystyrene) is reacted with an alkali metal hydrosulfide in the presence of a catalytic amount of an organic quaternary onium compound (e.g. tetrabutylammonium bisulfate).

6 Claims, No Drawings

Figure 11.1   Patent application.

1

METHOD OF MAKING
SULFHYDRYL-CONTAINING POLYMERS

BACKGROUND OF THE INVENTION

This invention pertains to a new process for making sulfhydryl-containing polymers whose backbones comprise units of the formula

$$+CH_2-CH+$$
$$|$$
$$C_6H_4-CH_2SH$$

Such polymers are known in the art and have been used, for example, as catalysts in various reactions and in complexing metal ions, such as copper and mercury.

Previous methods of preparing such sulfhydryl-containing polymers included the technique described by Okawara et al. (Chemical Abstracts 53: 5730d) wherein chloromethylated polystyrene was first reacted with thiourea and the intermediate thus formed was subsequently hydrolyzed with caustic to form the desired polymers.

SUMMARY OF THE INVENTION

A novel process for preparing sulfhydryl-containing polymers has now been discovered. The new process comprises reacting by contacting (a) a chloromethylated vinylaromatic polymer with (b) an alkali metal hydrosulfide in the presence of (c) a catalytic amount of an organic quaternary onium compound. The novel process is schematically represented below:

$$+CH_2-CH+_n + M^\ominus SHO \xrightarrow[\text{catalyst}]{\text{onium}} +CH_2-CH+ + M^\oplus Cl^\ominus$$
$$|\qquad\qquad\qquad\qquad\qquad\qquad |$$
$$C_6H_4\qquad\qquad\qquad\qquad\qquad C_6H_4$$
$$|\qquad\qquad\qquad\qquad\qquad\qquad |$$
$$CH_2Cl\qquad\qquad\qquad\qquad\quad CH_2SH$$

The results of this process were surprising both in the high rate of reaction and the high degree of nucleophilic displacement.

In practice, the chloromethylated vinylaromatic polymers are swollen or dissolved in an inert water-immiscible organic solvent (e.g. benzene, toluene, 2,4-dichlorobenzene, etc.) to form an organic phase. An aqueous phase is then formed by dissolving the alkali metal hydrosulfide in water. The two phases are combined in a suitable vessel and the quaternary onium catalyst(s) added to the reaction mixture. After a suitable reaction period, the organic phase containing the sulfhydryl-containing polymer is separated from the reaction mixture and the product recovered therefrom by flashing off the solvent or by other conventional means.

The reaction temperature may be varied to convenience but satisfactory reaction rates have been observed at temperatures of from about 20° to about 80°C (preferably 40°–60°C). The reaction rate is also increased by efficient blending (e.g. stirring) of the reaction mixture.

The reactants involved in the instant process are known classes of compounds.

The known class of alkali metal hydrosulfides includes lithium, sodium, potassium, rubidium and cesium hydrosulfide. Sodium and/or potassium hydrosulfide are normally used, based on their relative cost and commercial availability.

2

The chloromethylated vinylaromatic polymers are likewise well known. The polymer backbone of such polymers comprises units of the formula

$$+CH_2-CH+$$
$$|$$
$$C_6H_4-CH_2-Cl$$

Normally the polymers bear an average of one chloromethyl group per aromatic ring and it is in the meta and/or para ring position. The polymers are prepared by homopolymerizing ar-vinylbenzyl chloride (VBC) or by interpolymerizing VBC with other suitable vinyl monomers. Alternatively, the chloromethylated vinylaromatic polymers can be prepared by chloromethylating polystyrene or interpolymers of styrene with, for example, chloromethyl methyl ether. The polymerization techniques for VBC and the chloromethylation of polyvinylaromatics with chloromethyl methyl ether are, of course, well known. See, for example, Hoffenberg (U.S. Pat. No. 2,981,758) who describes chloromethylation with chloromethyl methyl ether and see Jones et al. C.A. 55: 17078i and C.A. 56: 10373d, Lloyd et al. C.A. 58: 8051h, Askarov et al. C.A. 76: 113565a and C.A. 77: 140572, McMaster (U.S. Pat. Nos. 2,992,544 and 3,022,253), Clarke et al. (U.S. Pat. No. 2,780,604), and Rassweiler et al. (U.S. Pat. No. 3,068,213) who describe some of the polymers of VBC. Examples of suitable such polymers therefore include chloromethylated polystyrene, poly(VBC), styrene-VBC interpolymers, acrylonitrile-VBC interpolymers, and the like. The polymers can be linear or lightly cross-linked (e.g. up to about 5 percent with divinylbenzene).

The stoichiometry of the reaction requires one mole of alkali metal hydrosulfide per chloromethyl equivalent in the polymer. A slight excess of alkali metal hydrosulfide is normally used, however.

The catalysts here used are organic onium salts of the elements of group 5(a) and 6(a) of the Periodic Chart and are known in the art as phase-transfer catalysts. Suitable such salts are therefore ammonium salts, phosphonium salts, sulfonium salts, etc. as described, for example, by Starks and Napier in British Patent Specification No. 1,227,144 and by Starks in J. Am. Chem. Soc. 93, 195 (1971). Preferred onium salts have from about 10 to about 30 carbon atoms. Additionally, the ammonium salts are currently preferred over the other onium salts and benzyltrimethyl-, benzyltriethyl- and tetra-n-butylammonium salts are most preferred. Suitable onium salts have a minimum solubility of at least about 1 weight percent in both the organic phase and the aqueous phase at 25°C. To further illustrate the type onium salts which can be used, suitable ammonium salts are represented by the formula $R_1R_2R_3R_4N^+A^-$, wherein R_1–R_4 are hydrocarbyl groups (e.g. alkyl, aryl, alkaryl, aralkyl, cycloalkyl, etc.) and R_1 can join with R_2 to form a 5- and 6-membered heterocyclic compound having at least one quaternized nitrogen atom in the ring and may also contain one atom of nitrogen, oxygen or sulfur within the ring. Typically, R_1–R_4 are hydrocarbyl groups of from 1 to about 16 carbon atoms. Similar formulas can be drawn for the phosphonium and sulfonium salts and the like. The neutralizing anion portion of the salt (e.g. A^-) may be varied to convenience. Chloride, bromide and bisulfate are preferred anions, but other representative anions include nitrate,

Figure 11.1 *(Continued)*

3,900,451

3

tosylate, acetate, etc. The following compounds are illustrative: tetraalkyl ammonium salts, such as tetramethyl-, tetraethyl-, tetra-n-butyl-, tetrahexyl-, methyltriethyl-, and trioctylmethyl-, hexadecyltriethyl- and tridecylmethylammonium chlorides, bromides, bisulfates, tosylates, etc.; aralkylammonium salts, such as tetrabenzylammonium chloride, benzyltrimethyl-, benzyltriethyl-, benzyltributyl- and phenethyltrimethylammonium chlorides, bromides, iodides, etc.; arylammonium salts, such as triphenylmethylammonium fluoride, chloride or bromide, N,N,N-trimethylanilinium chloride, N,N,N-triethylanilinium bromide, N,N-diethyl-N-ethylanilinium bisulfate, trimethylnaphthylammonium chloride, p-methylphenyltrimethylammonium chloride or tosylate, etc.; 5- and 6-membered heterocyclic compounds containing at least one quaternized nitrogen atom in the ring, such as N,N,N', N'-tetramethylpiperaziniumdichloride, N-methylpyridinium chloride, N-hexylpyridinium iodide, 4-pyridyltrimethylammonium iodide, 1-methyl-1-azoniabicyclo[2.2.1]heptane bromide, N,N-dibutylmorpholinium chloride, N-ethylthiazolinium chloride, N-butylpyrrolium chloride, etc., and the corresponding phosphonium and sulfonium salts, and the like. The onium catalysts are included in the reaction mixture in small but sufficient amounts to catalyze the reaction between (a) and (b) above. E.g. from about 0.01 to about 10 mole percent. In most instances, however, amounts of from about 0.2 to about 3 mole percent are preferred. The following example further illustrates the invention.

EXAMPLE 1

Resin beads of a lightly cross-linked styrene-VBC interpolymer (10 g.; 0.021 mole of VBC) were swollen in o-dichlorobenzene. To this was added a concentrated aqueous solution of sodium hydrosulfide (0.153 mole) and tetra-n-butylammonium bisulfate (0.003 mole). The mixture was stirred at 100°C for 24 hours and cooled to room temperature. The resin beads thus

4

treated were filtered from the mixture, washed with methylene chloride and dried in a vacuum oven at 60°C and 0.1 mm/Hg for 48 hours. Elemental analysis for sulfur in the polymer showed the conversion of

$$+CH_2-CH+ \quad\quad to \quad +CH_2-CH+$$
$$\underset{C_6H_4-CH_2-Cl}{|} \quad\quad\quad \underset{C_6H_4-CH_2-SH}{|}$$

to be greater than 95% complete.

Other polymers and catalysts as described above can be similarly used.

We claim:

1. A process for preparing sulfhydryl-containing polymers comprising reacting by contacting

   a. a chloromethylated vinylaromatic polymer whose backbone comprises units of the formula

$$+CH_2-CH+$$
$$\underset{C_6H_4-CH_2-Cl.}{|}$$

   said polymer being swollen or dissolved in an inert water-immiscible organic solvent, and

   b. an aqueous solution of an alkali metal hydrosulfide, in the presence of

   c. a catalytic amount of an organic onium salt.

2. The process defined by claim 1 wherein (c) has a total carbon content of from about 10 to about 30 carbon atoms.

3. The process defined by claim 1 wherein (c) is quaternary ammonium salt.

4. The process defined by claim 3 wherein (c) is benzyltrimethyl-, benzyltriethyl- or tetra-n-butylammonium salts.

5. The process defined by claim 1 wherein (b) is an aqueous solution of sodium or potassium hydrosulfide.

6. The process defined by claim 4 wherein (b) is an aqueous solution of sodium hydrosulfide.

* * * * *

Figure 11.1 *(Continued)*

broad use of any alkali metal hydrosulfide with any organic quaternary onium salt as a catalyst and ending with the more specific use of sodium hydrosulfide and one of three specific catalysts.

11.8 PATENTING AN INVENTION

The steps involved in obtaining a patent have been described from the viewpoint of the chemist[6,7] and the Patent Office examiner.[4] The references cited should be consulted for more detailed discussions of the procedures and documentation involved. The following fictionalized narrative should serve to illustrate the general sequence of events in patenting an invention.

While employed as research chemists, Ben Tefertiller and Jim Nevill jointly conceived the idea of synthesizing a new class of trifunctional alcohols—the hydroxyalkyl esters of 3,3,3 nitrilotripropionic acid. They believed that these compounds could be formed readily by the reaction of the corresponding hydroxyl alkyl acrylates with ammonia.

$$NH_3 + 3CH_2 = CH - \overset{\overset{\displaystyle O}{\|}}{C} - OR \longrightarrow N(CH_2 - CH_2 - \overset{\overset{\displaystyle O}{\|}}{C} - OR)_3$$

where R = monohydroxyalkyl.

They also recognized that one potential use for these compounds would be as cross-linking agents in polyurethanes. The rationale for this was the known utility of other distantly related polyols such as those based upon glycerine and pentaerythritol.

A search of the chemical literature revealed that this class of compounds had not been previously reported. The two chemists decided to synthesize the beta-hydroxyethyl ester by the reaction of ammonia with beta-hydroxyethyl acrylate. The reaction took place readily, and a sample of the product was submitted to a polyurethane development group for routine screening as a cross-linking agent. The results of these screening tests indicated that the compound did indeed act as a cross-linking agent.

At this point the three requirements for patentability—novelty, unobviousness, and utility appeared to have been satisfied. Since the initial conception by Tefertiller and Nevill encompassed not only the new family of compounds but also a utility and the general scheme for reducing the concept to practice, no other persons could be considered as co-inventors. Tefertiller and Nevill therefore described their work to the company patent department, using a Disclosure of Invention form (Fig. 11.2). The questions on the disclosure form are designed to determine just what the invention is, if it is a patentable invention, and what proof exists as to the dates of conception and reduction to practice.

In the course of completing the disclosure form the two inventors recognized that other similar compounds might well fit within the scope of their inventions and made plans to synthesize homologs of the beta-hydroxyethyl ester. These plans were discussed with the patent solicitor who had been assigned to the disclosure. The solicitor had evaluated the disclosure, considered it worthwhile, and encouraged them to complete this experimental work since it would broaden and strengthen the patent application. In addition, he suggested that the inventors give thought to broadening the scope of the proposed uses of these new compounds. Since the compounds were trifunctional alcohols, it was conceivable that they would also act as cross-linking agents in the formation of polyester-type polymers. Since the compounds were a type of alkanol amine, it was also possible that they would act as acid scavengers similar to the ethanolamines. Accordingly,

DISCLOSURE OF INVENTION TO PATENT DEPARTMENT

Please fill out carefully - The requested information is required for proper preparation
of a patent application. Call your Patent
Department contact if you have a question.

DO NOT WRITE IN THIS SPACE
Disclosure No.
Date:

Approval of Department Head

(IT IS SOMETIMES HELPFUL TO ANSWER QUESTIONS N THRU T BEFORE A THRU M,
GIVE ANSWERS ON SEPARATE SHEET WHERE INSUFFICIENT SPACE)

A	Title of invention or discovery.
B	To whom and when did you first disclose this invention?
C	What written records show this disclosure to others? (attach copies)
D	What other records exist that describe the concept of the invention?
E	When was the first work done on this invention, and where is such work described? Who did that work, and who else knew of it?
F	List available records that describe the work to date on this invention, in addition to those identified in answers above.
G	To your knowledge, has this invention been used for any commercial purpose in a plant process, by sale or by offer to sell? If so, please give dates and details.
H	Give dates and details regarding any sample or other information relating to this invention that has been given to persons not employed by
I	Do you know of , disclosures or patent applications that you believe tie in closely with this invention: if so, please identify.
J	Is further work planned as to this invention? Please outline.
K	
L	If this invention was made or completed in the course of work under a contract with any governmental agency or other outside party, please identify the party and the contract.
M	Are you aware of any activity? If so, please describe.

Originating Department or Laboratory:

Submitted by: (Sign & Print name in full)	Home Address:	Citizenship:	Date:
1			
2			
3			

Witness: (Sign & Print name in full)

I have read and understood the foregoing Disclosure and attachments.

Signed: Date:

Figure 11.2 Form for disclosure of invention to patent department.

samples were submitted for evaluation for these two utilities. Results were positive. The synthesis of homologs was complicated somewhat by the fact that both Tefertiller and Nevill had been assigned to other, higher priority projects. In a meeting with their laboratory director, it was decided that the suggested synthetic work should be pursued. Therefore, a lab technician was assigned to this phase of the program, to work under the general supervision of Tefertiller. The additional seven compounds were synthesized without difficulty and proved to possess cross-linking properties similar to, the beta-hydroxyethyl compound.

With the experimental work completed, the solicitor re-evaluated the potential value of the patent and the strength of the application. Considering them worthwhile, he proceeded to write the patent application. After several revisions and consultations with the inventors, an application resulted that was satisfactory to all parties. From the chemists viewpoint, it was scientifically accurate; from the solicitor's standpoint, it represented as strong and as broad an application as the data warranted. The application was typed in its final form. The inventors signed an oath that they believed they were the true inventors and assigned power of attorney to the solicitor (Fig. 11.3). They assigned to the company the ownership of any patent that might result from the application (Fig. 11.4). The completed application, with appropriate filing fee, was submitted to the Patent Office.

After acknowledging receipt of the application, the first action by the Patent Office was to reject all claims. This action was not unusual. Less than 20% of the applications are approved the first time they are examined.[4] The basis for this rejection was a prior patent on the use of certain nitrilotripropionate esters as plasticizers in polyvinylchloride formulations. This patent (Fig. 11.5) specifically covers the use of nitrilotripropionate esters where the three R groups of the ester are nonacetylenically unsaturated hydrocarbon groups. The similarities and the differences between the claimed substances of the patent application and the prior patent may be represented by the formula

$$N \underset{\displaystyle \diagdown CH_2CH_2COOR^3}{\overset{\displaystyle \diagup CH_2CH_2COOR^1}{-CH_2CH_2COOR^2}}$$

Prior patent   R = aryl, aralkyl, alkaryl, alkyl and alkenyl groups of 6 to 18 carbon atoms

Patent application   R = monohydroxylalkyl groups of 2 to 4 carbon atoms

It was the opinion of the patent examiner that the prior patent "generally teaches the compounds claimed" and that the "claimed compounds became obvious in view of the common utility."

Oath and Power of Attorney $\left(\substack{\text{JOINT} \\ \text{INVENTORS}}\right)$

Being duly sworn, We, __BEN A. TEFERTILLER and__

_____JAMES I. NEVILL_____

depose and say that we are citizens of the United States residing at_____

_____Midland, Michigan_____

respectively; that we have read the foregoing specification and claims and we verily believe we are the original, first, and joint inventors of the invention or discovery in

__TRIS(HYDROXYALKYL)3,3',3"-NITRILOTRIPROPIONATES__

described and claimed therein; that we do not know and do not believe that this invention was ever known or used before our invention or discovery thereof, or patented or described in any printed publication in any country before our invention or discovery thereof, or more than one year prior to this application, or in public use or on sale in the United States for more than one year prior to this application; that this invention or discovery has not been patented in any country foreign to the United States on an application filed by us or our legal representatives or assigns more than twelve months before this application; and that no application for patent on this invention or discovery has been filed by us or our representatives or assigns in any country foreign to the United States;

And we hereby appoint WILLIAM M. YATES, (Registration No. 14594), and __C. E. Rehberg_____(Registration No. __17,741__), and _____(Registration No._____), whose addresses are, P.O. Box 1967, Midland, Michigan 48640, our attorneys or agents, with full power of substitution and revocation, to prosecute this application and to transact all business in the Patent Office connected therewith. Please address all communications to William M. Yates.

We hereby subscribe our names to the foregoing specification and claims, oath and power of attorney, this _17th_ day of _October_ , 19 _72_

	Post Office Addresses:
(signature) Ben A Tefertiller	5401 Sunset Drive
(signature) James I Nevill	Midland, Michigan 48640
	1278 Poseyville Road, R#10
	Midland, Michigan 48640

STATE OF __MICHIGAN_____ ⎫
County of __Midland_____ ⎬ ss.

Before me personally appeared__BEN A. TEFERTILLER and__

_____JAMES I. NEVILL_____

to me known to be the persons described in the above application for patent, who signed the foregoing instrument in my presence, and made oath before me to the allegations set forth therein as being under oath, on the day and year aforesaid.

(SEAL)   *(signature) Nona Z. Reinsch*
                          Notary Public

FORM.NO. C-8720 PRINTED IN U.S.A. R-7-71

Figure 11.3   Oath and power of attorney exhibit.

ASSIGNMENT – JOINT INVENTORS

For One Dollar ($1.00) and other valuable consideration received of the hereinafter named assignee, We, ____BEN A. TEFERTILLER and____
____JAMES I. NEVILL____

of__Midland, County of Midland, State of Michigan____

United States of America, joint inventors and owners of the hereinafter described invention, do hereby sell, assign and transfer unto THE DOW CHEMICAL COMPANY, a corporation organized and existing under the laws of the State of Delaware, and having its executive offices at Midland, County of Midland, State of Michigan, United States of America, the entire right, title and interest, for the United States of America and Territories thereof and for all foreign countries, in and to the invention in____
__TRIS(HYDROXYALKYL)3,3',3"-NITRILOTRIPROPIONATES____
(Case No.__16,308____)
set forth in the application signed by us even date herewith for Letters Patent of the United States (and we do hereby authorize said assignee to insert here the Serial Number_299869_, and filing date _Oct 24, 1972_, of said application upon receipt of official notification thereof); and in and to said application and any reissue, divisional, continuing or substitute application based thereon, and any corresponding application filed in a country foreign to the United States of America, including all rights to claim priority under the International Convention in the name of said assignee, and in and to any Letters Patent that may be issued for said invention in the United States of America or in any foreign country; and we do hereby covenant and agree for ourselves and our legal representatives to assist the said assignee in the prosecution of the said application or applications and in any interference, conflict, opposition or litigation which may arise involving said invention or the patent or patents thereon, and to execute any reissue, divisional, continuing, substitute, or foreign application based on or corresponding to the above specifically identified application, as may be requested by said assignee; and we do hereby authorize and request the Commissioner of Patents to issue such Letters Patent to said THE DOW CHEMICAL COMPANY, in accordance herewith, this assignment being under covenant, not only that we have full power to make the same, but also that such assigned right is not encumbered by any grant, license or other right heretofore given.

Signed by us this_17th_day of_October_, 19_72_

Ben A Tefertiller

James I Nevill

Attested by:

David N. Field

Nona Z. Reinsch

STATE OF__MICHIGAN_____ ⎫
County of __Midland_____ ⎬ ss.
⎭

RECORDED
U. S. PATENT OFFICE

DEC 1 0 1973

Rene D. Tegtmeyer

ACTING COMMISSIONER OF PATENTS

On this_17th_day of_October_, 19_72_, before me personally appeared
__BEN A. TEFERTILLER and JAMES I. NEVILL_____

to me known to be the persons named in and who executed the foregoing instrument, and acknowledge that they executed the same as their free act and deed.

Nona Z. Reinsch

(SEAL)

NONA Z. REINSCH   Notary Public
Notary Public, Bay County, Michigan
Acting in Midland County, Michigan
My commission expires January 15, 1974

FORM NO. C-4481 R-7-71   PRINTED IN U.S.A.

Figure 11.4 Joint inventors' assignment exhibit.

1

3,278,478
VINYL RESINS PLASTICIZED WITH TRISUBSTI-
TUTED NITRILOTRIPROPIONATES
James E. Masterson, Jenkintown, and David H. Clemens,
Willow Grove, Pa., and Arthur W. Ritter, Jr., Haddon
Heights, N.J., assignors to Rohm & Haas Company,
Philadelphia, Pa., a corporation of Delaware
No Drawing. Filed Feb. 2, 1965, Ser. No. 429,887
4 Claims. (Cl. 260—31.8)

This application is a continuation-in-part of application Serial No. 296,461, filed July 22, 1963, now abandoned. which in turn is a continuation-in-part of application Serial No. 269,855, filed April 2, 1963, now abandoned.

This invention deals with trisubstituted nitrilotripropionates, processes for making these compounds, and resinous compositions. This invention also deals with disubstituted iminodipropionates and processes for making these compounds.

The trisubstituted nitrilotripropionates of the invention may be represented by the formula:

$$N \underset{\displaystyle CH_2CH_2COOR^3}{\overset{\displaystyle CH_2CH_2COOR^1}{—CH_2CH_2COOR^2}} \qquad (I)$$

in which

R^1 represents a non-acetylenically unsaturated hydrocarbon group. Typical thereof are the following: aryl, aralkyl, alkaryl, alkyl, and alkylene. Each of these groups generally has a minimum content of 6 carbon atoms and a maximum content of 18 carbon atoms;

R^2 represents a non-acetylenically unsaturated hydrocarbon group. Typical thereof are the following: aryl, aralkyl, alkaryl, alkyl, and alkylene. Each of these groups generally has a minimum content of 6 carbon atoms and a maximum content of 18 carbon atoms;

R^3 represents a non-acetylenically unsaturated hydrocarbon group. Typical thereof are the following: aryl, aralkyl, alkaryl, alkyl, and alkylene. Each of these groups generally has a minimum content of 6 carbon atoms and a maximum content of 18 carbon atoms.

It is preferred that the total carbon atom content of R^1, R^2, and R^3, taken collectively, be no less that 20 carbon atoms and no more than 30 carbon atoms.

The substituents R^1, R^2, and R^3 need not all be the same; they may all be different or two of the three substituents may be the same; all three may be the same.

Typical of the substituents represented by R^1, R^2, and R^3 are the following: 2-ethylbutyl, n-hexyl, sec-hexyl. n-octyl, isooctyl, 2-ethylhexyl, 2-ethylisohexyl, tert-octyl, isodecyl, n-decyl, stearyl, benzyl, palmityl, oleyl, monochlorobenzyl. dichlorobenzyl, nitrobenzyl, phenethyl, 6-phenylhexyl, and tetrahydrofurfuryl. Other substituents typical of R^1, R^2, and R^3 include mixtures of unsaturated and saturated substituents derived from the selective reduction of vegetable and animal fats and oils, such as tallow, soybean oil, safflower oil, linseed oil, and the like. They are commonly mixtures of palmityl, stearyl, oleyl, linoleyl, and linolenyl groups in various proportions, depending on the initial composition of the fat or oil and the degree of selectivity of the reduction.

Typical of the trisubstituted nitrilotripropionates of the invention are the following: trihexyl nitrilotripropionate, tri-2-ethylhexyl nitrilotripropionate, tri-n-octyl, nitrilotripropionate, triisooctyl nitrilotripropionate, tri-n-decyl nitrilotripropionate, diisodecyl nitrilotripropionate, tristearyl nitrilotripropionate, tripalmityl nitrilotripropionate; tribenzyl nitrilotripropionate, tritetrahydrofurfuryl nitrilotripropionate, triphenethyl nitrilotripropionate, dioctylmonobutyl nitrilotripropionate, distearylmonobutyl

2

nitrilotripropionate, tripentachlorobenzyl nitrilotripropionate, trioleyl nitrilotripropionate, trioctenyl nitrilotripropionate, n - octyl-n-decyl(1.5–1.5)nitrilotripropionate, dihexylmonoctyl nitrilotripropionate, and monohexyldioctyl nitrilotripropionate.

A particularly valuable group of the compounds of the invention is the trialkyl nitrilotripropionates in which the alkyl groups may be all the same or different. The process for preparing the trisubstituted nitrilotripropionates of the invention comprises reacting an acrylate ester of the formula

$$CH_2=CHCOOR^{1-3} \qquad (II)$$

with ammonia. In the above formula, R^{1-3} is as defined above. When a mixture of acrylate esters is employed, the resulting trisubstituted nitrilotripropionates of the invention have R^{1-3} substitutents which are dissimilar. In the reaction of the invention, the molar ratio of acrylate to ammonia should be at least 3 to 1. Generally, it is preferred to use an excess of acrylate ester, preferably not exceeding 10 mole percent, over the stoichiometric amount required, although higher excesses are not detrimental.

The temperature at which the reaction proceeds may vary widely. It may range broadly from about 0° C. to the temperature at which the formation of the amide from the acrylate predominates over the addition reaction across the double bond of the acrylate to form the ester. Generally, the temperature is in the range of 5° to 150° C., and more preferably in the range of 30° to 100° C. For best results, it is very advisable to continue heating the reaction mixture after essentially all of the ammonia has been consumed. Usually heating is carried out at a temperature higher than that at which the reaction between the ammonia and the acrylate ester proceeds. Such heating temperature ranges from 50° to 100° C. or higher. The progress of the reaction may be followed by the disappearance of any intermediate disubstituted iminodipropionates that may be formed during the reaction. Such iminodipropionates may be represented by the following formula

$$HN \underset{\displaystyle CH_2CH_2COOR^2}{\overset{\displaystyle CH_2CH_2COOR^1}{<}} \qquad (III)$$

wherein R^1 and R^2 are as defined above in conjunction with the nitrilotripropionates.

In accordance with the invention, the reaction between the acrylate ester and ammonia proceeds preferably in the presence of a solvent for the ammonia. Typical solvents are: methanol, ethanol, propanol, butanol, dimethyl sulfoxide, dimethyl formamide, tetrahydrofuran, ethylene glycol, dimethyl ether, 2-methoxyethanol, 2-butoxyethanol, and diethylene glycol.

It has also been found that for best results the reaction is carried out in the presence of a catalyst selected from the group consisting of salts of alkaline and alkaline earth metals (metals from Groups IA and IIA of the Periodic Table) and compounds of the generic formula

$$R^1—\overset{\displaystyle R^3}{\underset{\displaystyle R^4}{N^+}}—R^2X^- \qquad (IV)$$

where

R^1, R^2, R^3, and R^4 are hydrogen, alkyl, aryl, or aralkyl, and X is an anion whose conjugate acid has a dissociation constant greater than 10^{-9}.

Examples of catalysts are the following: ammonium formate, ammonium acetate, ammonium butyrate, ammonium chloride, monomethyl ammonium chloride, monomethyl ammonium acetate, tributyl ammonium sul-

Figure 11.5 Patent on nitrilotripropionate esters as plasticizers.

fate, trioctyl ammonium iodide, tetramethyl ammonium butyrate, tetrabutyl ammonium acetate, tetrabutyl ammonium chloride, lithium chloride, lithium acetate potassium chloride, sodium acetate, sodium propionate, sodium dichloroacetate, magnesium acetate, strontium trichloroacetate, and the like. Preferably, the catalyst is soluble in the reaction medium.

The catalyst is employed in catalytic amount, i.e., that amount which increases the speed of the reaction. Generally, an amount in the range of .05% to 5% of the total weight of the reactants is used.

The greatest benefits from the action of the catalyst are obtained by its presence in the reaction mixture from the time essentially all of the ammonia has been consumed. But, the catalyst may also be present during the reaction of the ammonia and the acrylate.

In a further embodiment of the invention, there is provided a transesterification process which comprises reacting trisubstituted nitrilotripropionates of the formula Ia

$$N \begin{cases} CH_2CH_2COO \text{ lower alkyl} \\ CH_2CH_2COO \text{ lower alkyl} \\ CH_2CH_2COO \text{ lower alkyl} \end{cases} \quad (Ia)$$

with at least one alcohol of the formula

$$R^4OH \quad (V)$$

In Formula Ia above, "lower alkyl" is taken to mean an alkyl group of 1 to 4 carbon atoms. Preferably, the lower alkyl group is an alkyl group of 1 to 2 carbon atoms. In Formula V, the R⁴ substituent may represent a nonacetylenically unsaturated aliphatic hydrocarbon group. Typical thereof are the following: aralkyl, alkyl, and alkylene. These groups generally have a minimum carbon atom content of 6 atoms and a maximum carbon atom content of 12 to 18 with a maximum of 12 carbon atoms being preferred. In this embodiment of the invention, the particular group which R⁴ represents is so selected as to be different and of substantailly higher molecular weight than the lower alkyl groups in Formula Ia.

Alcohols typical of R⁴OH are the following: n-hexyl alcohol, 2-ethylbutyl alcohol, n-octyl alcohol, isooctyl alcohol, 2-ethylhexyl alcohol, n-decyl alcohol, 2-ethylisohexyl alcohol, isodecyl alcohol, stearyl alcohol, palmityl alcohol, benzyl alcohol, tetrahydrofurfuryl alcohol, oleyl acohol, phenethyl alcohol, and others.

This embodiment of the invention provides a process for making, in essentially quantitative yields, trisubstituted nitrilotripropionates in which the alcohol moiety of the ester is of higher molecular weight than that of the acrylate ester by reacting ammonia and the acrylate ester and then reacting the nitrilotripropionate with an appropriate alcohol.

These alcohols may be used individually or in any mixture thereof. Where the alcohols are used in a mixture, the resulting trisubstituted nitrilotripropionates have R¹, R², and R³ which are different, the respective ratios R¹, R² and R³ depending therefor on the particular ratio of alcohols used.

At the end of the reaction, the product is purified by suitable methods, as by distillation, removal of excess reactants, such as the alcohols and the acrylate.

As illustrative of the invention are the following examples. All parts are by weight unless otherwise noted. The symbol Hg stands for mercury.

EXAMPLE 1

There are charged to a two-liter reactor 5 parts of ammonium acetate and 75.0 parts of methanol. There are added in portions over a two-hour period 451.0 parts of methyl acrylate and 27.2 parts of ammonia. The temperature is raised gradually to 70° C. over a one-hour period and held at that temperature for six hours. The product is stripped of volatile materials to a temperature of 100° C. at 0.15 mm. Hg. The product (441.8 parts)

is essentially pure trimethyl nitrilotripropionate having a neutralization equivalent of 273.73.

EXAMPLE 2

The procedure of Example 1 is followed but replacing the ammonium acetate by 4.3 parts of lithium chloride. The same product is obtained.

EXAMPLE 3

The procedure of Example 1 is followed but the catalyst is omitted. The same product is obtained after a longer reaction time.

EXAMPLE 4

A stirred two-liter reactor equipped with a dropping funnel and gas inlet tube is charged with 4.30 parts of lithium chloride. The flask is evacuated to 5 mm. of Hg and 75 ml. (60 parts) of methanol is charged followed by a total of 500 parts of ethyl acrylate and 27.1 parts of ammonia which are added in portions over about a six-hour period while maintaining the temperature between 18° and 38° C. The reaction mixture is allowed to stir for 16 hours at room temperature. An additional 4.3 parts of lithium chloride is added and the reaction mixture heated at a temperature of 70° C. for 3 hours and then stripped of methanol and excess ethyl acrylate at a temperature of 100° C. and a pressure of 0.35 mm. of Hg. There remains 480 parts of almost pure triethyl nitrolotripropionate having a neutralization equivalent of 315 compared to theory of 317.4.

EXAMPLE 5

The procedure of Example 4 is followed but omitting the lithium chloride. The same product is obtained.

EXAMPLE 6

The procedure of Example 4 is followed but 9 parts of lithium chloride are added only after all the ammonia is consumed.

EXAMPLE 7

A stirred two-liter reactor equipped with a dropping funnel and gas inlet tube is charged with 4.30 parts of lithium chloride. The flask is evacuated to 5 mm. of Hg and 75 parts of methanol charged, followed by a total of 430.1 parts of n-butyl acrylate and 21.4 parts of ammonia which are added in portions over about a 7-hour period while maintaining the temperature between 28° C. and 35° C. The reaction mixture of di-n-butyl iminodipropionate and tri-n-butyl nitrilotripropionate is allowed to stir for 13 hours at room temperature. An additional 4.30 parts of lithium chloride is added and the reaction mixture heated at a temperature of 70° to 75° C. for 3 hours and then stripped of methanol and excess butyl acrylate at a temperature of 100° C. and a pressure of 0.25 mm. of Hg. The product which is obtained is tri-n-butyl nitrilotripropionate.

EXAMPLE 8

The procedure of Example 7 is followed omitting the lithium chloride. The same product is obtained.

EXAMPLE 9

The procedure of Example 7 is followed but 9 parts of lithium chloride are added only after all the ammonia is consumed.

EXAMPLE 10

A stirred two-liter reactor equipped with a dropping funnel and gas inlet tube is charged with 4.30 parts of lithium chloride. The flask is evacuated to 5 mm. of Hg and 150 ml. (120 parts) of methanol charged, followed by a total of 430.1 parts of 2-ethylhexyl acrylate and 13.1 parts of ammonia which are added in portions over about a 6-hour period while maintaining the temperature betwen 28° and 35° C. The reaction

Figure 11.5   *(Continued)*

5

mixture is allowed to stir for 22 hours at room temperature. An additional 4.0 parts of lithium chloride are added and the reaction mixture heated at a temperature of 70° C. for 4½ hours and then stripped of methanol and excess 2-ethylhexyl acrylate at a temperature of 185° C. and a pressure of 0.3 to 0.4 mm. of Hg. The product which is obtained is tri-2-ethylhexyl nitrilotripropionate.

EXAMPLE 11

The procedure of Example 10 is followed but omitting the lithium chloride. The same product is obtained.

6

EXAMPLE 14

The procedure of Example 13 is followed replacing the catalyst by zinc acetate. The same product is obtained.

EXAMPLE 15

The procedure of Example 13 is followed replacing the catalyst by aluminum triisopropoxide. The same product is obtained.

Examples 16 to 27 illustrate the preparation of various higher molecular weight nitrilotripropionates from the corresponding lower ester.

Table I.—Transesterification of nitrilotripropionates

Examples	Ester Type	Alcohol		Catalyst, Percent	Time (Hrs.)	Pressure, mm.	Product		
		Type	Moles per Mole of COOR				Yield (Percent)	N.E.	
								Found	Calcd.
16	E	Isodecyl	1.50	STA, 1.0	4 plus 2½	97–110 18–21	95.13	642.7	653
17	E	Mixture of 55% n-octyl alcohol, 45% n-decyl alcohol.	1.50	STA, 2.0	5¼ plus 3½	70–100 15	95.15	603.4	607
18	M	Isooctyl	2.00	STA, 0.75	4 plus 5	90–100 18–22	97.2	562.9	599
19	E	Complex mixture of branched and straight chain alcohols averaging to 7 to 9 carbon atoms.	1.33	STA, .06	3 plus 3½	95–103 47–24	98.3	546.5	549.7
20	E	A mixture of n-C₈, C₉, and C¹⁰ alcohols, average composition of C₉ H₁₉.₄	1.33	STA, .03	8 plus 3	97–120 17–55	102	589.4	573.8
21	E	Benzyl	2.0	STA, 0.33	10 (Toluene azeotrope)	Atm.	80.6		
22	E	Tetrahydrofurfuryl	2.0	STA, 0.6	14½	Atm.	77	486.2	486.6
23	E	Cetyl/Stearyl averaging to 17.5 carbon atoms.	1.0	STA, 0.2	4 plus 7	85–100 50–65	98	979.2	969
24	E	2-ethylhexyl	1.33	Aluminum triisopropoxide, 0.34.	9 plus 2	95–100 30–50	98+	554	569
25	E	2-ethylhexyl	1.50	Zn(OAc)₂.2H₂O, 0.17	8½	100	95+	550.9	569
26	M	Oleyl	1.50	STA, 0.33	10	100	97	975	983
27	M	Allyl	1.50	STA, 0.33	12	200	98	350	353

¹ N.E. stands for neutralization equivalent.
² E stands for triethyl nitrilotripropionate; M stands for trimethyl nitrilotripropionate.
³ STA stands for sodium hydrogen titanium alkoxide.
Note 1.—Temperature 100°–135° C.

EXAMPLE 12

The procedure of Example 10 is followed but 9 parts of lithium chloride are added only after all the ammonia is consumed.

EXAMPLE 13

There are charged to a 500 ml. reactor arranged for distillation 98.6 parts of trimethyl nitrilotripropionate, 186 parts of 2-ethylhexanol, and 0.5 part of a sodium hydrogen titanium mixed alkoxide. The pressure is reduced to 99–106 mm. of Hg and the reaction mixture heated to 92°–116° C. After three hours, the pressure is reduced to 41 to 50 mm. of Hg and the reaction mixture maintained at 118°–120° C. for two hours. The pressure is then reduced to 24 mm. of Hg with the temperature at 105°–117° C. The pressure is returned to atmospheric and 0.20 part of acetic acid and 0.40 part of water are added. The mixture is heated at 80°–95° C. for two hours and stripped of volatile materials at a temperature of 178° C. and a pressure of 0.15 mm. of Hg. The product (202.8 parts) is tri-2-ethylhexyl nitrilotripropionate and has a refractive index of 1.4295 at 25° C.

The esters of the invention are valuable plasticizers for polyvinyl halide resins. The term "polyvinyl halide resin" refers to polymers containing a predominant quantity, that is, a quantity greater than 50%, generally over 60%, by weight of the monomer as vinyl halide units. This includes the homopolymers of the vinyl halides as well as the copolymers and interpolymers of a vinyl halide and an unsaturated monomer copolymerizable therewith. Other monomers that may be copolymerized with the vinyl halide include the vinyl type monomers such as, for example, those having a single $CH_2=C=$ group, such as vinylidene chloride, vinyl chloroacetate, chlorostyrene, chlorobutadienes, etc., and those copolymers of such vinyl compounds and other unsaturated materials copolymerizable therewith, for example, copolymers of a vinyl halide, such as vinyl chloride, with such materials as vinylidene chloride, vinyl esters of carboxylic acids, for example, vinyl acetate, vinyl propionate, vinyl butyrate, vinyl benzoate; esters of unsaturated acids, for example, alkyl acrylates, such as methyl acrylate, ethyl acrylate, propyl acrylate, butyl acrylate, allyl acrylate and the corresponding esters of methacrylic acid; vinyl aromatic compounds, for example, styrene; esters of

Figure 11.5 (Continued)

7

α,β-unsaturated carboxylic acids, for example, the methyl, ethyl, butyl, amyl, hexyl, octyl esters of maleic, crotonic, itaconic, fumaric acids and the like. Further useful copolymers are those obtained by copolymerization of vinyl chloride with an ester of an α,β-unsaturated dicarboxylic acid, such as diethyl maleate or other esters of maleic, fumaric, aconitric, itaconic acid, etc., in which 5 to 20 parts by weight of diethyl maleate or other analogous esters are used for every 95 to 80 parts by weight of vinyl chloride.

When the esters of the invention are employed as plasticizers for polyvinyl halide resins, they are ordinarily incorporated into the vinyl halide polymers by mixing the powdered resin with the liquid plasticizer followed by mixing and/or kneading and then by curing the mix at an elevated temperature, for example, within the range from 150° to 200° C., on hot rolls or in a heated mixer, such as a Werner-Pfleiderer or Banbury mixer. The proportion of esters that may be employed may vary over a great range since it is dependent on the particular esters of this invention which is selected, the specific polyvinyl halide resin to be plasticized, and the final degree of plasticization desired in the resin, this factor in itself being dependent on the ultimate application intended for the resin. With these facts in mind, one skilled in the art may use the esters in a "plasticizing amount" for most purposes, this being from about 5 to 100 parts, and more commonly from 20 to 60 parts, of esters per 100 parts of resin. In amounts from 60 to about 150 parts of ester per 100 parts of polyvinyl chloride resin, as from 60 to 90 parts, the esters of the invention are more commonly suitable for use in organosols and plastisols. One or more esters may be used in the polyvinyl halide resin.

In accordance with the invention there may be employed one or more esters of this invention in polyvinyl halide compositions; also the esters of the invention may be employed as the sole plasticizer; or they may be employed in conjunction with conventional plasticizers, such as alkyl phthalates, alkyl phosphates, monomeric or polymeric epoxides, and other plasticizers known in the art.

With the polyvinyl halide resin, there may be incorporated various stabilizers, fillers, dyes, pigments, and the like.

The value of the esters of the invention is further demonstrated by the following illustrations.

A standard resinous composition is made up from the following ingredients.

Table II.—Standard formulation

	Parts
Polyvinyl chloride	60
Plasticizer	40
Barium cadmium laurate	1.0

The resinous compositions are evaluated in accordance with tests further described hereinafter.

Table III

Compound	Tests				
	1	2	3	4	5
Tri-2-ethylhexyl nitrilotripropionate	71	−50	1.5	4.2	0.6
Triisooctyl nitrilotripropionate	77	−46	1.6	7.3	1.0
Triisodecyl nitrilotripropionate ¹	78	−40	5.6	6.4	0.3
Tri-n-octyl/n-decyl nitrilotripropionate ¹	77	−45	5.7	7.4	0.4

¹ Data are for a mixture of the plasticizer with 50% of di-(2-ethylhexyl)-phthalate. It is evident from the data that the polyvinyl resinous compositions are well plasticized, soft, pliable, and flexible. These properties are retained even at very low temperatures. The plasticizers also remain integrated in the resin even under high temperature conditions; they are resistant to soap extraction and highly compatible.

8

The above formulations are modified by replacing polyvinyl chloride by copolymers of

(A) 87 parts vinyl chloride : 13 parts vinyl acetate
(B) 80 parts vinyl chloride : 20 parts vinylidene chloride
(C) 80 parts vinyl chloride : 20 parts methyl acrylate
(D) 95 parts vinyl chloride : 5 parts vinyl isobutyl ether

The resinous compositions are tested as described above. All compositions are supple and flexible, and they exhibit good low temperature flexibility allied with good resistance to high temperature conditions.

The plasticized polyvinyl chloride compositions of the invention are useful for many outdoor uses, such as greenhouse glazing, pond liners, ditch liners and silo covers. The compositions are well suited for many automotive applications, such as door panels, crash pad covers, and gasketing; for jacketing for electrical cables and many others.

The nitrilotripropionates of the invention are, furthermore, unique in their performance in plastisols in combining excellent gelation and fusion behavior with outstanding viscosity stability.

A typical plastisol composition comprising 100 parts of finely divided polyvinyl chloride, 3 parts of a Ba/Cd/Zn complex stabilizer and 85 parts of tri-2-ethylhexyl nitrilopropionate was mixed for 15 minutes in a Hobart mixer and then deaerated at 28 in. of Hg for 10 minutes. The viscosity of the plastisol was 17 poises at 60 r.p.m. after aging one day at 77° F. A comparative composition with di-(2-ethylhexyl) phthalate (DOP) had an 18 poise/60 r.p.m. viscosity. After 21 days at 77° F., the viscosity of the composition with DOP had risen to 29. The viscosity of the nitrilotripropionate composition had decreased to 12 poise. In another experiment, at 104° F., the viscosity for the nitrilotripropionate after one day was 7 poises/60 r.p.m. The composition with DOP was 20. After 21 days at 104° F., the nitrilotripropionate composition was 5; the DOP had a viscosity which increased to 56. Thus, while the conventional DOP composition showed gradual and perceptible gelation, the plastisol of the invention showed remarkable viscosity stability. The term "plastisol" is well known in the art. See Modern Plastics, vol. 29, page 87 (1951), and U.S. Patent No. 3,050,412, col. 3, lines 48–67, for instance.

TESTS

Test 1—Shore hardness.—A Shore "A" durometer, under a weight of 3 pounds, is applied to the test specimens. A recording is made at once and after 10 seconds; and the hardness is expressed by the two values, of which the first recording is the higher.

Test 2—Torsional modulus at low temperatures.—A 1¼" x ¼" sample is cut and mounted in a Tinius-Olsen stiffness tester, which measures the torsional modulus of plastic at various temperatures. The temperature at which a specimen has a torsional modulus of 135,000 lbs./sq. in., known as T_f of $T_{135,000}$, is determined. This roughly corresponds to the "brittle point" obtained by cantilever apparatus.

Test 3—Activated carbon volatility.—2" squares of weighed specimens are placed between 2-inch layers of activated carbon in sealed glass jars, which are maintained at 90° C. for 24 hours. The specimens are removed, dusted free of carbon and reweighed.

Test 4—Soapy water extraction.—3" squares of weighed specimens are immersed in a 1% aqueous solution of Ivory soap at 90° C. for 24 hours, after which they are thoroughly washed, dried, and reweighed.

Test 5—Compatibility.—Weighed, conditioned duplicate samples, 4 inches by 4 inches by 0.010 inch, are placed between two sheets of cardboard, which have been conditioned at least 15 hours. The specimens are in contact with the white, coated side of the cardboard. The cardboard-specimen sandwiches are placed between 5 inches by 5 inches by 1 inch wood blocks under a 3-

Figure 11.5 (Continued)

3,278,478

9

kilogram weight. After seven days, the specimens are removed from the stack, conditioned, and reweighed. Percent plasticizer loss is calculated. The cardboard sheets are examined qualitatively for evidence of plasticizer or plasticizing stabilizer stains.

We claim:

1. A resinous vinyl chloride composition comprising a vinyl chloride polymer selected from the group consisting of homopolymers of vinyl chloride and copolymers of 50 to 95 percent by weight vinyl chloride and from 50 to 5 percent by weight of a monoethylenically unsaturated monomer copolymerizable therewith, said polymer being plasticized with a trisubstituted nitrilotripropionate of the formula

$$\begin{array}{l} CH_2CH_2COOR^1 \\ N-CH_2CH_2COOR^2 \\ CH_2CH_2COOR^3 \end{array}$$

in which each of R^1, R^2 and R^3 is an alkyl group of 8 carbon atoms.

2. The resinous composition of claim 1, wherein the

10

nitrilotripropionate is tri(2 - ethylhexyl)nitrilotripropionate.

3. The resinous composition of claim 1, wherein the nitrilotripropionate is triisooctyl nitrilotripropionate.

4. A vinyl plastisol composition comprising a vinyl chloride resin selected from the group consisting of homopolymers of vinyl chloride and copolymers of 50 to 95 percent by weight vinyl chloride and from 50 to 5 percent by weight of a monoethylenically unsaturated monomer copolymerizable therewith, and tri(2-ethylhexyl)nitrilotripropionate plasticizer.

References Cited by the Examiner

UNITED STATES PATENTS

2,296,331   9/1942   Bogemann et al. _____ 260—482

OTHER REFERENCES

Mellan: The Behavior of Plasticizers; 1961,; Pergamon Press; pages 27, 30, 31.

MORRIS LIEBMAN, *Primary Examiner.*

L. T. JACOBS, *Assistant Examiner.*

Figure 11.5 *(Continued)*

Upon receipt of this rejection, the solicitor met with the two inventors to determine the nature of their response. It was agreed that the rejection was unjustified on the basis of (1) the vast difference in functionality of the chemical structures and (2) the basic difference in function between a cross-linking agent and a plasticizer. The solicitor thereupon went to Washington and discussed the matter with the Patent Office examiner. Following this visit, a written response was submitted.

Shortly thereafter a letter was received from the examiner that all claims of the application would be allowed. This was followed by a formal Notice of Allowance. After submitting an additional fee, the patent (Fig. 11.6) was issued. The patenting process took about 16 months from date of application to date of issue. This time is probably normal for cases that are not overly complex.

11.9   THE CHEMIST'S RESPONSIBILITIES

A chemist in a research and development group has been hired to invent. The inventions or innovations that result from his work naturally and legally belong to the company. Some of these inventions will be patentable. In this regard the company has certain expectations.

It expects a prompt and full disclosure of the invention including a clear description of how it differs from the prior art. Since the inventor usually has the best knowledge of the overall area of technology, he is expected to

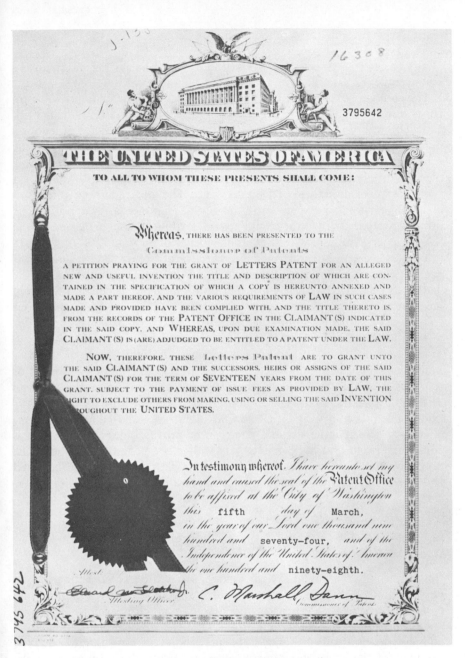

Figure 11.6 Patent exhibit.

assist the patent solicitor with the patent application so that the best possible patent coverage is obtained.

The chemist is expected to advise the patent solicitor before disclosing the invention or sampling the product outside the company.

The company expects the chemist to maintain complete and provable records of his work. The primary record in research and development work is the laboratory notebook. Each company has specific rules for notebooks, but they are almost identical because their objectives are essentially the same. In the legal world it seems that the terms "proof" and "provable records" are only semispecific. What constitutes legal proof is what is accepted as sufficient in that particular instance. The rules regarding notebooks are designed to make them most readily acceptable as legal proof. The principal requirements are: completeness, legibility, date, signature of the researcher, signature of witnesses, and evidence of originality.

Witnesses are important. Whenever legal proof of an act or a record is needed, witnesses may be required to testify as to their knowledge of the matter to be proved. An inventor can testify in his own behalf, but it will have no value as proof unless corroborated by others. Ideal witnesses are fellow workers who are familiar with the work and have personal knowledge of what is done. Joint inventors cannot serve as witnesses.

The notebook itself should furnish evidence of originality. It should be a bound notebook with numbered pages so that pages cannot be inserted or removed. It should be filled consecutively with a dated signature at the end of each day's entry. Blank spaces compromise the authenticity of the notebook since it is conceivable that additional entries could be added at some later time. The genuineness of a page can be proven quite easily when it is completely filled and fits consecutively with other pages in the book. The notebook should be reviewed frequently by a witness and should be dated and signed by the witness under a notation that it has been read and understood.

11.10 BENEFITS OF PATENTS

Philosophically the patent system was established to promote the progress of science and useful arts. Practically, patents make it worthwhile for private enterprise to invest in research and development. Research is expensive, and the outcome is frequently uncertain. There would be little incentive for a company to gamble this money if its competitors were waiting to copy the products of the successful research projects. From the standpoint of the company, the greater the prospects for a good patent position, the greater the justification for spending on research.

Products and processes that are protected by patents generate higher profits. This is the principal benefit of patents. There are other benefits. Patents also act to defend a company's technology. Thus a patent on an alternate process, even if it is not used by the company, can be of value by preventing competitors from entering a market. Patents also have been used to give a company a trading position to enter fields previously denied by competitors' patents. The exchange of patent rights is not uncommon. Last, patents can be licensed to others to collect royalties.

11.11 FURTHER READING

An excellent series of papers on chemical patents, written for the chemist, is contained in *Advances in Chemistry Series* Volume 46, entitled, "Patents for Chemical Inventions," edited by E. J. Lawson and E. A. Godula, and published by the American Chemical Society, Washington, D.C.

11.12 REFERENCES

1. D. H. Fifield, private communication.
2. E. J. Lawson and E. A. Godula (Eds.) "Patents for Chemical Inventions," *Advances in Chemistry Series*, Vol. 46, American Chemical Society, Washington, D.C., 1964.
3. Choate and Francis, *Cases and Materials on Patent Law*, West Publishing Co., St. Paul, Minnesota, 1973.
4. A. P. Kent, "Current Patent Office Procedures," *Chem. Tech.* **2**, pp. 599–605, 1972.
5. Judge Clarence Newcomer, "Mueller Brass vs Reading Industries," 176 United States Patent Quarterly 361 at 372, 1972.
6. J. H. Saunders, *Careers in Industrial Research and Development*, Dekker, New York, 1974.
7. F. W. Billmeyer, Jr. and R. N. Kelley, *Entering Industry*, Wiley, New York, 1975.

12

ECONOMICS

FINANCIAL ANALYSIS OF THE PROCESS

Probably the single most important task of the management function is the making of decisions on the investment of money in order to make more money. The future of the company is directly affected by the skillfulness of its managers in choosing profitable ventures. Although a company of some size and diversity might withstand a certain number of bad decisions, the cumulative effects of a string of losers can be just as devastating in the chemical business as at the race track.

One of the inherent problems is that these decisions involve predictions of the future and therefore are based upon uncertainties. Another problem is that in any well run, dynamic company there are many proposed projects and a limited amount of money to invest. The question is not simply whether a project is deemed worthwhile, but which project is *most* worthwhile from the standpoint of risks and anticipated economic return. The ultimate objective of an economic analysis program is to provide management with the best possible basis for its investment decisions. It should do this in a manner that minimizes the risks of uncertainties and that provides some scale for comparing the relative worth of various projects or proposals.

Many companies have evolved systematic methods of evaluating the economics of proposed investments. Although the details of these methods may vary, there are certain common considerations and concepts. In this chapter we develop and describe two analysis procedures similar to those currently being used in the chemical industry.

12.1 THE ECONOMIC STUDY

An economic study is an analysis of the financial aspects of a problem or a proposed venture for use in making investment decisions.

There are three aspects to the economic study: the problem, the analysis, and the use to be made of the study. The economic study first clearly defines the problem or proposal. It must then identify and quantify the economic consequences of pursuing a particular course. Last, it must result in some criteria of value or worth so that they can be used by management in reaching a decision.

12.2 THE APPROACH

In order to fairly and equitably compare the economic value of various proposed projects, the analytical procedure must be standardized. The differences in the results of two economic studies must reflect differences in the economics of the two proposals—not differences in the evaluation methods.

Frequently the basic approach in defining the problem is to compare the economic consequences of pursuing a proposed project with *not pursuing* it. The alternative to investing capital is simply not investing capital. If each project is evaluated relative to this common base and each analysis is done in a standard manner, then the results can be used to compare the relative worth of a number of projects.

In defining the problem in terms of alternatives, it is important that the two *best* alternatives be compared. As described above, these alternatives are frequently *risking* capital and *not risking* capital. It sometimes happens that not investing capital is a completely unrealistic alternative. For example, if a plant is down because of a pump failure, the "do-nothing" alternative cannot be considered. An economic study in this case might compare spending $6000 for a replacement pump with spending $800 for the repair of the existing pump.

After defining the true alternatives the differences in these two courses of action must be analyzed. Even in a relatively simple problem, probing analysis frequently reveals a very large number of variables. Most of these variables can be expressed in terms of dollars and brought into some semblance of order by reducing them into two categories: dollars of cost and dollars of capital. The cost dollars can then be subtracted from the sales revenue dollars to give a measure of the earnings or profit from the project. Unfortunately, the simple dollars of profit is not a very good criterion of the value of a proposed project. By itself it does not give a measure of how much time is required to yield the profit; or how many capital dollars must be invested to generate the profit dollars.

12.3 COST FACTORS

The total expenditures associated with the production and marketing of a product may be divided into capital monies and expense monies. *Capital* consists of money that is invested at the initiation of a project or at certain intervals during the course of the project. These are nonrecurring disbursements. *Expenses*, on the other hand, are continuing costs that recur month after month throughout the life of the production facility.

The total expenses associated with the operation of a chemical company are made up of a large variety of individual cost items. Some of these—such as raw materials costs—can be directly associated with a particular product. These are termed *direct costs*. Others—such as the costs of administration and public relations—cannot be tied to one specific product. These are *indirect costs*. Both are, of course, *real* costs and the total revenue from all of the products must be sufficient to cover the total costs of the company. The indirect costs must be allocated or apportioned out among the various products by some systematic though arbitrary formula. Some indirect costs, which appear to be mainly associated with personnel, might be allocated on the basis of the number of employees involved with a particular product. Others that relate more closely to the volume of business might be allocated on the basis of the annual sales revenue of each product. The particular allocation formulas vary from company to company, but one way or another the total costs of the company are borne by the only source of revenue—the production and marketing of products and services.

Costs can also be classified as *fixed* or *variable*. The distinction here is the relation of the cost to the production level of the product. The test is to imagine what would happen if, on a short-term basis, the production rate dropped to zero. Some costs, such as the purchase of raw materials, would drop to zero. Raw materials costs vary directly with production rate, therefore, they are *variable costs*. Other costs, such as rent or depreciation of buildings and equipment, would continue unchanged even though no product was being produced. These are termed *fixed costs*. Many cost items would not drop to zero and would not remain unchanged, they would simply decrease to a lower level. The operating labor force could be decreased by layoff or transfer. But probably a few key trained operators would be retained. Maintenance costs would be lower, but would not drop to zero. Even a shutdown plant requires some routine maintenance. These cost items are handled by considering them to be partially variable and partially

Table 12.1 Typical Cost Factors

Cost Item	Description	Basis for Estimate	% Var.	% Fixed
Raw materials	Itemized cost of all raw materials including solvents, catalyst, etc.; current delivered prices	Process data (unit ratios and prices)	100	0
Labor	Includes factors for vacations, pensions, payrolls taxes, etc.	Engineering estimate	80	20
Maintenance	Includes labor, materials shop expenses	Engineering estimate	50	50
Utilities	Steam, power, process water, cooling water, waste disposal charges	Engineering calculations	100	0
Factory overhead	An overhead cost associated with expenses outside battery limits (control lab, safety department, roads, lights, etc.)	Estimate based on operating labor	65	35
Depreciation	Distribution of investment costs over life of project; method and time period variable	Capital estimate	0	100
Taxes & Insurance	Taxes and insurance, real estate, buildings, etc.	2% of direct fixed capital per year	0	100
Packaging	Cost of bags, drums, paks., tank car	Cost schedule	100	0
Research	Allocation of cost of total research effort	14% of production cost at 100% capacity	0	100
Selling expense	Varies with product, market characteristics	Marketing estimate	20	80
General & Admin.	Allocation of administrative costs	3% of production cost	100	0

fixed. The maintenance costs, for example, might be considered to be 50%
variable. The specific factors used will vary for each company. Table 12.1
lists a number of operating cost items and typical assigned factors for the
variable and fixed components.

12.4   CAPITAL

Capital expenditures can be classified into three categories that are estimated
separately in the economic study. *Direct fixed capital* is the money required
to build and equip the production facility. The cost is estimated by the
design engineering group upon completion of the design work. *Indirect fixed*
capital or "back-up" capital includes the prorated cost of all facilities or
plants that supply raw materials, utilities, or services to the particular production plant. Indirect capital is usually calculated by considering the money
invested in the back-up facility and the percentage of the total output utilized
by the particular project being evaluated. *Working capital* is the fund of
money required to do business. There is a time lag from the purchase of raw
materials through the production of the product, sale, and delivery to the
customer and, finally, to payment of the invoice by the customer. The
amount of money required to operate the business venture is related to the
overall volume of business and is frequently estimated as a certain fraction
of the estimated annual sales revenue. One-sixth of the annual sales—which
represents the money required for a 60 day lag—is a reasonable estimate
for the working capital requirements. This money is considered to be invested
or encumbered at the time of plant start-up.

The investment in fixed capital, both direct and indirect, can be considered
as money loaned by the company to the production unit. The repayment of
this loan over an extended time period is termed the *depreciation* of the
capital assets. There are various accounting procedures for depreciation.
It can be spread evenly over the time period—straight-line depreciation.
There also are a number of formulas in which the depreciation rate is initially
high and gradually decreases over the time period—sum of the years,
declining balance, double-declining balance. Since depreciation is a cost
and therefore directly affects taxable profits, the government has regulations
that specify the depreciation rate and the time period. These regulations
vary with various categories of facilities or equipment and tend to fluctuate
through the years. Regardless of the depreciation method used for accounting
and tax purposes, the method used in the economic study is established by
the practice or philosophy of the individual company. Table 12.2 compares
the depreciation rates calculated by three different methods.

Table 12.2   A Comparison of Several Common Depreciation Methods

% Depreciated

Year	Straight Line	Sum of Years Digit	Double Declining Balance
1	10.00	18.18	20.00
2	10.00	16.37	16.00
3	10.00	14.55	12.80
4	10.00	12.73	10.24
5	10.00	10.91	8.19
6	10.00	9.09	6.55
7	10.00	7.27	5.24
8	10.00	5.46	4.20
9	10.00	3.64	3.36
10	10.00	1.80	2.64
TOTAL①	100.00	100.00	89.22

①In the straight line and sum of years methods, the amount depreciated
is the original investment less estimated salvage value at the end of
the depreciation period; in the double declining balance method, the
amount depreciated is the original investment, the salvage value is
assumed to be 10.78%.

12.5 SALES FORECASTS

In the typical economic study the capital investments and the anticipated
costs are estimated and treated in considerable detail. In contrast to this,
the forecasts of sales revenues appear simple and straightforward and perhaps of secondary importance. This is, of course, deceptive. The final calculations in any economic study involve the difference between total revenues
and total expenditures. The accuracy of the completed study therefore can
be no better than the accuracy of the market forecast. In addition to this
direct effect, there are more subtle effects due to the interrelation of many
factors. Thus the projected sales volume not only affects the sales revenues
but also sets the size of the proposed plant, which in turn affects the manufacturing costs as well as the capital requirements.

Forecasting the sales volume and selling price of a product is a highly
complex art. Although marketing specialists have evolved some systematic
approaches, the key to reliable forecasting is a detailed familiarity with the
specific market in question. These forecasts originate within the marketing
function of the company. The job titles of the individuals involved may vary
depending upon the organization of the particular company. Market forecasts or marketing plans may cover a wide time range. Thus a forecast may

include price and volume projections for 1, 2, 5, and 10 years. The accuracy of these forecasts decreases as the time span increases.

Although sales revenues are the simple product of sales volume times selling price, the situation is immediately complicated by the effect of price upon volume. Frequently, the customer demand for a product will increase as the selling price is decreased. Economists term this the *elasticity of demand*. The objective of marketing strategy is not to establish the highest possible selling price but to maximize the total sales revenues and long-term profitability of a product.

Pricing policy is a very important part of marketing strategy. There are a number of factors that influence pricing policy. It is interesting that cost is not considered to be the most important of these. The value of the product to the customer is the major factor. This value is perhaps more difficult to define than cost, in terms of cents per pound. In the final analysis, however, the product must satisfy the customer's needs or wants at a cost that he considers to be economically advantageous to him. Otherwise there is no sale and the "price" is merely an "asking price" not a "selling price."

Where then does the manufacturing cost enter into the pricing picture? In a sense it sets the lower limit or floor on the price. In the short run prices might sometimes be set lower than manufacturing costs—in order to establish a position in a market or to fill out a product line. On some occasions, such as situations where industry-wide production capacity far exceeds the total market demand, prices might be set below costs simply to minimize losses. In spite of such special cases, a firm must routinely set prices above costs in order to stay in business.

The matter of costs setting the floor on price is complicated somewhat by two facts. The first is that the floor is obscured by the difficulty in determining the real cost of a product. In a highly integrated chemical complex, it is virtually impossible to calculate the true total cost of one product among the dozens or hundreds of products being produced. This is in spite of rather sophisticated cost accounting procedures and is due to the many intraplant transfers of materials and the inherent inequalities in allocating the costs of shared utilities, services, and functions. The second complication is that cost, as the price floor, tends to move. The total manufacturing cost is made up of fixed costs plus variable costs. Variable costs vary with the production rate. Since the production rate depends upon sales volume and sales volume varies with selling price, there is no simple solution to the cost/price relationship.

If the price floor is somewhat hazy, what about the price ceiling? This could be considered to be that price that maximizes current profits. Setting prices at this level would certainly appear to optimize the overall profit picture of the company. However, it seems to make better sense over the long term to establish prices somewhat below this maximum. The reasons for this may involve relations with customers, the government, labor unions, and stockholders. The principal reason however is competition. High prices and uncommonly high profits invite competitors to enter the market. The construction of competing production capacity invariably leads to lower prices and drastically lower profits.

12.6 CRITERIA OF PROFITABILITY

The owners of a company have entrusted funds to the management with the expectation of making a profit on their investment. Management's function is to utilize these monies to generate more money. The future of the company depends upon channeling funds into projects that will generate high earnings and improve the economic stance of the company.

Relatively large sums of capital may be involved in these projects. Chemical plants are expensive to build; since they are usually very specialized, their salvage value is low. Once the authorization to proceed has been made and the money has been committed, there is no convenient way to back out. The effects of a poor investment decision therefore can depress the firm's profit picture for a long time. It is essential that each proposed project be evaluated very carefully.

The end result of an economic study is expressed in terms of some financial yardstick or commonly recognized criteria that allow comparisons between various proposals. Actually there are several commonly used criteria. We shall consider payout, return on investment, and net present value.

A small percentage of the money available to management for investment comes from the sale of stock. The major sources are retained income, depreciation, and long-term loans. A project that recovers the invested capital quickly so that it can be reinvested in other projects is beneficial to company growth. This gives rise to the profit criteria known as *payout*, which is simply the number of years required to recover the money that has been invested in the project.

If payout were used as the only criterion, low investment, short-term projects would tend to be favored over long-term, high investment projects. Another criterion is needed that covers a longer time interval and considers the profit dollars generated per capital invested. This is the return on investment concept. *Return on investment* is defined as the average annual profit divided by the total capital investment. Choosing a realistic time interval for the economic study is very important since it can affect the calculated return on investment.

Payout and return on investment are both valuable profit criteria; they are simple, easily calculated, and widely understood. They do have some shortcomings, however, and sometimes do not give the best answer when used to compare the relative value of competing proposals. This usually occurs when the projects differ in the timing of cash expenditures and cash income. Payout only considers time for a short period. Return on investment considers only dollars expended and dollars returned over the total designated project life.

The discounted cash flow method is an evaluation technique based upon the fact that money has a value that is time related. Two profit criteria that evolve from the discounted cash flow method are the net present value and the discount rate. The significance of these two criteria is discussed later in this chapter.

12.7 BASES FOR THE STUDY

One of the first and most important tasks in conducting an economic analysis is establishing the bases for the study. The bases used in a particular study are the variables that have been fixed or the assumptions that have been made for the particular study. For the results of a study to be significant and valid, the bases must be realistic. There are three main areas of concern:

1. The proposed project itself should represent the best possible choice consistent with realistic technological and marketing considerations.
2. The alternative to the proposal should be the most realistic alternative.
3. The time period considered in the study should be chosen carefully.

The information to be used in the study is normally supplied by a number of different groups within the company. A preliminary review should establish that it is realistic and mutually consistent. For example, the process engineering group might base its design for a new plant on a production capacity of 1 million lb/month because this size is considered necessary to achieve the economies of scale. If the marketing department is projecting sales of 0.5 million lb/month, there is an obvious problem, and some adjustments will be needed to reconcile the discrepancy. On the other hand, the sales forecast might be for 0.5 million lb, increasing to 1 million lb/month by the fifth year of operation. This is a different situation, and the capacity chosen for the design work might or might not be realistic. Perhaps the study should be based upon the initial construction of a 0.5-million lb plant followed by a later expansion. A number of other variations might be considered. The point is simply that not only should the bases be consistent with all of the information on hand, but also they should represent the optimum conditions for the project studied.

12.8   ESTABLISHING THE STUDY BASES

As described above, one of the most critical stages in an economic analysis is establishing the bases for the study. The decisions and value judgments made here are frequently not discussed in the economic study report, but they often can swing the outcome of the study one way or the other. The following example, which describes the determination of a raw material cost, illustrates the type of analysis used in establishing the bases.

The proposed project to be studied is the production of a new agricultural chemical. Field testing and the gathering of data for government registration is being completed and appears most encouraging. An economic study is required to back up the request for authorization to build a production plant.

One of the starting materials for the new product is 2-chloro-4-nitrophenol (CNP), which is currently being manufactured by the same company. The existing CNP plant has a production capacity of 400,000 lb/month. Sales of CNP have been averaging 200,000 lb/month so the plant has been operating at 50% of rated capacity. Production of the new agricultural chemical would utilize 200,000 lb/month of CNP. It would appear that there is a good fit and the proposed project would have a secondary beneficial effect of bringing the CNP plant up to capacity. One of the questions that must be answered in establishing the bases for the study of the proposed project is: What cost should be charged for the 200,000 lb/month of CNP?

There are a number of answers to this question. Among the possibilities that might appear to be reasonable are:

1. The difference in total costs between operating the CNP plant at 400,000 and 200,000 lb/month.
2. One-half the total cost of operating the CNP plant at 400,000 lb/month.
3. The estimated cost of manufacture in an imaginary new plant with a capacity of 200,000 lb/month.
4. A prorated portion of the estimated cost of manufacture in a new or expanded CNP plant; for example, one third of the estimated cost in a 600,000 lb/month plant.
5. The current market price, less selling and shipping costs.

Which one of these five possibilities is most reasonable? The answer is that it all depends! It depends on what would be most likely to happen to the

CNP plant in the future, in the event that the proposed project is *not* approved. Using this philosophy then:

1. Would be the best choice if no change is expected in the CNP market and it is judged that there is no possibility of finding any other use for the material.
2. Would be more reasonable if no change is foreseen in the CNP external market, but it is considered likely that another internal use will develop in the future.

3-4. Would be considered if it is felt that the CNP market will grow and commitment of this material to this project now will require construction or expansion of production facilities sometime in the future.

5. Would be the choice if it is known that new facilities would never be built, but the CNP market will most likely grow. The commitment of raw material to the project will thus eventually result in lost sales.

The point of this example is to illustrate the penetrating analysis that is frequently required in using the concept of alternatives. The proposed project is the manufacture of a new product. The alternative is to *not* manufacture it. What are the true consequences of *this* alternative? The selection of the most likely course of events in the future is not as simple and straightforward as it might first appear.

12.9 TYPES OF ECONOMIC STUDIES

In order to compare objectively the economic values of various proposals, it is necessary to standardize the evaluation procedures. However, there is such a variety of proposals and questions that one single standardized procedure would be impractical. A study made to determine the optimum inventory levels for ethylene glycol would be inherently different from a study to consider the value of constructing a plant to manufacture a new textile preservative.

The problem is to achieve some measure of standardization while assuring that the treatment used is pertinent and realistic for each proposal or problem. Many companies have evolved a limited number of types of studies. Which type to use in a particular situation depends to a great extent on the type of decision that will be made on the bases of the completed study. Some of the kinds of decisions that might be involved are:

1. Deciding to invest capital in a new plant to manufacture a new product.
2. Deciding to expand the capacity of an existing plant.
3. Selecting a process for a new product.
4. Establishing a price for a new product.
5. Changing the price of an existing product.
6. Discontinuing an existing product.
7. Selecting equipment.
8. Determining optimum inventory levels.
9. Selecting a plant site.

In considering the kind of decision that will be made, several factors are important. Is the expenditure of capital involved? What is the time interval of the decision? A pricing study on an existing product involves no investment of capital. The time interval of the decision is indeterminate but short-term from the standpoint that even though undesirable, any bad decisions can be corrected by another price change. In contrast, the new product study involves capital investment and a long time period. Once the investment is approved and the new plant is built, there is no way to undo a bad decision.

In this chapter we limit our concerns to studies used in evaluating new product proposals. Two types of studies are considered—the return on investment (ROI) study and the discounted cash flow (DCF) study.

The ROI study is a long-term study that compares the alternatives of investing capital and making a product with not investing capital and not making the product. It is designed so that the profit criteria of payback and return on investment are directly comparable with studies of other proposed products. To this end the direct capital, indirect capital, depreciation schedule, plant size, sales volume, and selling price are treated in a standardized manner. The profit criteria are widely recognized and easily understood. The evaluation procedure is simple and straightforward.

A DCF study method can be used on a number of types of problems. When used on the analysis of a proposed new product, it is similar to the ROI study from the standpoint that it allows comparisons between various proposals. It differs in that it is based upon the time value of money and therefore considers in detail the timing of all cash expenditures and cash income. It yields a more sophisticated and more accurate analysis than is possible with the ROI study approach.

12.10 THE RETURN ON INVESTMENT STUDY

ROI studies should answer the question: Is the return on investment from this project high enough to commit the company to making this product on a long-term basis? These studies are based upon optimum sized plants with capacity and sales volume in balance. Sales revenue is based upon long-term average selling price. Capital requirements, both direct and indirect, are expressed in current dollars. Existing facilities are thus estimated in terms of replacement value. Depreciation is for a 10-year period and is based upon the straight-line method.

Example 12.1   Return on Investment Study—Fire Retardant 647

A convenient worksheet for a ROI study is depicted in Fig. 12.1a. As an example of the method, consider the evaluation of a proposed new product, Fire Retardant 647.

Step 1. In Fig. 12.1b the various bases, estimates, and projections have been entered on the worksheet. The marketing department has furnished the projections for the long-term sales volume and selling price, the packaging requirements, and the estimated selling expenses. The process engineering group has completed a process design and provided estimates for capital, raw material and utility requirements, maintenance costs, and labor usage. The unit costs for raw materials, labor, and packaging are based upon current quoted prices or cost accounting data.

Step 2. In Fig. 12.1c, the raw material requirements and costs have been extended and totaled. The *factory overhead* has been estimated as 28% of the *labor cost. Depreciation* has been calculated on the basis of a 10-year straight-line depreciation of the $1,200,000 *direct fixed capital. Taxes and insurance* are estimated on the bases of 2% of the *direct fixed capital* per year. *Production costs* are then totaled. *Research costs* and *general and administrative costs* are then estimated on the bases of 14% and 3% of the production costs. These costs are then totaled to yield the *total cost* of manufacture.

Step 3. The *total cost* of manufacture calculated at this point has been based upon a production rate of 100% of capacity. In order to separate this total cost into fixed cost and variable cost elements, it is convenient to recalculate the costs at a production rate of 0% of capacity. Imagine that the plant is shut down temporarily. Raw material, packaging, and utility costs would decrease to zero. Other costs would decrease by factors based upon the accounting history of the particular company. Typical factors for the fixed/variable components of cost items are included in Table 12.1. This calculation is shown in Fig. 12.1d.

Step 4. The completed worksheet is shown in Fig. 12.1e. The difference between the *total cost* and the *fixed cost* is the *variable cost*. The net sales of

PROJECT	

BASES

(1) CAPACITY :   / YR   PROD'N RATE :   % CAP

(2) SELLING PRICE :

(3) RAW MATERIALS :   UNIT RATIO   UNIT COST

      A

      B

COST ESTIMATE	$ / YR
RAW MATERIALS	
TOTAL RAW MATERIALS	
LABOR	
MAINTENANCE	
UTILITIES	
FACTORY OVERHEAD	
DEPRECIATION	
TAXES & INS.	
PACKAGING	
PRODUCTION COST	
RESEARCH	
SELLING EXPENSE	
G & A	

TOTAL COST		$ / LB
Fixed Cost		
Fixed Cost - Dep'n.		
Variable Cost		

CAPITAL ESTIMATE	$ M
DIRECT FIXED	
INDIRECT FIXED	
TOTAL FIXED	
WORKING	
TOTAL	

SUMMARY

NET SALES	
COST	
PROFIT Before Tax	
After Tax	
DEPRECIATION	

PROFIT CRITERIA

	Before Tax	After Tax
PAYBACK		
R.O.I.		
N.P.V.		

(a)

Figure 12.1   *(a)* A worksheet for a ROI study.

PROJECT	FIRE RETARDANT 647

BASES

(1) CAPACITY: 5,000,000 lbs/YR PROD'N RATE: 100 %CAP

(2) SELLING PRICE: $ 0.430/lb

(3) RAW MATERIALS:

	UNIT RATIO	UNIT COST
A	0.700 lbs A/lb 647	$0.075/lb
B	0.600 lbs A/lb 647	$0.064/lb

COST ESTIMATE	$/YR	$/LB
RAW MATERIALS		
TOTAL RAW MATERIALS		
LABOR	120,000	
MAINTENANCE	99,000	
UTILITIES	56,000	
FACTORY OVERHEAD		
DEPRECIATION		
TAXES & INS.		
PACKAGING @ $0.015/lb		
PRODUCTION COST		
RESEARCH		
SELLING EXPENSE	200,000	
G & A		
TOTAL COST		
Fixed Cost		
Fixed Cost - Dep'n.		
Variable Cost		

CAPITAL ESTIMATE	$M
DIRECT FIXED	1,200
INDIRECT FIXED	-0-
TOTAL FIXED	1,200
WORKING	
TOTAL	

SUMMARY

NET SALES	
COST	
PROFIT Before Tax	
After Tax	
DEPRECIATION	

PROFIT CRITERIA

	Before Tax	After Tax
PAYBACK		
R.O.I.		
N.P.V.		

(b)

Figure 12.1 (*b*) Base estimates and projections for Fire Retardant 647 project.

PROJECT	FIRE RETARDANT 647

BASES

(1) CAPACITY: 5,000,000 lb/YR PROD'N RATE: 100 %CAP

(2) SELLING PRICE: $0.430/lb

(3) RAW MATERIALS:

	UNIT RATIO	UNIT COST
A	0.700 lb A/lb 647	$0.075/lb
B	0.600 lb B/lb 647	$0.064/lb

COST ESTIMATE	$/YR	$/LB
RAW MATERIALS		
A(5,000,000)(0.700)(0.075)	262,500	
B(5,000,000)(0.600)(0.064)	192,000	
TOTAL RAW MATERIALS	454,500	
LABOR	120,000	
MAINTENANCE	99,000	
UTILITIES	56,000	
FACTORY OVERHEAD	33,600	
DEPRECIATION	120,000	
TAXES & INS.	24,000	
PACKAGING @ $0.015/lb	75,000	
PRODUCTION COST	983,000	
RESEARCH	137,620	
SELLING EXPENSE	200,000	
G & A	29,490	
TOTAL COST	1,350,110	
Fixed Cost		
Fixed Cost - Dep'n.		
Variable Cost		

CAPITAL ESTIMATE	$M
DIRECT FIXED	1,200
INDIRECT FIXED	-0-
TOTAL FIXED	1,200
WORKING	
TOTAL	

SUMMARY	
NET SALES	
COST	
PROFIT Before Tax	
After Tax	
DEPRECIATION	

PROFIT CRITERIA

	Before Tax	After Tax
PAYBACK		
R.O.I.		
N.P.V.		

(*c*)

Figure 12.1 (*c*) Raw material costs for Fire Retardant 647 project.

PROJECT	FIRE RETARDANT 647

BASES

(1) CAPACITY: 5,000,000 lb/YR PROD'N RATE: 0 %CAP

(2) SELLING PRICE: $0.430/lb

(3) RAW MATERIALS:

	UNIT RATIO	UNIT COST
A	0.700 lb A/lb 647	$0.075/lb
B	0.600 lb B/lb 647	$0.064/lb

COST ESTIMATE	$/YR	$/LB
RAW MATERIALS		
TOTAL RAW MATERIALS	-0-	
LABOR 20%	24,000	
MAINTENANCE 50%	49,950	
UTILITIES 0%	-0-	
FACTORY OVERHEAD 35%	11,781	
DEPRECIATION 100%	120,000	
TAXES & INS. 100%	24,000	
PACKAGING 0%	-0-	
PRODUCTION COST	229,731	
RESEARCH 100%	137,620	
SELLING EXPENSE 80%	160,000	
G & A 0%	-0-	
TOTAL COST	527,351	
Fixed Cost	527,351	
Fixed Cost - Dep'n.		
Variable Cost		

CAPITAL ESTIMATE	$M
DIRECT FIXED	1,200
INDIRECT FIXED	-0-
TOTAL FIXED	1,200
WORKING	
TOTAL	

SUMMARY	
NET SALES	
COST	
PROFIT Before Tax	
After Tax	
DEPRECIATION	

PROFIT CRITERIA	Before Tax	After Tax
PAYBACK		
R.O.I.		
N.P.V.		

(d)

Figure 12.1 (*d*) Fixed/variable component costs.

BASES

① CAPACITY: 5,000,000 lb/YR   PROD'N RATE: 100   % CAP

② SELLING PRICE: $0.430 /lb

③ RAW MATERIALS: UNIT RATIO   UNIT COST
   A   0.700 lb A / lb 647 $0.075/lb
   B   0.600 lb B / lb 647 $0.064/lb

COST ESTIMATE	$/YR	
RAW MATERIALS		
A (5,000,000) (0.700) (0.075)	262,500	
B (5,000,000)(0.600) (0.064)	192,000	
TOTAL RAW MATERIALS	454,500	
LABOR	120,000	
MAINTENANCE	99,000	
UTILITIES	56,000	
FACTORY OVERHEAD	33,600	
DEPRECIATION	120,000	
TAXES & INS.	24,000	
PACKAGING @ $0.015/lb	75,000	
PRODUCTION COST	983,000	
RESEARCH	137,620	
SELLING EXPENSE	200,000	
G & A	29,490	$/LB
TOTAL COST	1,350,110	0.270
Fixed Cost	527,351	0.105
Fixed Cost - Dep'n.	407,351	—
Variable Cost	822,759	0.165

CAPITAL ESTIMATE	$M
DIRECT FIXED	1,200
INDIRECT FIXED	– 0 –
TOTAL FIXED	1,200
WORKING	358
TOTAL	1558

SUMMARY

NET SALES	2150
COST	1350
PROFIT Before Tax	800
After Tax	416
DEPRECIATION	120

PROFIT CRITERIA

		Before Tax	After Tax
PAYBACK	YRS	1.9	3.7
R.O.I.	%	51.3	26.7
N.P.V.		—	—

(e)

Figure 12.1   (e) Completed worksheet.

$2,150,000 is the product of sales of 5,000,000 lb/year and the projected long-term selling price of $0.430/lb. The working capital required is estimated as $\frac{2}{12}$ or 16.7 % of the annual sales volume. The difference between *net sales* revenue and cost is *profit before tax*. At a tax rate of 48 % the *profit after* tax is 52 % of this figure. The *payback* period is calculated by dividing the total fixed capital investment by the annual profit. The return on investment is calculated by dividing the profit by the total capital investment.

Payback and *return on investment* are important profit criteria but the bases used in the study must be kept in mind. One important assumption was that production rate and sales volume would be equal to plant capacity. Any sales volume less than this level would reduce the profitability of the proposal. A *break-even chart* gives some measure of the profitability at less than optimum conditions. Such a chart, for this study, is shown in Fig. 12.2. The data from the worksheet used to construct this chart are: total cost, fixed cost, and net sales. The chart indicates a break-even point at 40.5 % of capacity or a production and sales volume of 2,025,000 lb/year if the selling

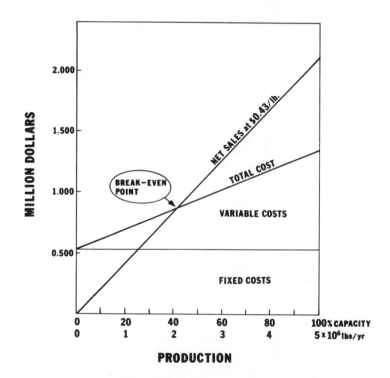

Figure 12.2 A break-even chart.

price is the projected $0.43/lb. This is the lowest production rate at which the operation would just cover costs. It should be noted that it is *not* the point at which the plant would be shut down. Operation below the break-even point, say at 20% of capacity, would result in a loss

$$5,000,000 \text{ lb} \times 0.20 \times \$0.43/\text{lb} = \$430,000 \text{ net sales}$$

$$\$(82,759 \times 0.20) + \$527,351 = \underline{\$691,903} \text{ total cost}$$

$$\$261,903 \text{ net loss}$$

But if the plant were shut down, the net loss would be greater than this because the fixed costs of $527,351 would continue.

It is of course quite unlikely that the selling price of $0.43/lb would be maintained with the plant operating at 20% of capacity. It is more probable that the price would be lowered in an attempt to increase the sales revenue by increasing the sales volume. The relation of sales volume to selling price is the elasticity of demand for the product and depends on the particular market and the competitive situation. If in our example sales could be increased to 80% of capacity, 4,000,000 lb/year, by dropping the price to $0.30/lb, then

$$4,000,000 \text{ lb} \times \$0.30/\text{lb} = \$1,200,000 \text{ net sales}$$

$$\$(822,759 \times 0.80) + \$527,351 = \underline{\$1,185,558} \text{ total cost}$$

$$\$ \quad 14,442 \text{ net profit before tax}$$

The sensitivity of return on investment to changes in the selling price is depicted in Fig. 12.3. This chart is based upon production at 100% of

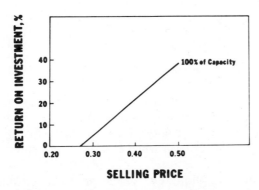

Figure 12.3   Return on investment versus selling price.

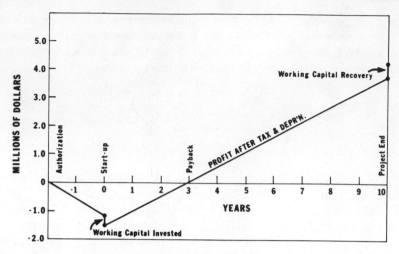

Figure 12.4   Cash flow diagram.

capacity. The return on investment at lower production levels could be indicated by a family of lines corresponding to 80, 60, 40 % of capacity.

A diagrammatic view of the cash flow into and out of this project is depicted in Fig. 12.4. This is a somewhat simplistic diagram owing to the basic assumptions of the ROI study. Because we assumed a static situation with regard to production rate, selling price, and costs, each year is the same in regard to profitability. The diagram shows a 2-year construction phase and the commitment of the working capital at the time of plant start-up. Over the 10-year project life, the investment is paid off and a profit generated. The rate is equal to the annual depreciation charges and the annual net profit. At the end of the 10-year period, if operations are shut down, the working capital is returned to the company capital fund.

12.11   THE DISCOUNTED CASH FLOW STUDY

The concepts of return on investment and payout have been used as criteria for evaluating the economics of investment proposals for many years. The long-term growth and profitability of many chemical companies attest to the effectiveness of these criteria when tempered with shrewd management judgment.

Payout measures the "liquidity" of the proposed venture, or the time required to recover the investment capital. Consideration of payout alone neglects the overall estimated project life and the amount and timing of the project income after the initial investment has been paid off. The return on investment is the percentage relation between the average annual profit earned by the project and the capital investment required. The ROI criteria do not consider the timing of the cash outlays and the cash incomes. Since they are based upon the *average* rate of return over the life of the project, they fail to distinguish between projects that have a high initial return that then declines from those that have a constant return or those that have low initial returns that gradually increase. Also these criteria fail to account for the amount of capital that is recovered each year over the life of the project. Since the amount of unrecovered capital is steadily decreasing each year, a rate of return calculated on the basis of the initial investment is unrealistically low.

The *discounted cash flow* (DCF) approach is based upon the proposition that money makes money. Money that is on-hand right now can be invested in a profitable venture and thus earn more money. Money that is to be received sometime in the future cannot be invested until it is received. It has no earning power until it is available. The value of money varies with time. Note that the concern here is the earning power of the money. Any effects of inflation or investment risks are not directly considered. Note also that this decrease in the value of money with time applies to expenditures as well as income. The factor used in calculating this decrease in value is the discount factor. It is essentially an interest rate.

In the discounted cash flow method, a time value is placed upon all flows of cash into or out of the company or, more specifically, the project being studied.

The discounted cash flow method as used in the chemical process industry has been adapted from methods used by financial institutions such as banks and insurance companies. The method requires somewhat more calculating time and effort, but the basic information needed is similar to that required in a ROI study. Thus a project life period must be defined; projected sales volumes and selling prices must be determined; cost factors and capital requirements must be estimated. More detail is required in some of these estimates. Sales revenue in the ROI study is based upon production (and sales) at optimum capacity and long-term average selling price. In the DCF study sales and selling price projections are required for each year of the project. Generally in a DCF study, long-term averages are not sufficient, the timing as well as the amount of every project expenditure and income is important.

It might seem reasonable that an approach that takes into account the difference between near dollars and distant dollars gives a more accurate picture of profitability. However, we have not yet described the calculations that yield the results described. Let us reexamine the Fire Retardant 647 project. The method we use will yield a *net present value* for the project. The net present value simply represents the total after tax profit dollars adjusted to current value by the appropriate discount factors.

The underlying philosophy of the method is the thought that the capital investment is being loaned to the project by the company. This investment will be returned to the company during the project life through depreciation charges. The use of this capital is not free, however. The company's capital resource is not a huge fund of idle cash ready for investment in a worthwhile project. The company obtained its capital in the real world and it has a cost. It might come from long- or short-term loans, sale of bonds, or stock issues. In practice the capital comes from all of these sources, each with an associated cost. An overall long-term cost of capital can be determined, however, and this is used as the basis for the discount factors used in the net present value calculation. In our example, we shall assume that the company has a cost of capital corresponding to an annual interest rate of 10%.

Example 12.2   Discounted Cash Flow Study—Fire Retardant 647

The sales forecast for Fire Retardant 647 is presented in Table 12.3. It is projected that sales will build to a maximum corresponding to plant capacity and then gradually decrease as competitive products are introduced into the market. The selling price, also responding to competitive pressures, is forecasted to decrease over the 10-year period.

Table 12.3   Sales Forecast

Year	Price $/lb	Volume lbs	Revenue $
1	0.55	1,000,000	550,000
2	0.55	2,500,000	1,375,000
3	0.53	4,000,000	2,120,000
4	0.47	4,500,000	2,115,000
5	0.44	5,000,000	2,200,000
6	0.42	5,000,000	2,100,000
7	0.40	5,000,000	2,000,000
8	0.34	4,500,000	1,530,000
9	0.30	4,000,000	1,200,000
10	0.30	3,500,000	1,050,000

**Table 12.4   Depreciation Schedule—Method: Sum of
Years Digits; Fixed Capital: $1,200,000**

Year	Factor	Amount
1	18.18	218,160
2	16.37	196,440
3	14.55	174,600
4	12.73	152,760
5	10.91	130,920
6	9.09	109,080
7	7.27	87,240
8	5.46	65,520
9	3.64	43,680
10	1.80	21,600
	TOTAL	1,200,000

The capital requirements for the project are the same as previously estimated: $1,200,000 for direct fixed capital and $358,000 for working capital. In the ROI study depreciation was handled by the straight-line method. In the DCF study the depreciation method actually to be used by the company must be used in the study. In this example the sum of years method is used for the direct fixed capital (Table 12.4). This money is assumed to be invested one year before plant start-up. The working capital is assumed to be committed at the time of start-up and to be returned at the end of the 10-year project. The discount factors for the 10% cost of capital are listed in Table 12.5.

Step 1.   Fig. 12.5 depicts a worksheet for the net present value calculation. The figures in Column A are the sales revenue projections from the sales forecast. The total cost of manufacture varies because the amount of product produced and sold varies from year to year.

Step 2.   The variable and fixed components of the total cost are listed in Columns B and C. These components were calculated at 100% capacity in the ROI study. See Fig. 12.1d. Note that the fixed cost component $407,351 does not include depreciation, which is handled separately in the DCF method. The variable costs listed in Column B are the products of the variable cost at 100%, $822,759 multiplied by the appropriate production rate expressed as percent of capacity. It should be noted that projected increases in costs due to inflation are not normally considered. The rationale is that selling prices

Table 12.5 Discount Factors

$$\text{Present Value} = \frac{1}{(1 + r)n}$$

n = years
r = interest rate

Year	10% Factor
-1	1.100
0	1.000
1	0.909
2	0.826
3	0.751
4	0.683
5	0.621
6	0.564
7	0.513
8	0.466
9	0.424
10	0.385

will increase at a comparable rate. However, if it is known through contracts or agreements that costs such as labor or raw materials will be increased, these increases should be considered in the study.

Step 3. The cash earnings listed in Column D result from revenue (Column A) less costs (Columns B and C). Column E lists the depreciation amounts from Table 12.4.

Step 4. The taxable income in Column F is the cash earnings, Column D, less the depreciation, Column E. Column G lists the annual amount of tax.

Step 5. The cash flow, Column H, is calculated by subtracting the annual income tax amount, Column G, from the cash earnings, Column D.

Step 6. The annual discounted cash flow, Column I, is calculated by multiplying each annual cash flow by the appropriate discount factor in Table 12.5. The net present value of the project is the total of these discounted cash flows.

The net present value of $756,640 indicates that the project would be a worthwhile venture. The project could borrow the required direct and working capital at the 10% interest rate, return the investment, and show a profit. The total dollars profit throughout the 10-year project is $2,365,443. Profit expressed in terms of today's dollars is $756,640.

PROJECT: FIRE RETARDANT 647
Depreciation Method: Sum of Years, 10 Year Period
Discount Rate: 10 %
Tax Rate: 48 %

Net Present Value
+ $756,640

Year	A Revenue	B Variable Cost	C Fixed Cost	D Cash Earnings	E Depreciation	F Taxable Income	G Tax 48%	H Cash Flow	Discounted Cash Flow
-1	—	—	—	—	—	—	—	-1,200,000	-1,320,000
0	—	—	—	—	—	—	—	-358,000	-358,000
1	550,000	164,552	407,351	-21,903	218,160	-240,063	-115,230	93,327	84,834
2	1,375,000	411,380	407,351	556,269	196,440	359,829	172,718	383,551	316,813
3	2,120,000	658,207	407,351	1,054,442	174,600	879,842	422,324	632,118	474,720
4	2,115,000	740,483	407,351	967,166	152,760	814,406	390,915	576,251	393,579
5	2,200,000	822,759	407,351	969,890	130,920	838,970	402,706	567,184	352,221
6	2,100,000	822,759	407,351	869,890	109,080	760,810	365,189	504,701	284,651
7	2,000,000	822,759	407,351	769,890	87,240	682,650	327,672	442,218	226,858
8	1,530,000	740,483	407,351	382,166	65,520	316,646	151,990	230,176	107,262
9	1,200,000	658,207	407,351	134,442	43,680	90,762	43,566	90,876	38,531
10	1,050,000	575,931	407,351	66,718	21,600	45,118	21,677	45,041 358,000	155,171
							TOTAL	2,365,443	756,640

Figure 12.5   A worksheet for the net present value calculation.

Another popular method of expressing the worth of a project is based upon a slightly different calculation. In this approach the discount factor or interest rate that the project could afford to pay and still fully recover the capital investment at the end of the 10-year period is determined. This discounted cash flow rate of return is then the financial yardstick.

12.12   REFERENCES

1. J. Happel and D. G. Jordan, *Chemical Process Economics*, 2nd ed., Dekker, New York, 1975.

2. M. N. Geragosian (Ed.), *Chemical Marketing Research*, Reinhold, 1967.

3. Herbert Popper and staff of Chemical Engineering, Editors, *Modern Cost Engineering Techniques*, McGraw-Hill, New York, 1970.

4. F. A. Lowenheim and M. K. Moran, *Faith, Keyes and Clarke's Industrial Chemistry*, 4th ed., Wiley-Interscience, New York, 1975.

5. R. E. Kirk and D. F. Othmer, *Encyclopedia of Chemical Technology*, 2nd Ed., Wiley, New York, 1969.

12.13 PROBLEMS

These problems are included to provide practice in the calculations used in economic studies. The cost factors provided are intended to be realistic but are not based upon any real or proposed production facility. It is suggested that the reader use the cost factors included in Table 12.1 and assume a 10-year project life. Working capital requirements may be estimated to be $\frac{2}{12}$ of the annual sales revenue. In the net present value calculations use the sum of years depreciation method (Table 12.4) and assume a cost of capital of 10% (Table 12.5).

1. It is proposed that a 6,000,000 lb/year plant be built to produce ANTI-OXIDANT Z-100. The direct fixed capital requirements are estimated at $1,000,000. There are no indirect fixed capital requirements. The raw material requirements are:

Raw Material	Unit Ratio (lb/lb)	Unit Cost ($/lb)
A	0.690	0.0737
B	0.369	0.1000
C	0.460	0.0200

   Estimated costs, at capacity production, include:

Labor	$115,000/year
Maintenance	$100,000/year
Utilities	$55,600/year
Selling costs	$210,000/year
Packaging	$0.015/lb

   (Other costs should be estimated using Table 12.1.)

   Calculate the total cost of production, the fixed cost, and the variable cost. Estimate the break-even point at a selling price of $0.38/lb. Plot the relation between the break-even point and selling price over the range $0.25 to $0.50/lb.

2. Styrene is produced by the dehydrogenation of ethyl benzene in a vapor-phase continuous reactor. Using the ROI approach, evaluate the economics of a proposed 500,000,000 lb/year plant. The yield of styrene is 92.66% of theory; the cost of ethyl benzene is $0.080/lb. Assume a negligible cost for catalyst replacement and no credit for the by-product, hydrogen. The capital estimates are: direct fixed, $35,000,000; indirect fixed, $23,000,000.

Labor	$650,000/year
Maintenance	$1,700,000/year
Utilities	$6,550,000/year
Selling cost	$300,000/year
Packaging	$500,000/year

The projected selling price of the product is $0.18/lb. Calculate the break-even point, the payback period, and the return on investment, both before and after taxes. Using prices ranging from $0.14 to 0.22/lb, calculate and plot the effect of selling price upon return on investment. Determine the sensitivity of return on investment to errors in various estimates by considering the effect of the following:

(a)   Underestimating labor costs by 10%.
(b)   Underestimating capital costs by 10%.
(c)   Overestimating sales volume by 10%.
(d)   Overestimating selling price by 10%.

3. Two commercially important processes for the manufacture of acrylic acid are the air oxidation of propylene and the reaction of acetylene with

Bases	Propylene Process	Acetylene Process
Capacity	150,000,000 lb/year	150,000,000 lb/year
Selling price	$0.14/lb	$0.14/lb
Raw material costs		
Propylene	$0.03/lb	—
Acetylene	—	$0.08
Carbon monoxide	—	$0.002
Air, water	N.C.	N.C.
Raw materials unit ratio		
Propylene	0.88	—
Acetylene	—	0.42
Carbon monoxide	—	0.36
Capital estimates		
Direct fixed	$10,100,000	$9,500,000
Indirect fixed	$6,800,000	$8,300,000
Cost estimates		
Labor	$495,000/year	$405,000/year
Maintenance	$600,000/year	$300,000/year
Utilities	$1,680,000/year	$1,200,000/year
Selling expense	$320,000/year	$320,000/year
Packaging	$150,000/year	$150,000/year

carbon monoxide and water. (*a*) Using the bases and estimates listed, calculate the payback period and return on investment for each process; (*b*) assume that the cost of all raw materials is doubled. Calculate the effect upon the return on investment for each process.

4. There are two important commercial processes for the production of propylene oxide from propylene. Many existing plants are based upon the old chlorohydrin route. Ethylene oxide was originally manufactured by this process. With the advent of the direct oxidation route to ethylene oxide, many of these plants were converted to the production of propylene oxide. The market for this product continues to grow at a good rate, principally due to the growth in polyurethane polymers. A more recent route to propylene oxide is the Oxirane process which is an indirect oxidation involving isobutane. The co-product in this reaction sequence is isobutylene or tert-butyl alcohol. It is suggested that the reader review the descriptions of the technology and chemistry of these processes.[4,5]

Complete an economic study comparing these two processes on the basis of return on investment and new present value.

Market projections (average selling price $0.23/lb):

Year	Sales Volume	Selling Price
1	100×10^6 lb/year	$0.20/lb
2	120	0.21
3	150	0.22
4	180	0.23
5	200	0.24
6	200	0.24
7	200	0.23
8	200	0.22
9	200	0.21
10	200	0.20

Capital estimates—capacity 200,000,000 lb/year:

	Chlorohydrin	Oxirane
Direct fixed	$1,120,000	$20,600,000
Indirect fixed	$1,000,000	$6,000,000

Cost estimates:

Raw Materials	Chlorohydrin		Oxirane	
	Unit ratio (lb/lb)	Unit cost ($/lb)	Unit ratio (lb/lb)	Unit cost ($/lb)
Propylene	0.98	0.08	0.80	0.08
Chlorine	1.43	0.04	—	—
Lime	1.28	0.02	—	—
Isobutane	—	—	2.50	0.023
Air	—	—	No charge	
Isobutylene credit	—	—	2.00	(0.023)

	Chlorohydrin	Oxirane
Labor	$800,000/year	$500,000/year
Maintenance	200,000	1,200,000
Utilities	1,100,000	2,300,000
Selling expense	120,000	120,000
Packaging	$0.005/lb	$0.005/lb

13

RESEARCH

WHERE ARE WE GOING?

Research and development (R&D) have been described and discussed by a number of writers.[1,2,3] There are many different concepts of these two activities as they are practiced in the chemical industry. Our concern here is not a rigorous definition of the terms, but a consideration of the research function from the standpoint of its relation to the overall goals of the company.

13.1 RESEARCH SPENDING

The chemical industry, which is highly competitive and characterized by rapid technological changes, spends heavily for its research and development efforts. In 1975 this outlay had risen to over \$1.3 billion/year.[4] This level of spending is considerably higher than that for all other manufacturing industries when considered as a percentage of sales.[5]

It is frequently of interest to compare the research budgets of various industrial firms. This is most often done on the basis of percentage of annual sales. A superficial review, however, can be misleading. In some firms, for example, the costs of the patent department are included in research; in others, it is a separate budget item. Similar inconsistencies exist in the handling of the costs of technical support to the sales department and to the construction and operation of pilot plants. A compilation of data reflecting the R&D spending as a percentage of sales volume thus shows a wide diversity among individual firms. In 1974 companies with annual sales over \$500 million spent between 0.9 and 5.0% of sales. For smaller companies, with sales between \$50 million and \$500 million, the range was 0.3 to 5.6%. In addition to differences in accounting methods and corporate philosophies, these wide ranges also reflect differences in the type of business in which the firms were engaged. Companies at the lower end of the expenditure range tended to be heavily engaged in large volume production of commodity products: fertilizers, industrial gases, and basic raw materials. Those listed as spending the highest amounts on research appear to be more involved in specialty products such as pharmaceuticals, pesticides, and food additives.

13.2 RESEARCH BUDGETS

There are a number of opinions regarding guidelines to use in setting the size of a research budget. Annual sales volume, profits, research staffing, competitors' R&D efforts, and the estimated costs of the research projects have all been suggested.[6]

Building up a budget by considering the costs and the merits of each individual research project would seem to be a rational, businesslike approach. Some of the practical problems that arise when this is attempted are the lack of valid evaluation criteria for the projects and the very real possibility of rather wide swings in budgeting levels from year to year. To be effective over the long term, a research organization needs a considerable amount of continuity. It is very difficult to expand the effort quickly and even more difficult to reduce it on short notice. Either action results in a marked decrease in the efficiency and productivity of the research group. For this reason, many firms tend to establish their budgets on a historical foundation; considering the present staffing, allowing for salary increases, any planned additions to the staff, and other increases in costs. The chemical industry generally has been increasing the size of research staffs by 3 to 4% per year. Manpower is the major item of R&D costs, accounting for 60 to 70% of the total budget.

Sales have long been used as a yardstick for determining the amount to spend on research. A number of companies budget a fixed percentage of annual sales or use this as a guide in establishing a ceiling on research spending. One advantage to this approach is that it automatically relates the research budget to some measure of the funds available. The major advantage, however, is that total sales volume tends to follow a reasonably steady growth curve, thus avoiding large swings in the research effort. To further smooth out possible fluctuations, a running three-year average of sales has sometimes been used.

Philosophically, there are a number of questions that can be raised regarding the use of sales as a guideline. Current sales, after all, reflect past research; what then is the logical relationship to current research? At a time when sales are decreasing, should this not be the very time that the research effort should be increased rather than decreased? If research spending is determined solely by factors outside the control of research management, has it not lost a voice in one of the most important decisions affecting research—the level of effort? In the practical application, however, the advantages of using a guideline that assures continuity of effort appear to override the apparent logic of such arguments.

In the final analysis, perhaps, no one factor establishes the research budget. The number of promising projects, the personnel and facilities available, overall business conditions, research efforts of close competitors, and anticipated income for the year are all reflected, in varying degrees, in the final budgeting decisions.

13.3 THE RESEARCH AND DEVELOPMENT FUNCTION

The research and development function of the typical modern chemical company includes a number of different activities. Following are some of those that commonly are the responsibility of the research department:

1. Process improvement.
2. Process development.
3. Product development.
4. Applications development.
5. Technical support to sales.
6. Background research.
7. Recruiting and training.
8. Miscellaneous services.

13.4 PROCESS IMPROVEMENT

This category includes a number of activities that represent technical support to the production department on an existing process. In severe operating problems, this can be an emergency situation and there is a great deal of pressure to solve the problem quickly and resume production. In other cases the time element is less compelling and the main objective is to improve the profits derived from the process.

When there is an increasing demand for a product, process improvement might involve "debottlenecking"—analyzing a process to determine how to most economically increase plant capacity. Since it is seldom that all of the components of the plant are in balance, this might involve identifying the capacity-limiting steps in the process and recommending equipment modifications or replacement. Occasionally, a change in operating procedures can shorten the time required for a processing step.

Many times the basic objective is to improve the profitability of the plant by decreasing operating costs. The specific approach to be taken might result from a consideration of the chemistry of the process or from a review of the cost accounting data for the particular plant. Such reviews might suggest which processing steps or which cost items are most susceptible to improvement. The problems encountered might typically include: attempts to increase yields by optimizing reaction conditions, utilizing lower cost raw materials, or investigating changes in operating procedures that would lower the energy requirements of the process.

Another type of process improvement study has as its objective the improvement of product quality in order to increase or maintain the sales volume. This might be related to establishing a new use for an existing product or simply attempting to capture a greater share of an existing market.

Probably the most important process improvement studies are those that concern the environments—both those of the individual operators in the plants and those of the surrounding community. All responsible chemical manufacturers are concerned with these problems and are allocating large portions of their research expenditures to them. Our knowledge of air pollution, water pollution, and industrial hygiene is changing very rapidly. These considerations are tied so closely to the economic foundations of our culture that we cannot afford to make decisions based upon platitudes, suppositions or just plain bad research. It is anticipated that in future years an increasing portion of our research effort—in terms of both magnitude and quality—will be addressed to these very real problems.

One of the satisfying aspects of process improvement assignments is that the value in terms of the environment, costs, or profits can generally be calculated with reasonable certainty and the payoff usually comes quickly.

13.5 PROCESS DEVELOPMENT

Process development assignments include devising a new process for an existing product as well as developing a process for a new product. In the first instance, the existing product may or may not be currently produced by the firm. But it is being produced and sold by someone. The market is thus usually quite well defined in terms of use pattern, price structure, and product specifications. The objectives of the effort in relation to these points can be established with some degree of confidence. The motivation for initiating a process development project for an existing product might be the availability of new technology, a change in a raw material situation, or a marked change in sales volume which changes the economics. Examples of the successful development of new processes for existing products are the oxychlorination process for vinyl chloride from ethylene, the cumene route to phenol, and the direct oxidation route to ethylene oxide.

The development of a process for a new product is tied very closely to the development of the product itself. There are a number of unknowns and uncertainties. The product that is acceptable to the potential customer must be defined. The purity required and the sales specifications must be determined. The product might be marketed as a formulation that contains additives or inhibitors. At the start the projected sales volumes and the expected selling price are unknown. The potential customers who might be evaluating the experimental product are vitally concerned with the anticipated selling price as well as the assurance that there will be a dependable source of supply. Every factor appears to depend upon several other unknown factors and no rational, definite answers are possible at the beginning of the project. The projections regarding the product, the selling price, and the sales volume become increasingly firm as the sales development program progresses. There are two main points here. The first is that projects of this type require an unusual degree of communication and cooperation between the process development, the applications research, and the market development groups. The second is the necessity of supplying larger than laboratory quantities of the experimental product for customer evaluation. This might require making a limited number of runs in a full scale production facility or operating a pilot plant. There is an obvious reluctance to constructing a full-scale plant before the sales potential of the product has been fairly well established.

Seeking uses for by-products of an existing process is an activity that can be considered to be related to process development. The upgrading of waste materials to saleable products can result in a marked increase in profitability for the plant. Usually, additional processing operations must be devised for the isolation and purification of the by-product material. It should be noted that the pricing and marketing of such a product requires unusual care and judgment. It sometimes happens that the growth of the new product outstrips the growth of the principal product, and there is no plant or process technology available for increasing the supply to meet the new demand.

13.6 PRODUCT DEVELOPMENT

A customer buys a chemical for a particular purpose. He is paying for the effect produced by the product, not the chemical compound per se. In many cases this is a simple effect. Thus a customer may purchase lime to neutralize an acid. Many soluble alkaline materials might be used for this purpose. If the customer could reduce his costs by using sodium hydroxide or potassium hydroxide, he would do so. His real interest is not in buying lime but hydroxide ions or any other suitable species to neutralize the acid.

Ethylene glycol is used extensively as an antifreeze in automotive cooling systems. The depression of freezing point is a general effect produced by any soluble substance. Why then is ethylene glycol used almost exclusively? Consider its properties—complete water solubility, good heat transfer characteristics, very low volatility, noncorrosive, thermally stable, and low molecular weight. The latter is important because freezing point depression is a colligative property and the lower the molecular weight, the greater the effect per pound of product. It would be difficult to conceive of a substance that would combine all these desirable properties in a more economical manner. Ethylene glycol dominates the antifreeze market because of technical and economic considerations not because of clever advertising campaigns.

Other products have more complex effects and have been described as "functional products" because they perform a function—such as preventing oxidation by a chelating action, stabilizing a plastic by ultraviolet absorption, or emulsifying by surface active properties.

The area encompassed by the term product development is so vast that no attempt will be made to describe the activities involved. There are certain generalizations that can be made, however. First, the ultimate purpose is continued long-term profit to the company. Many firms keep a close watch on the performance of new products since it is a good indicator of the effectiveness of their research effort. Many report data relating the percent sales and percent profit derived from "new products," which frequently are arbitrarily defined as products introduced within the past 10 years.

It is generally true that the newer the product, the more profitable it is. The typical product has a life cycle as depicted in Fig. 13.1. After introduction a new product commonly enjoys increasing sales and profits as it is accepted by the market. At some point the profitability frequently reaches a maximum and then declines. This may be due to competing companies introducing the same product, in which case the sales and profits may plateau at some lower value corresponding to the company's share of the total market for that product. In other cases the product may be displaced by a superior, competing product and profit falls so low that the product is discontinued.

In order to maintain reasonable corporate profits, the product mix must be continuously updated—introducing new items and terminating obsolete products.

There are several other characteristics of new product development. One is the close liaison required between research and sales. It is essential that

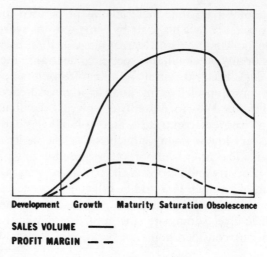

Figure 13.1   Typical product life cycle. *Source*:   Reference 7.

the producer completely understand the problems or the effect desired by the potential customer. Only when there is this close interaction with the customer can critical mistakes be avoided. Mistakes such as developing a product that cures *the* problem, but causes others; developing a product that produces the desired effect but at a price that is unrealistic; developing products that perform well in the evaluation tests, but are not effective in practice because the test method itself was ill chosen.

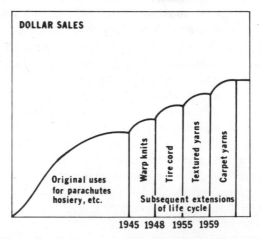

Figure 13.2   Extension of life cycle through development of new applications—nylon. *Source*: Reference 7.

13.7 APPLICATIONS DEVELOPMENT

This is a particularly worthwhile effort since it directly affects the product life cycle. The periodic discovery of new use areas can prolong the profitability of a product as illustrated for nylon in Fig. 13.2. Another aspect is that even if sales are not declining, the growth resulting from new uses can increase the scale of production, which leads to lower unit costs and greater profit. The effect might be termed autocatalytic from the standpoint that with the lowering of production costs the number of potential new use areas increase.

13.8 TECHNICAL SUPPORT TO SALES

Many chemical products are sold to perform specific functions for the customer. Examples include chlorinated solvents purchased for degreasing by a metal fabricating company; polystyrene granules purchased for extrusion by a toy company; or polypropylene fiber purchased by a textile mill. In each case the sale of the chemical is dependent upon providing technical service to the customer. In many instances there is a line of formulated chemical products that might vary as to stabilizer systems, plasticizers, and other property modifiers. Continued sales require that the customer is satisfied with the performance of the product. This requires close support by technical people who understand both the characteristics of the various products and the customer's application.

13.9 BACKGROUND RESEARCH

This category includes the activities frequently termed "basic research." Unfortunately, there are so many different opinions as to what constitutes basic research that the term itself has lost its meaning. In the chemical industry it may involve research studies aimed at increasing the understanding of certain chemical and physical processes. The economic payoff to the company results when this knowledge can be used to develop a new technology that can result in new products or processes.

13.10 RECRUITING AND TRAINING

This function simply reflects the fact that, in many chemical companies, research and development acts as an employment agency and manpower pool for the rest of the company. Very often chemists and chemical engineers are initially hired into R&D positions. Some elect to continue in these areas for their entire careers. Others, after working on a number of projects, transfer to other parts of the company such as production, sales, or administration.

13.11 MISCELLANEOUS FUNCTIONS

Because it represents a reservoir of technical and scientific talent, the research department is frequently called upon to provide services that do not fit into the categories described above. The services may involve long-term commitments of manpower or may be on a consultant basis. In some research groups, certain routine analytical procedures may be run for the production plants. In others, research personnel have become deeply involved in the industrial hygiene and safety programs. Wherever a particular technical expertise is needed—in patent matters, in problems concerning government regulations, or in evaluating raw material or energy sources—the research department supplies these needs.

13.12 RESEARCH PROJECT EVALUATION

Once the problem of determining how much money to spend on research has been answered, the next problem is determining how to spend it. Research managers must select projects and set priorities. Naturally, there is generally no lack of advice, suggestions, and perhaps pressure from other groups such as marketing and production. The final responsibility, however, rests with the research managers; indeed, evaluating the projects and establishing the overall program for the research effort constitutes their major role.

There is no magic formula for selecting or ranking research projects. Experience, intuition, and a keen awareness of the overall goals and direction of the company are necessary. There have been a number of attempts to establish models that systematize or quantify the selection process.[6,7] It is instructive to review some of these formulas:

Fred Olsen, former vice-president for research at Olin, devised a formula for the value of a research project that was based upon an "index of return." This index of return was based upon setting an arbitrary value upon the savings or sales resulting from the proposed research. This value was then modified by a factor representing the chances of success compared with the estimated cost of the research.

$$\text{Value of research} = \frac{(\text{index of return})\,(\text{probability of success})}{(\text{estimated research cost})}$$

where index of return = value of process savings for one year or sales value of new product for 5 years or sales value of improved product for 2 years.

Carl Pacifico at Alcolac used a "project number" which was the ratio of estimated profit during the life of the project, modified by estimated risks, to the total costs of commercialization.

$$\text{Project number} = \frac{R_c \times R_T \times (P - C) \times V \times L}{TC}$$

where R_c = chance of commercial success (0.1 to 1.0)
R_T = chance of technical success (0.1 to 1.0)
P = product selling price ($/lb)
C = product cost ($/lb)
V = sales volume (lb/yr)
L = life of project (yrs)
TC = total cost of project including costs of research, engineering, plant, market development, etc.

Soloman Desman of Abbott Labs developed a formula that indicated the maximum amount of money that should be spent on a project. This is then compared to the estimated cost of the project in order to establish priorities.

$$\text{Project priority} = \frac{MEJ}{\text{total R\&D costs}}$$

$$MEJ = R_c \times R_T \times \frac{I_1}{(1 + r)} + \frac{I_2}{(1 + r)^2} \cdots \frac{I_n}{(1 + r)^n}$$

where MEJ = maximum expenditure justified
R_c = risk of commercial success
R_T = risk of technical success
r = rate of return desired
n = commercial life of project
I = estimated net income for year $1, 2, \ldots n$

Many research directors scoff at the use of such standardized formulas, relying instead upon "mature judgment." In the exercise of this judgment, however, it is very probable that common elements are considered. All of the formulas, very naturally, include some measure of return expected from the project, if successful—this may be expressed in terms of annual sales, cost savings, profit, or even discounted cash flow. Many formulas modify this expected return by factors that express the degree of risk or probability of success. Most include consideration of the cost of the project. Even research managers who claim to operate by intuition undoubtedly include these same elements—return, risk, and costs in their considerations. Usually other factors must be considered since most formulas do not differentiate between big, long-range and small, short-term projects; or between high-risk, high-profit and low-risk, low-profit proposals. Whether a company employs one of these formulas or depends on managerial judgment, a number of significant factors regarding a research proposal must be considered before committing a significant effort to a project. To keep these considerations consistent from project to project, many companies employ checklists. These checklists are used, not to rate a proposed project in a quantitative manner, but simply to assure that *all* pertinent factors have been considered in the evaluation.

13.13   MARKET FORECASTS

The evaluation of any proposed research project that involves a product requires some estimate of the expected sales revenue from the product. Generally, estimates of sales volume and selling prices are not as reliable as estimates of manufacturing costs. Market forecasts require the judgment and perhaps the intuition of persons who have a thorough knowledge of the particular market involved.

With a product that is currently being sold by a competitor, the selling price and market volume have been and are being established by the market itself. However, the forecaster cannot blindly use the current market price as the forecast price. The market is not static. Entering an existing market may well disrupt that market. Consideration must be given to the economics of the current producers as well as to the economics of the proposed project. In particular, the question that must be answered is: "What is the lowest price industry can tolerate during periods of overcapacity and other competitive pressures?" This would be based upon the out-of-pocket costs of the lowest cost producer. Some rather accurate estimates of the competitor's cost situations are needed.

The factor of prime importance in determining the selling price of a new product is simply the value of the product to the customer. Regardless of the manufacturing costs or the persuasiveness of the salesman, the product will have a certain, finite economic value to the customer. This value might be known precisely or it might be based upon a somewhat vague judgment; but this value exists and it is the single most important factor in forecasting a selling price.

Realistic market forecasts are based upon an awareness of the needs of the potential customers. The proposed new product may be designed to perform some function for the customer. It is quite probable that some other product is presently performing this function. The proposed product may well be vastly superior to the product presently being used. But the pricing of the existing product will be a determining factor in price forecasts of the proposed product. Occasionally the new product price is based algebraically upon the market price of the currently available competitive product. For example, the price of a new rigid polyurethane foam formulation used for insulation was related to the market price of less efficient polystyrene foam by a simple ratio of their relative insulating efficiencies. The motivation for a customer to switch to the new product then was based upon secondary benefits such as lower fabrication costs or the savings and increased appeal of refrigerators with thinner walls. It is more common, however, that the prices cannot be related in such a direct logical manner. In a more typical case a new product that might be 60% better than a current product will do well to command a 20% premium in price.

A similar consideration would show that in those cases where a competitive product does not exist, the selling price must still relate to the intended function or performance. The customer will buy the product if it can be demonstrated to have an economic value. The value of a new antioxidant for salad dressing, for example, could be related to the current situation where the lack of such an antioxidant leads to a dollar loss due to a shortened shelf life.

In theory, perhaps, it should be possible for a person to invent a new product that is so superior to current products that it will displace these products even at a significantly higher selling price. In reality this seldom happens. The introduction of biodegradable detergents where displacement of the current products was assisted by government regulations did not result in drastic price increases in the new intermediates that were required. Industry competition held the price structures and profits to about the same level as before. This generalization is illustrated well by Herman Zabel's "Exclusion Charts," which show that the total market for a chemical product tends to be inversely proportional to its price.[8] In spite of improved properties, a high priced product does not enjoy high volume sales.

The foregoing discussion emphasized the necessity for a thorough knowledge of the market, the competitive products, and the competitors' processing economics. In addition to knowledgeable estimates as to a competitors' probable manufacturing costs and investment, it is also required to predict what that competitor might consider to be a reasonable return on his investment. Although there may be special circumstances that cause exceptions, there is a good general guideline. Competition enters the picture here also. The most logical sales return is that which nets an after tax return on investment of 12 to 15%. If a corporation does not, over the long term, average in this range, it will not be in a position to maintain and acquire the capital necessary for growth.[9]

This average reflects the fact that some plants are operating below capacity, others are producing outmoded products that must settle for a lower sales price to maintain sales volume. Proposed new projects therefore must be justified on the basis of an above average return on investment.

From the standpoint of the overall product mix, new and old, the company should earn about 20 cents before taxes on each sales dollar. This nets 10 cents after taxes. It pays out 5 cents in dividends and retains 5 cents in the business. It also generates about 5 cents in depreciation. There is a cash flow then of 10 cents to invest in expansion. This is sufficient to maintain a 10% expansion. In order to maintain this level of earnings, new products earning 15% must be continually introduced to balance the older products whose earnings have slipped to 5%.

13.14 NEW PRODUCT DECISIONS

The decision to commercialize a new product is an exceedingly important one in the chemical industry. A healthy company must continually add to its product line to maintain its economic position. But each decision involves risks. Once the development monies have been spent and the capital monies invested, it is too late for questions that should have been considered earlier.

One method of assuring that the appraisal of a new product proposal considers all of the key characteristics of the product is through the use of the New Product Profile Chart.[10] This chart, which was developed at Monsanto, graphically illustrates a product's merits and shortcomings with respect to 26 specific criteria. This is done on a simple four-level scale of desirability. The emphasis however is upon the visual indication of how the product meets each of the criteria, not upon the overall numerical score or the total score of one proposed product compared to another. The criteria have been grouped into four areas: financial aspects, production and engineering aspects, research and development aspects, and marketing and product aspects. This is illustrated in Fig. 13.3. The specific criteria and the rating system for this particular chart were developed for use at Monsanto. At another company the criteria and the rating system might be altered to reflect different corporate capabilities and objectives.

The basic point here is that the success of a new product cannot be predicted upon a single criterion such as return on investment. There are many different individual considerations involved. A valid appraisal method must

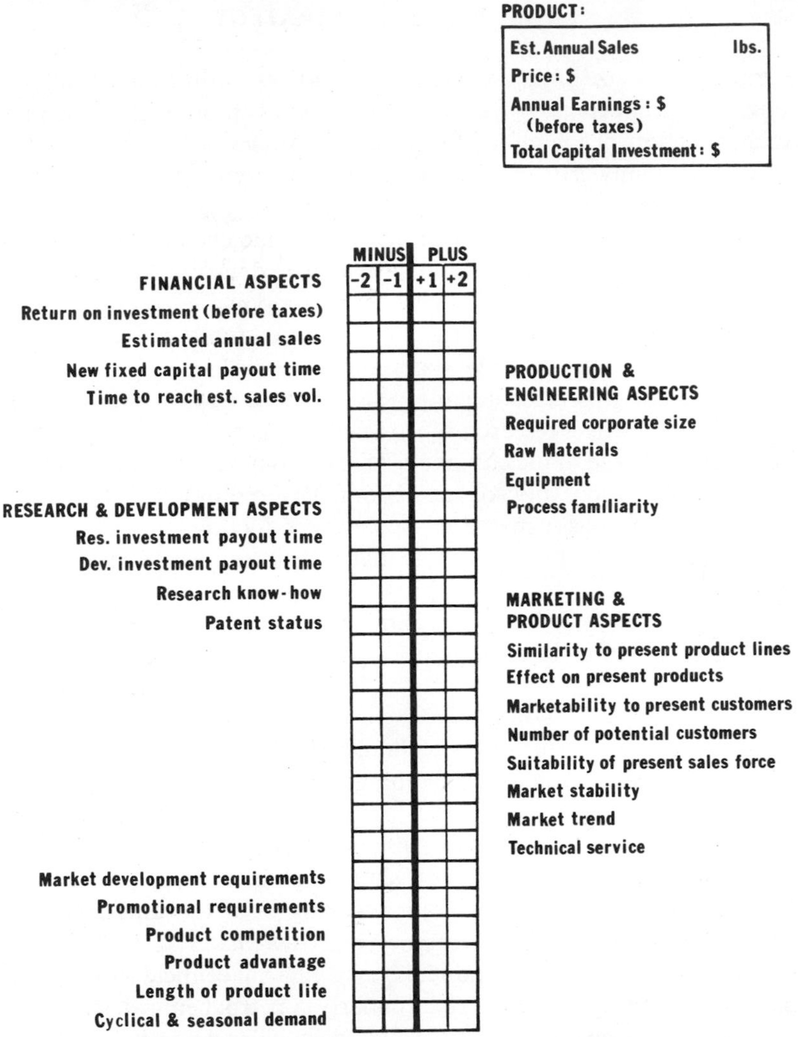

PRODUCT:

Est. Annual Sales	lbs.
Price: $	
Annual Earnings: $ (before taxes)	
Total Capital Investment: $	

	MINUS		PLUS	
FINANCIAL ASPECTS	-2	-1	+1	+2
Return on investment (before taxes)				
Estimated annual sales				
New fixed capital payout time				
Time to reach est. sales vol.				

PRODUCTION & ENGINEERING ASPECTS
Required corporate size
Raw Materials
Equipment
Process familiarity

RESEARCH & DEVELOPMENT ASPECTS
Res. investment payout time
Dev. investment payout time
Research know-how
Patent status

MARKETING & PRODUCT ASPECTS
Similarity to present product lines
Effect on present products
Marketability to present customers
Number of potential customers
Suitability of present sales force
Market stability
Market trend
Technical service

Market development requirements
Promotional requirements
Product competition
Product advantage
Length of product life
Cyclical & seasonal demand

Figure 13.3   New product profile chart. *Source:*   Reference 9.

consider and evaluate each of these aspects individually. Further, there is no method of averaging or combining these individual evaluations into one overall weighted value that is a measure of the attractiveness of the proposal. The product profile chart recognizes this and presents the results of the individual evaluations in a simple, graphical technique that clearly distinguishes the advantages and disadvantages of a particular new product.

13.15 RESEARCH STRATEGY

Determining the size of the research effort and evaluating individual projects is obviously important. Another matter of vital importance is setting the overall direction of the effort. Since it will provide the technological base for future company growth, it must be integrated with the overall corporate goals, plans, and capabilities.

It is the optimistic philosophy of many dynamic chemical firms that they can do *anything* they want to do—provided it does not violate any legal, moral, or physical laws. But no company can do *everything* it wants to do. No company, no matter how large, has unlimited funds available to pursue every project that appears to be economically sound.

In a thriving, growing company research does not operate in an ivory tower. To be effective, research must develop technology that results in an economic advantage to the company. This normally requires that ultimately a product be produced and sold at a profit. This requires the efforts of other functional groups beside research—such as production and marketing. It is the whole organization plus capital that translates new technology into a profitable product.

In order to truly evaluate a firm's research effort over the long term, the following factors must be appraised:[11]

1. The technological output.
2. The economic value of the technology produced.
3. The degree to which the technology supports company goals.

In this context, then it is very possible for a project—or, indeed, a whole research program—to be a failure even though technically or scientifically successful. Thus a research program resulting in a new and economically attractive process for a large volume basic raw material might well be a waste of time and effort if conducted by a pharmaceutical firm that is incapable of exploiting the discovery. Similarly, a petrochemical firm probably would not be in a position to successfully develop and commercialize a new drug.

It is the responsibility of research management to assure economic relevance. The research program must be integrated with the goals of management—and research must interface with engineering, production, marketing, legal, and other functions. This need has been recognized in a number of firms. In some, project review groups that consist of representatives of each pertinent functional area have been established. In many companies that have organized around the business team concept the product managers are involved in establishing the direction of the research efforts.

In a truly productive research group this two-way communication takes place not just at the management level but throughout the organization. The process development chemist should have the freedom and the responsibility to interact with the process design engineer and with the production supervisors. The chemist engaged in product development should develop close ties with the market development or technical sales personnel. In addition, the efficient and timely transfer of new technology from the laboratory to the plant can frequently be improved by assigning production personnel to the research program as it enters the development stages and by transferring research personnel with the project as it proceeds from the laboratory to the pilot plant to full production.

We have been discussing the key points in developing a strategy for the overall corporate research program. Strategy is also important in planning the individual research project.[7] The so-called "systems" approach to problem solving involves the following:

1. A clear statement of the objective or definition of the goal.
2. A listing of alternate solutions.
3. Development of an analytical or evaluation scheme to measure the effectiveness of a solution.
4. Experimentation.
5. Measurement of results and comparison with the goal.
6. Possible modification of Steps 1 to 3 and further experimentation.

It might seem that this systematic, stepwise procedure merely represents a pretentious description of an approach that should be second nature to a well-trained, rational scientist. In practice, however, many of us, particularly when under pressure, tend to seek quick, easy solutions to problems. Any project that is worth initiating deserves the benefit of a well-thought-out project plan.

Every member of a research group contributes to some extent to the development of a research strategy. To the individual scientist these considerations can be summarized as follows:

1. Know what business you are in.
2. Know the relevant science and technology.
3. Reexamine goals and objectives frequently.
4. Use a systems approach in project planning.

13.16 REFERENCES

1. J. H. Saunders, *Careers in Industrial Research and Development*, Dekker, New York, 1975.
2. A. Baines, F. R. Bradburg, and C. W. Suckling, *Research in the Chemical Industry*, American Elsevier, Amsterdam, New York, 1969.

3. F. W. Billmeyer and R. N. Kelly, *Entering Industry—A Guide for Young Professionals*, Wiley, New York, 1975.

4. *Chem. Eng. News*, June 2, 1975, p. 63.

5. American Chemical Society, *Chemistry in the Economy*, Washington, D.C., 1973.

6. *Chem. Eng. News*, March 23, 1964, p. 88.

7. T. Holzman, *Chem. Tech.*, February, 1972, p. 81.

8. R. Williams, *Chem. Tech.*, October, 1973, p. 592.

9. A. Jonnard, "Chemical Economics and Price Forecasting" in N. H. Giragosian (Ed.), *Chemical Marketing Research*, Reinhold, New York, 1967.

10. J. S. Harris, *Chem. Tech.*, September, 1976, p. 554.

11. J. B. Quinn, *Harv. Bus. Rev.*, March–April, 1960, pp. 69–80.

14

DEVELOPMENT OF A PROCESS

THE MANUFACTURE OF UREA
A CASE STUDY

The primary purpose in concluding this book with a case study is to show students how the chemistry they have learned is used in the solution of a real industrial problem. We have tried to do this by considering the stages in the development of a process and showing how an understanding of chemical principles is involved at each stage and, indeed, is necessary to carry out the project to a successful conclusion. This case study deals with a viable and important process of the modern chemical industry.

In order to maintain some continuity in the case study, we have adopted the concept of following the history of the product from the stage at which a chemical reaction is first conceived as a reasonable possibility, through the stages by which it becomes viable as a process, to its final fruition as a full-scale operation.

The development of a process progresses by a series of question-and-answer steps. Initially the questions are simple: "Is there a demand for this particular product?" "Can it be made from available starting materials in a small number of reaction stages?" "Is it required in one-ton or thousand-ton quantities?" The questions increase rapidly in number and complexity as the investigation proceeds, but it is generally possible to recognize three key questions that are of prime importance and can be answered only in terms of basic chemical concepts:

1. A reaction to make a desired product from available starting materials is proposed; will such a reaction occur to an appreciable extent?
2. If reaction does occur, is the rate of reaction adequate?

3. Since the desired product will be mixed with unchanged reactants and by-products, how can it be isolated in the required purity?

These basic questions are considered in turn in the urea case study that follows.

14.1 INTRODUCTION

In 1773 Rouelle separated a white crystalline substance from the urine of an animal; he named it urea, after urine. Some years later the chemical formula was found to be NH_2CONH_2. But it was in 1828 that urea really caught the attention of chemists, for it was in that year Wohler prepared urea from ammonium cyanate, an inorganic compound.

$$NH_4CNO \xrightarrow{\text{HEAT}} NH_2CONH_2$$

Ammonium   Urea
cyanate

The excitement came from the fact that this was the first time anyone had ever made an organic compound in the laboratory. The synthesis of urea is considered to be the beginning of synthetic organic chemistry. Before 1828 many chemists believed that only the living processes could produce organic compounds such as urea.

But urea was not to remain merely a laboratory curiosity marking the beginning of a new chemical era. In 1862 Basarov learned how to make urea from ammonia and carbon dioxide which, because of the simple raw materials, brought it into consideration as a product that might be made economically on a large scale. The reaction is illustrated by the following equation:

$$2\,NH_3 + CO_2 \rightleftharpoons NH_4COONH_2 \rightleftharpoons NH_2CONH_2 + H_2O$$

Ammonium   Urea
carbamate

Although the reaction dates back to 1868, it did not attain commercial importance until 1920, when the I. G. Farbenindustrie developed a commercial urea process. The first major production of urea in the United States was pioneered by DuPont in the early 1930s at Belle, West Virginia. After World War II the use of urea in the United States expanded rapidly. Production facilities for the manufacture of urea now number approximately 50 in the United States alone. Urea production in the United States during 1972 was 3,200,000 metric tons. This production figure serves to point out the importance of urea as a product in the modern chemical market.

The phenomenal rate of expansion of urea production is due to the increasing use of urea as a nitrogenous fertilizer, as a cattle feed supplement, and as a raw material in the manufacture of plastics. The most important use of urea is as a nitrogenous fertilizer. In fact, it is rapidly becoming the leading solid nitrogenous fertilizer.

Urea is the most concentrated solid nitrogenous fertilizer available, having a nitrogen content of about 46%, which is 30% higher than that of ammonium nitrate and more than 100% higher than ammonium sulfate. In addition, under some field conditions, urea nitrogen is not so easily leached from the soil as the nitrogen in ammonium nitrate. Consequently, it is the recommended fertilizer in regions with abundant rainfall or for irrigated lands.

Urea can be applied in a solid form, as a concentrated solution with ammonia, or mixed with phosphorus and potassium salts. It can also be applied by spraying plants with a solution containing 0.5 to 6.0% urea together with other agricultural chemicals such as pesticides, fungicides, and micronutrients. On acid soils that are not treated with lime, the efficiency of urea nitrogen is higher than that of nitrogen from ammonium nitrate. Because of the higher nitrogen concentration, the transportation and application costs are 10 to 20% less for urea than for ammonium nitrate.

The use of urea in vineyards, tobacco fields, hop fields, root plant cultures, and so on, results not only in a crop increase but also ensures an improvement in their quality. When used in hardy fruit growing and market gardening, it speeds up ripening by about two or three weeks.

In recent years, in addition to its application as a mineral fertilizer, urea is used in stock feeding as a substitute for albuminoid fodder for beef cattle because the nitrogen in 1 kg of urea is equivalent to between 2.6 and 2.8 kg of protein, to 6.0 kg of oil-bearing seeds, or to between 22 and 25 kg of oats. In addition, 1 kg of urea produces an increase between 1.0 and 1.2 kg of meat or 6.8 kg of milk.

The foregoing discussion has served to answer the first question that must be dealt with in developing a new chemical process—"Is there a demand for the particular product?" In the case of urea, the answer was not difficult to obtain, for it had long been recognized that urea could function as an excellent solid nitrogenous fertilizer. Thus, a tremendous market existed for urea, if a process could be found for producing urea at a price that would make it competitive with other solid nitrogenous fertilizers.

Before we discuss the development of the various steps in the manufacture of urea and the factors affecting each step, it would be beneficial to consider a simplified flow sheet. In so doing, the purpose behind each stage in the development of the process can be put into perspective more easily.

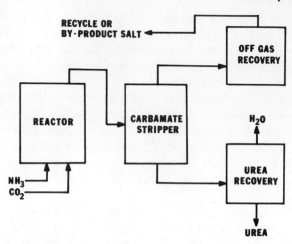

Figure 14.1   Simplified flow sheet for urea manufacture.

14.2   THE PROCESS

The process for synthesis of urea from carbon dioxide and ammonia is illustrated in Fig. 14.1.

Ammonia and carbon dioxide are compressed and fed continuously into a reactor that is maintained at a sufficiently high temperature and pressure to form a melt of ammonia, carbon dioxide, ammonium carbamate, urea, and water. A continuous stream of the melt flows from the reactor, through a pressure let-down valve, to the carbamate stripper where the melt is separated into a liquid phase of urea and water (and perhaps varying amounts of ammonium carbonate and ammonia) and a gaseous phase of ammonia, carbon dioxide, and small quantities of water. This gas is recycled to the process or used in the production of by-product salts such as ammonium nitrate or ammonium phosphate. The liquid product of the carbamate stripper can be used in the preparation of urea-ammonia solutions or it can be processed further in an evaporator to produce a solid urea.

We can now study the effect of several process variables and the problems associated with the development of each of the four steps in the manufacture of urea.

14.3   THEORETICAL BASIS FOR UREA SYNTHESIS

In order to develop a commercial process for manufacturing urea from CO_2 and NH_3 a number of questions must be answered. The first question that we must answer is, "Will the reaction of CO_2 and NH_3 to form urea occur to an appreciable extent?" Second, we must determine if the rate of reaction is adequate, and since the desired product will be mixed with unchanged reactants and by-products, how can it be recovered in the required purity?

Since the first step in the synthesis of urea is the reaction of CO_2 and NH_3 to form ammonium carbamate, the first order of business in our process development program will be to study this reaction.

14.3.1   The Formation of Ammonium Carbamate

Ammonium carbamate is formed on simple contact of anhydrous CO_2 and NH_3. When these two compounds are dry, carbamate alone is formed regardless of the relative proportions of the two constituents.

$$CO_2 \text{ (g)} + 2\,NH_3 \text{ (g)} \quad \rightleftharpoons \quad NH_2CO_2NH_4 \text{ (s)} \qquad (14.1)$$

The reaction is an equilibrium reaction, and the system is divariant. That is, the pressure depends upon the temperature and the composition of the gas phase. When the initial product is carbamate, this composition is fixed and the dissociation pressure depends only upon the temperature.

We would expect Eq. 14.1 to occur in the forward direction to an appreciable extent only at such temperatures that the pressure is below the dissociation pressure of ammonium carbamate, or conversely, the pressure on the reaction system must be maintained above the vapor pressure of ammonium carbamate at any given operating temperature.

The Dissociation Pressure of Ammonium Carbamate.   The dissociation pressure of ammonium carbamate at various temperatures was measured in the laboratory,[1] and the data are given in Table 14.1. The vapor from solid ammonium carbamate, as determined by analysis, is practically all NH_3 and CO_2, since ammonium carbamate vapor dissociates very rapidly. Therefore, the "vapor pressure" of ammonium carbamate is also the dissociation pressure. The vapor pressures enclosed in brackets at 180°C and 200°C are extrapolated values obtained from the plot of $\log P$ versus $1/T$ shown in Fig. 14.2. The reason for the extrapolated values being higher than the experimental values is that the pressure measured at high temperatures is complicated by the simultaneous decomposition of ammonium carbamate into urea and subsequent decomposition of urea into biuret and ammonia.

The plot of $\log P$ versus $1/T$ as shown in Fig. 14.2, strictly speaking, is not a straight line. It is a curve with a slope of increasing steepness as the

Table 14.1   Dissociation Pressure of Ammonium Carbamate

Temp. °C	Diss. Pressure, Atm.
40	0.31
60	1.05
80	3.1
100	8.5
120	21
140	46
160	98
180	150,(190)
200	200,(360)

temperature rises. But, if we do make the approximation that the experimental points fall on a straight line, we can derive Eq. 14.2.

$$\text{Log } P = \frac{-2748}{T} + 8.2753 \tag{14.2}$$

where P is in atmospheres and T in degrees Kelvin. This equation can be used to calculate the dissociation pressure for any given temperature. Since the melting point of ammonium carbamate is 145°C, pressures calculated

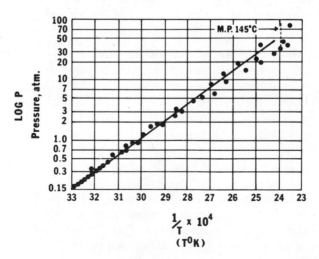

Figure 14.2   Dissociation pressure of ammonium carbamate. *Source*:   Reference 1.

below this temperature will be for the dissociation of the solid and for temperatures above 145°C for liquid ammonium carbamate.

The dissociation pressure data for ammonium carbamate can be used to calculate the equilibrium constant for its dissociation. These calculations are shown in the next section.

Equilibrium Constant for Ammonium Carbamate Dissociation. The equilibrium constant for the dissociation of carbamate to form ammonia and carbon dioxide can be derived in the following way

$$NH_2CO_2NH_4 \ (s) \ \rightleftharpoons \ 2NH_3 \ (g) + CO_2 \ (g) \qquad (14.3)$$

$$Kp = P_{CO_2} \times P_{NH_3}^2 \qquad (14.4)$$

P_{CO_2} and P_{NH_3} are the partial pressures of CO_2 and NH_3, respectively. In the case of synthesis from the stoichiometric ratio of carbon dioxide and ammonia, or in the case of heating ammonium carbamate in vacuum, K_p may be calculated from the total pressure P_T as shown in the following equations:

$$P_{NH_3} = \frac{2P_T}{3} \qquad (14.5)$$

$$P_{CO_2} = \frac{1P_T}{3} \qquad (14.6)$$

$$K_p = P_{CO_2} \times P_{NH_3} = \frac{4P_T^3}{27} \qquad (14.7)$$

The calculated equilibrium constants from various dissociation pressure measurements are listed in Table 14.2.

Table 14.2   Equilibrium Constants for Ammonium Carbamate Dissociation

Temp. °C	K_p, Atm^3
20	9.11×10^{-5}
40	4.91×10^{-3}
80	4.64
100	8.23×10
130	3.68×10^3
150	3.75×10^4
200	6.88×10^5

This phase of the study shows that since ammonium carbamate has a high dissociation pressure, it is formed most readily under high pressure. From these considerations, therefore, it will be advantageous in plant operation to use a high pressure to obtain a high yield of ammonium carbamate, which in turn will produce a high yield of urea.

Furthermore, in a continuous flow manufacturing process, the applied pressure in the reactor must be substantially higher than the equilibrium dissociation pressure of ammonium carbamate corresponding to the operating temperature in the reactor. Otherwise the reaction

$$NH_2CO_2NH_4 \rightleftharpoons 2NH_3 + CO_2$$

will more or less cancel the reaction in which ammonium carbamate is formed, thus decreasing the ammonium carbamate concentration and hence decreasing the yield of urea.

On the other hand, ammonium carbamate can be removed from the system by decomposing it to NH_3 and CO_2 under a low applied pressure and high temperature. This property will form the basis for developing a method for separating ammonium carbamate from urea.

The heat of formation of ammonium carbamate at constant volume can also be calculated from the dissociation pressure of ammonium carbamate by use of the van't Hoff equation:

$$\frac{d \ln K_p}{dT} = \frac{-\Delta E}{RT^2} \tag{14.8}$$

Calculations using the slope of the plot given in Fig. 14.2 yield a value of 37.8 kcal of heat evolved for every mole of ammonium carbamate formed at constant volume. The following equation can be used to convert the value at constant volume into the value at constant pressure.

$$\Delta H = \Delta E - \Delta nRT \tag{14.9}$$

ΔH is the heat of reaction at constant pressure, ΔE is the heat of reaction at constant volume, Δn is the change in the number of moles of gas involved in the reaction per mole of ammonium carbamate formed. Using 3 for the value of Δn and a temperature of 25°C, Eq. 14.9 gives a value of -39.6 kcal/mole for the heat of formation of solid ammonium carbamate.[4]

Using a value of 3.6 kcal/mole for the heat of fusion of ammonium carbamate, the heat of formation of molten ammonium carbamate can be calculated as follows:

$$2NH_3 \text{ (g)} + CO_2 \text{ (g)} \longrightarrow NH_2CO_2NH_4 \text{ (s)}; \quad \Delta H = -39.6 \text{ kcal} \tag{14.10}$$

$$NH_2CO_2NH_4 \text{ (s)} \longrightarrow NH_2CO_2NH_4 \text{ (l)}: \quad \Delta H = 3.6 \text{ kcal} \tag{14.11}$$

adding Eqs. 14.10 and 14.11 gives

$$2\,NH_3\,(g) + CO_2\,(g) \longrightarrow NH_2CO_2NH_4\,(l); \qquad \Delta H = -36.0\,\text{kcal}$$
$$(14.12)$$

When liquid ammonia is the reactant instead of gaseous ammonia, the heat of reaction is given by Eq. 14.14.

$$2\,NH_3\,(g) + CO_2\,(g) \longrightarrow NH_2CO_2NH_4\,(l); \qquad \Delta H = -36.0\,\text{kcal}$$
$$(14.12)$$

$$2\,NH_3\,(l) \longrightarrow 2\,NH_3\,(g); \qquad \Delta H = 9.5\,\text{kcal}$$
$$(14.13)$$

$$2\,NH_3\,(l) + CO_2\,(g) \longrightarrow NH_2CO_2NH_4\,(l); \qquad \Delta H = -26.5\,\text{kcal}$$
$$(14.14)$$

Thus, when ammonia and carbon dioxide react to form ammonium carbamate, a great deal of heat is evolved. This heat must be removed in the process of manufacture to prevent the reaction temperature from reaching that level at which the carbamate dissociation pressure becomes equal to the pressure being maintained in the reactor.

The next step in our investigation is to study the rate of the formation of ammonium carbamate.

The Rate of Formation of Ammonium Carbamate. Laboratory studies show that the rate of formation of ammonium carbamate from gaseous NH_3 and CO_2 is very much influenced by the pressure.[2] All other conditions being constant, the rate increases about proportionately to the square root of pressure.

The rate of formation of ammonium carbamate likewise increases with the temperature. The curves in Fig. 14.3 summarize this behavior. At constant

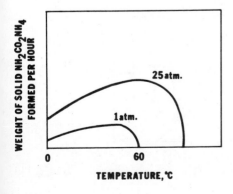

Figure 14.3   Rate of formation of ammonium carbamate.

pressure the amount of ammonium carbamate formed per unit time at first increases slowly, passes through a maximum, then decreases very rapidly and becomes zero at temperature T, which corresponds to that for which the dissociation pressure is equal to the reaction pressure.

In industrial production, therefore, it is advisable to operate at the highest pressure possible and at the highest temperature compatible with this pressure. The reaction of CO_2 with NH_3, which is rather slow at ordinary pressure and temperature, becomes almost instantaneous at pressures of the order of 100 to 150 atm and at temperatures of about 150°C. Under these reaction conditions, equilibrium data from the previous section indicate that the formation reaction for ammonium carbamate goes almost to completion.[3]

The next step in the reaction to produce urea is the dehydration of ammonium carbamate to form urea and water according to Eq. 14.15.

$$NH_2CO_2NH_4 \rightleftharpoons NH_2CONH_2 + H_2O \qquad (14.15)$$

14.3.2   The Conversion of Ammonium Carbamate into Urea

If ammonium carbamate is heated in a closed vessel, a part of it dissociates; and if the degree of filling of the apparatus is sufficiently high, the pressure produced by dissociation prevents the decomposition of the remaining ammonium carbamate, which dehydrates spontaneously to yield urea.[5] This dehydration occurs solely in the solid or liquid phase and not in the gas phase. The dehydration reaction is an endothermic process with $\Delta H_{25°C} = 6.8$ kcal.

As with most organic ammonical salts, the dehydration with the formation of the amide is never complete, but rather is an equilibrium phenomenon. The phase diagram for the CO_2-NH_3 reaction mixture is given in Fig. 14.4.[6]

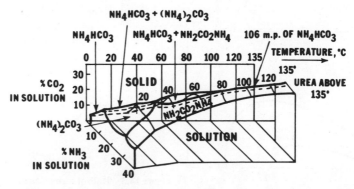

Figure 14.4   Phase equilibrium model of the system NH_3—CO_2—H_2O. *Source:*   Reference 6.

The diagram shows that ammonium carbamate reacts with the water of dehydration to form ammonium carbonate. Up to 60°C five components exist in the system. They are NH_4HCO_3, $(NH_4)_2CO_3 \cdot H_2O$, $(NH_4HCO_3)_2 \cdot (NH_4)_2CO_3$, $NH_4HCO_3 \cdot NH_2CO_2NH_4$, and $NH_2CO_2NH_4$. Ammonium carbamate is relatively ineffective in urea synthesis. However, the tendency for ammonium carbonate to form is very much reduced when both the temperature and pressure are raised to a high level, or when there is an excess of NH_3 in the system. Above 60°C the solute is predominantly ammonium carbamate, which begins to change to urea as the temperature is increased — particularly when it is raised above 130°C under superatmospheric pressure.

Thus at synthesis temperatures of 150 to 190°C the dehydration of ammonium carbamate is an equilibrium reaction in which a complex homogeneous mixture is obtained, consisting of water, urea, unchanged ammonium carbamate, and ammonium carbonate produced by the reaction of water with ammonium carbamate.

Effect of Temperature on the Conversion Yield of Ammonium Carbamate to Urea. In order to define the relationship between conversion yield and temperature, solid ammonium carbamate was loaded into sealed glass tubes to a loading density of 1 g/cc. The pressure in the tubes was not measured except that the tube was at atmospheric pressure and room temperature when it was closed and before the temperature was raised. After equilibrium had been achieved at the test temperature, the tubes were broken open and analyzed for the urea content. The data from these experiments are shown in Table 14.3 and are plotted in Fig. 14.5.[7] It can be seen that the conversion of ammonium

Table 14.3 Experimental Yields of
Urea from the Dehydration
of Ammonium Carbamate
at Various Temperatures

T°C	Yield, %
135	42.7
146	45.0
150	45.3
160	47.0
170	48.9
180	49.7
190	52.1
200	50.8

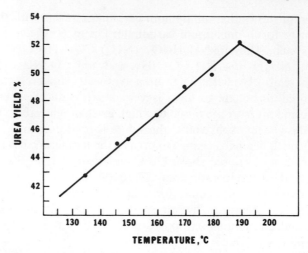

Figure 14.5 Effect of temperature on the conversion of ammonium carbamate to urea. *Source*: Reference 7.

carbamate to urea gradually increased as the reaction temperature increased until it reached approximately 52% at 190°C. The conversion then suddenly decreased with increasing temperature.

The decrease in the degree of conversion of ammonium carbamate to urea above 190°C may be explained by the fact that the degree of decomposition of ammonium carbamate into its initial components, CO_2 and NH_3, also increases with rising temperature. Therefore, although the yield of the conversion reaction to form urea is improved by increasing temperatures, the pressure must also be increased to prevent the reagents from becoming gaseous. Consequently, the working pressure in the reactor must be higher than the dissociation pressure of ammonium carbamate at the reaction temperature to avoid poor conversion yields.

Effect of Pressure on the Conversion Yield of Ammonium Carbamate to Urea. The effect of pressure on the conversion of ammonium carbamate to urea is shown in Table 14.4. As can be seen, the yield of urea increases significantly when the pressure increases.[8] If the process were to operate at 160°C, then the pressure in the reactor must be greater than 130 atm to obtain a yield greater than 46%. As higher temperatures are used, the pressure in the reactor must also be increased. For example, the maximum yield at 180°C is obtained at a pressure of 210 atm.

Effect of Excess Components on the Yield of Urea.   Thus far in our studies of those factors that affect the yield of urea, we have been working with stoichio-

Table 14.4   Effect of Pressure on the Conversion of
Ammonium Carbamate to Urea at 160°C

Pressure, Atm	Urea Yield, %
69	36.2
76.5	42.2
89	44.2
107.5	45.3
136.5	46.2

metric amounts of reactants and products. That is, according to the equation

$$2\,NH_3 + CO_2 \rightleftharpoons NH_2CO_2NH_4 \rightleftharpoons NH_2CONH_2 + H_2O \tag{14.16}$$

the molar ratio of NH_3/CO_2 is $2:1$. Now according to the law of mass action, an excess of ammonia must increase the degree of conversion of CO_2 to urea, and an excess of CO_2 must increase the degree of conversion of NH_3 to urea.

Data on the conversion of ammonium carbamate in the presence of excess ammonia into urea are given in Table 14.5.[9] The data were obtained by

Table 14.5   The Effect of Excess Ammonia on the Conversion of
Ammonium Carbamate into Urea

Starting Materials	Excess NH_3, %	Conversion of Carbamate into Urea, %
Carbamate	0	43.5
Urea + H_2O	0	44.0
Carbamate + NH_3	23.0	49.5
Urea + H_2O + NH_3	23.0	50.2
Carbamate + NH_3	45.2	58.2
Urea + H_2O + NH_3	45.2	58.7
Carbamate + NH_3	76.5	67.0
Urea + H_2O + NH_3	82.5	67.0
Carbamate + NH_3	222	81.5
Urea + H_2O + NH_3	279.0	85.2

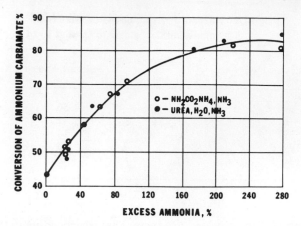

Figure 14.6 Effect of excess ammonia on ammonium carbamate-urea equilibrium at 155°C.
Source: Reference 9.

approaching equilibrium from both directions. As can be seen, the percent of conversion is independent of the direction from which equilibrium is established. The data in Table 14.5 are plotted in Fig. 14.6.

Part of the favorable effect that excess ammonia has on the conversion of ammonium carbamate to urea may be accounted for by its dehydrating activity, which reduces the water content of the molten reaction mixture, thus preventing its reaction with urea and thereby shifting the equilibrium toward the urea side.

Increasing the amount of excess ammonia in a manufacturing plant in order to increase the conversion yield of urea is economical only to a certain limit. Beyond this, any further excess becomes unprofitable because of rising utilities, electrical energy and steam requirements necessary for removing the excess ammonia from the product urea. In addition, an increase in the $NH_3 : CO_2$ ratio results is an increase in the specific volume of the molten mass in the reactor. This results in a corresponding reduction in rate of urea production. The most favorable excess of ammonia is that which allows the minimum consumption of utilities and the highest production of urea for the given synthesis and plant gas separation equipment. This value is estimated from pilot plant data in a later section.

As the data in Table 14.6 show, excess CO_2 has no appreciable effect on the conversion of ammonium carbamate to urea.[9] A plausible reason why an excess of carbon dioxide does not have a large effect on the degree of conversion is because excess carbon dioxide remains in the gaseous phase in the reactor, while the formation reaction of urea proceeds in the liquid phase.

Table 14.6   Effect of Excess Carbon Dioxide on the Conversion of Ammonium Carbamate into Urea

Starting Materials	Excess CO_2, %	Conversion of Carbamate into Urea, %
Carbamate + CO_2	97.0	44.3
Urea + H_2O + CO_2	100.0	44.0
Carbamate + CO_2	61.2	44.3
Urea + H_2O + CO_2	61.2	45.7

When ammonium carbamate is converted into urea, one mole of water is produced along with each mole of urea. According to the law of mass action, extra water entering with the reactants would have an adverse effect on the yield of urea. The data shown in Fig. 14.7 confirms this relationship.[11,14] In addition to reducing the conversion of ammonium carbamate to urea, extra water would take up room in the reactor and thereby reduce the productive capacity. Extra water also dilutes the urea solution from the reactor, thereby increasing the load on the evaporators, and a resulting increase in energy consumption.

Now that we have studied all of the variables that affect the conversion of ammonium carbamate into urea, the question may be asked, "How can we calculate the yield of urea for any given set of operating conditions?" The next section deals with answering this question.

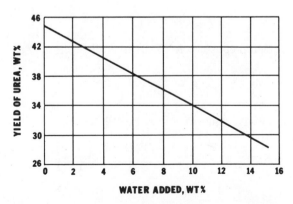

Figure 14.7   Effect of water on yield of urea at 150°C. *Source*:   References 11 and 14.

Calculating the Urea Yield.   Our studies have established that the quantity of urea produced from the reaction of ammonia and carbon dioxide involves the equilibrium of a system comprising five components: NH_3, CO_2, H_2O, ammonium carbamate, and urea; these along with temperature and pressure, make seven variables, each of which has an effect upon the others. To calculate the equilibrium yield of urea from such a system involves a number of as-yet-unknown quantities, such as the activities of the components under high temperature and high pressure conditions. To get around this difficulty we use an approximation calculation based upon the overall reaction.

$$2NH_3 + CO_2 \rightleftharpoons NH_2CONH_2 + H_2O \tag{14.17}$$

starting with 2 moles of NH_3 and 1 mole of CO_2 and obtaining an equilibrium yield Y of urea in moles.

Since there are a total of 3 moles of reactants the mole fraction of urea in the reactor is

$$N_{urea} = \frac{Y}{(3 - Y)} \tag{14.18}$$

likewise, the mole fraction of water is

$$N_{H_2O} = \frac{Y}{(3 - Y)} \tag{14.19}$$

The mole fractions of unreacted NH_3 and CO_2 are

$$N_{NH_3} = \frac{(2 - 2Y)}{(3 - Y)} \tag{14.20}$$

$$N_{CO_2} = \frac{(1 - Y)}{(3 - Y)} \tag{14.21}$$

Then the equilibrium constant K of the overall reaction (Eq. 14.17) is

$$K = \frac{(N_{urea})(N_{H_2O})}{(N_{NH_3})^2(N_{CO_2})} \tag{14.22}$$

Substituting the above equations into Eq. 14.22 gives

$$K = \frac{[Y/(3 - Y)][Y/(3 - Y)]}{[(2 - 2Y)/(3 - Y)]^2[(1 - Y)/(3 - Y)]}$$

$$K = \frac{Y^2(3 - Y)}{4(1 - Y)^3} \tag{14.23}$$

Since we have already determined Y as a function of T, the data given in Fig. 14.5 establishes a series of values for Y as a function of temperature in the range 130 to 190°C.

$$Y = 1.7 \times 10^{-3}T + 0.198 \tag{14.24}$$

where T is in degrees Celsius. Equation 14.24 is valid in the temperature range of 130 to 190°C. Therefore, Y may be calculated from the temperature and then K from Y.

For a nonstoichiometric mixture, the nonstoichiometric reaction can be expressed as follows:

$$aNH_3 + bH_2O + CO_2 \quad \rightleftharpoons$$
$$XNH_2CONH_2 + (a - 2X)NH_3 + (b + X)H_2O \tag{14.25}$$

where X is the extent of conversion of CO_2 to urea expressed as a fraction of the CO_2 in the total feed; a is the molar ratio of NH_3/CO_2 in the total feed; and b is the molar ratio of H_2O/CO_2 in the total feed. The total feed is fresh feed plus any recycle feed to the reactor. The equilibrium expression for Eq. 14.25 is as follows:

$$K = \frac{X(b + X)}{(1 - X)(a - 2X)^2} \tag{14.26}$$

Equation 14.26 can be solved for X after the value for K has been obtained from Eq. 14.23.

Now that we can calculate equilibrium conversion yields of urea, it is necessary to study the rate at which equilibrium is established.

14.3.3 The Rate of Formation of Urea

Studies on the rate of formation of urea are important and essential to the successful development of the urea synthesis process. For one thing, the design of the reactor in a continuous flow process for manufacturing urea will be determined by the residence time required for the urea reaction mixture to reach equilibrium.

Studies presented in the first part of our development program indicated that under conditions that give a high yield of urea, high temperature and pressure, the formation of ammonium carbamate from CO_2 and NH_3 is almost instantaneous. Therefore, we need only to study the rate of dehydration of ammonium carbamate to establish the overall rate of formation of urea.

The effect of time and temperature on the conversion of ammonium carbamate to urea is given in Fig. 14.8.[14] The data show that the velocity of the dehydration of ammonium carbamate is strongly affected by temperature. By examining the shape of the reaction velocity curves in Fig. 14.8,

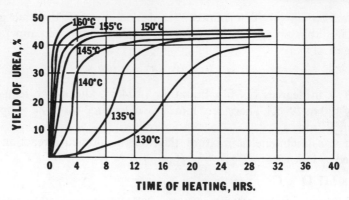

Figure 14.8   Rate of conversion of ammonium carbamate into urea. *Source*: Reference 14.

it can be seen that at temperatures below 145°C, the melting point of ammonium carbamate, the shape of the velocity curves is similar to the shape of curves for autocatalytic reactions. A reasonable explanation for this is that the reaction velocity is slower in the solid phase than in the liquid phase. It may also be considered that the water formed during the dehydration reaction depresses the melting point of the reaction product, thereby acting as a catalyst.

The effect of excess NH_3 on the rate of conversion of ammonium carbamate to urea is shown in Fig. 14.9.[15] The data indicate that approximately the same amount of time is required to reach equilibrium as was required for the stoichoimetric mixture.

14.3.4   Summary of Process Variables Obtained from Batch Reactor Data

The laboratory data that we have gathered thus far in our development of a urea synthesis process have served to establish the following process guidelines: The first stage in the synthesis is the combination of ammonia and carbon dioxide to produce ammonium carbamate.

$$2NH_3 + CO_2 \rightleftharpoons NH_2CO_2NH_4$$

This reaction is followed by dehydration of ammonium carbamate to give urea and water:

$$NH_2CO_2NH_4 \rightleftharpoons NH_2CONH_2 + H_2O$$

Both reactions are reversible and occur simultaneously in the reactor. The first reaction is exothermic and goes rapidly to completion under the processing conditions of high temperatures and pressure; the second is endothermic

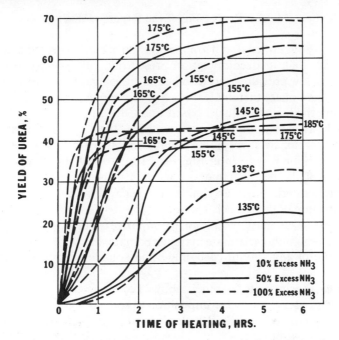

Figure 14.9 Conversion of ammonium carbamate to urea in the presence of excess NH_3.
Source: Reference 15.

and, depending on the process conditions adopted, goes to 40 to 80%
completion. Reaction pressures are within the range 130 to 260 atm, and
temperatures within the range 160 to 200°C. At these temperatures and
pressures the urea, ammonium carbamate, and components in the reaction
mixture are in the liquid form. Excess NH_3 over the stoichiometric ratio of
2 : 1 will increase the conversion yield of urea, while on the other hand,
excess water will decrease the yield.

A reaction time of approximately 1 hr is required in order to approach
the equilibrium yield of urea. The effluent from the reactor consists of a
gaseous mixture of unreacted ammonia and carbon dioxide, water vapor,
and an aqueous solution of urea, ammonia, and ammonium carbamate.

The next stage in our urea synthesis project deals with the development
of a method for recovering urea in a pure usable form from the reactor
effluent. In addition, the economics as well as environmental considerations
of the process require that the unreacted ammonia be recovered. Therefore,
a method will be developed for recycling the unconverted reactants back
to the reactor or diverting them to other uses, such as the manufacture of
coproducts.

14.3.5   Processing of the Reactor Effluent

We have already learned that if ammonium carbamate is heated at a pressure that is lower than its decomposition pressure it will decompose into NH_3 and CO_2. At low pressures and with heating, these gases can then be stripped from the liquid phase thereby leaving a solution of aqueous urea. A method for treating the reactor effluent based upon these principles is to allow the urea synthesis melt to flow from the reactor, through a pressure let-down valve, to a carbamate stripper. You may recall that the flow scheme for this type of separation was given in Fig. 14.1. The product stream will be cooled as a result of the flashing that occurs. Consequently, the temperature of the material entering the carbamate stripper will be a function of the operating pressure of that vessel.

Additional heat must be added to vaporize soluble ammonia and decompose ammonium carbamate in the carbamate stripper in order to decrease the quantity of ammonium carbamate and soluble ammonia in the liquid product.

However, if temperatures much above 120°C are used in the carbamate stripper it is observed that decomposition of some of the urea to undesirable biuret will occur. We will need to look more closely at this decomposition in our next phase of studies.

Theoretically, 1 mole of water is produced per mole of urea; and, therefore, if all the ammonium carbamate were dissociated in the stripper, its liquid product would be a urea-water solution having a urea concentration of approximately 75 wt %. In order to form a solid soluble product, the urea-water solution can be spray dried, evaporated, and crystallized, or it can be evaporated and prilled. Prilling is a process whereby molten urea is sprayed into a tower. The drops of molten urea are cooled in a countercurrent stream of air introduced at the bottom of the tower by fans, whereupon they solidify in the form of spherical prills. During this processing, the length of time that the material is held at temperatures above 110°C must be kept at a minimum in order to lessen the decomposition of urea to biuret.

Our earlier studies demonstrated that the conversion of carbon dioxide to urea in the absence of excess ammonia is only about 50% per pass. Therefore, excess ammonia will normally be used. This increases the conversion of carbon dioxide but decreases the conversion of ammonia. Consequently recovery of the off-gases, especially ammonia, from the carbamate decomposer is mandatory from an economic standpoint. Accordingly, the resulting NH_3–CO_2 gaseous mixture must either be recycled to the reactor or else used for the synthesis of some other product (this is the so-called "once-through" process)—an alternative that may not always be economical.

Recycling of the off-gas, however, is not a simple operation. Initial attempts to recycle the off-gases from the decomposer by direct compression (without separation) were found to be impractical because of precipitation of solid ammonium carbamate in the compressor system. (Remember that high pressure favors the formation of ammonium carbamate.) Therefore, some other method must be found for recycling the NH_3–CO_2 mixture.

There are several feasible ways of handling this recycle problem. One method is to conduct the carbamate decomposition at two different pressure levels. If the first level is at about 25 atm, the product gas will be mostly NH_3, and this can be recycled to the reactor without any difficulty. The reason for this is that at 25 atm the ammonium carbamate will not decompose at the temperature maintained. Consequently, only the excess NH_3 in the solution will be evolved and this will be free of CO_2. The product gas from the second-stage decomposer (about 1 atm pressure) will be a mixture of NH_3 and CO_2, and can be used in a coproduct fertilizer plant. This is called the "partial-recycle process."

Another method for recycling the gases is to remove the carbon dioxide from the off-gas stream of the carbamate decomposer before it is returned to the reactor. This could be done by passing the NH_3–CO_2 mixture through a monoethanoloamine solution whereupon the CO_2 would be selectively absorbed and the purified NH_3 could be condensed and pumped back into the reactor. If the absorbing solution is a urea-nitrate mixture instead of ethanoloamine the ammonia gas will be selectively absorbed. Both of these methods are referred to as "gas separation total-recycle" processes, since all of the ammonia is recovered.

Another method for recycling the NH_3–CO_2 mixture is to absorb the gas mixture in water and recycle the liquid solution back to the reactor. The additional water that will be introduced into the reactor will tend to reduce the urea yield; but if the amount of water is carefully controlled, the reduction will not be too severe. This method for handling the off-gases is referred to as a "total solution recycle" process.

Each of the methods that we have proposed for handling the NH_3-CO_2 gas mixture is summarized in Fig. 14.10. We will direct most of our attention to the development of the once-through process. The reason being that it is the simplest process and can serve as a model for developing the other more complicated processes.

Before progressing into the pilot plant stage with our urea project, it is necessary that we look more closely at the problems associated with the thermal decomposition of urea. In attempting to concentrate the aqueous urea solution from the carbamate decomposer, we have observed that some biuret is formed. We must find some way of minimizing this biuret formation. This phase of the investigation is handled in the next section.

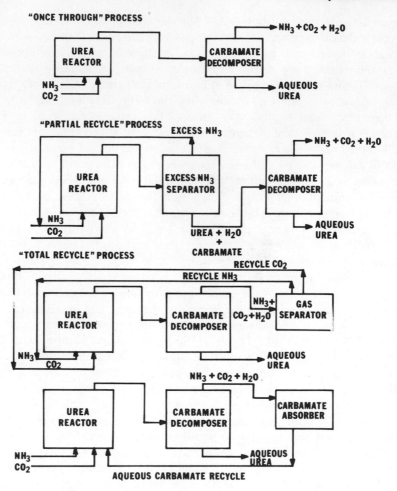

Figure 14.10   Various processes for handling NH_3—CO_2 off-gas mixture.

14.3.6   Formation of Undesirable Impurities During Urea Processing

As has already been noted, the urea rich product solution will be subjected to heat during the stripping, evaporation, prilling, and drying steps in the process of producing solid urea. It is important to determine if this heating will result in the decomposition of the urea product with subsequent production of undesirable impurities.

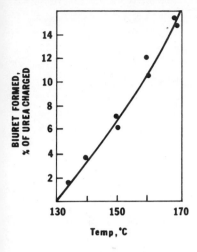

Figure 14.11 Biuret formation in dry urea versus temperature (per hour). *Source*: Reference 13.

Laboratory studies[13] show that when dry urea is heated to temperatures near, or above, its melting point (132°C) it slowly decomposes into biuret:

$$2\,NH_2CONH_2 \rightleftharpoons H_2NCONHCONH_2 + NH_3 \qquad (14.27)$$

      Urea   Biuret

Biuret has been found to be toxic and harmful to some crops. Its presence in most fertilizer-grade urea should be kept below 1%. Therefore, it is necessary to determine and define the conditions where biuret formation is at a minimum.

The effect of temperature on the rate of formation of biuret in anhydrous molten urea was studied first, and the results shown in Fig. 14.11. The data were obtained for an unstirred melt of pure urea, maintained at the indicated temperatures for 60 min.[13] The data show that temperature is a very important factor in biuret formation.

The data plotted in Fig. 14.12 show the effect that water has on biuret formation in urea melts at 140°C. As can be seen, low percentages of water have little effect on the initial rate of biuret formation; but after several hours, the two curves separate, and after 7 hr, only two thirds as much biuret is formed in the presence of water as is produced in the control run.

A comparison of the rate of biuret formation in urea melts at 140°C in contact with air and in an ammonia atmosphere of 1 atm is shown in Fig. 14.13. The data show that biuret formation is only about 60% as great in the presence of ammonia as compared to air.

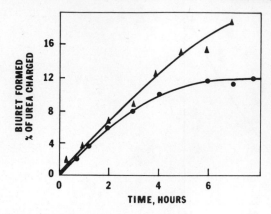

Figure 14.12   Biuret formation at 140°C (▲ pure dry urea at 140°C; ● pure urea plus 5% water at 140°C). *Source*:   Reference 13.

In summary we can say that biuret formation in the synthesis reactor is minimal because of the presence of a large excess of ammonia in the reaction mixture. However, during the final concentration of the urea solution where excess ammonia is not present, significant amounts of biuret may form. Biuret formation at this stage in the process can be minimized if the solution is concentrated under a vacuum at the lowest possible temperature. Where prills are preferred to crystals, the urea crystals should be melted and prilled in such a way that the time interval in which urea stays above 130°C is kept to a minimum.

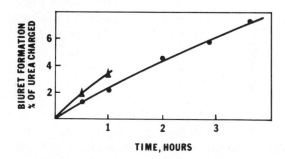

Figure 14.13   Effect of ammonia on biuret formation in dry urea at 140°C and atmospheric pressure (▲ urea in air; ● urea in ammonia atmosphere). *Source*:   Reference 13.

14.4 THE PILOT PLANT STAGE

Now that the bench scale studies have been completed and a successful commercial process for the production of urea appears possible, a decision has to be made about the next stage in the development project. This may be the construction of the full-scale plant if there is sufficient confidence in the "scale-up" of the process. If this is not the case, it may be decided to build a "pilot plant" which is a model, say of one-twentieth or one-tenth the capacity, of the proposed full-scale plant. As the object is to study the problems likely to be encountered, the design has to be studied carefully so that the data obtained will be of maximum value. The scale of the pilot plant will vary according to the type of problem to be investigated. Only a portion of a process may be studied if enough is known about the remaining sections.

It should be emphasized, however, that pilot plant experimentation should be considered carefully and, if deemed necessary, carried out on the smallest scale that is feasible. At the pilot plant stage, the development costs of a process increase steeply. Because of their size, pilot plants are much more expensive than laboratory apparatus. Moreover, to be fully useful, they have to be heavily instrumented, including automatic control instruments, and 24-hr operation will often be necessary, requiring a team of plant operators. As a result, the costs increase immediately by an order of magnitude over laboratory-scale operations. A further drawback of pilot plant work is the time-lag involved, for the plant has to be designed, the equipment brought together and erected, and finally commissioned.

The need for extensive pilot plant experimentation is not as great as it once was. Improved methods of chemical analysis, together with better mathematical techniques for correlating the results, have led to a much better understanding of the processes. Also, as a result of the development of more reliable scale-up procedures, the chemical engineer can have more confidence in the design of a plant item tens or hundreds of times larger than the laboratory equipment.

Nevertheless, information gained only from laboratory work is often insufficient to permit reliable scale-up, and the construction of a full-scale plant would involve uncertainties that might require much time and money to rectify if difficulties arose during start-up. Pilot plant development would minimize this problem. Yet, the importance of having some production from the plant as soon as possible may justify taking the risk that problems will arise when the plant is first put into production. Under these conditions, there will be a corresponding saving in time and effort at the process development stage, and it may very well be that the scale-up from a pilot plant to the full-scale unit will not in itself be devoid of risk.

The decision is never an easy one to make. It involves a careful analysis of the problems that may arise on scaling up each stage of the process.

In the development of the urea process the next logical stage of development is to gather some pilot plant data where the reactor is operated under continuous-flow conditions. The reason for going to the pilot plant stage is owing to the uncertainties in scaling up a laboratory size *batch* unit to a full plant size *continuous-flow* operation. The smart thing to do is to study the process in a small-scale continuous-flow plant and then scale up from this unit.

The laboratory batch process studies have generated enough knowledge about the synthesis of urea so that a minimum amount of time need be spent in the pilot plant stage. In a real situation, a considerable amount of economic analysis would be done at this point in order to justify the large pilot plant expenditures.

A small continuous-flow unit capable of producing 250 lb of urea/day was constructed. A flow diagram for the unit is given in Fig. 14.14. Liquid CO_2 is used for convenience only, since in the full scale plant operation, gaseous CO_2 will be charged to the reactor. In the pilot plant the anhydrous liquids are introduced separately into the base of the lead-lined reactor by means of single-acting hydraulic-type pumps activated by a common crankshaft. Adjustment of the NH_3/CO_2 ratio is effected by changing the stroke length of the pumps. Both reactants enter the reactor where reaction and conversion to urea take place. After leaving the reactor the charge, now consisting of a liquid mixture of urea, water, and ammonium carbamate, enters a still in which the dissociation of ammonia carbamate is accomplished by heat. The urea leaves the still as an aqueous solution. The ammonia and carbon dioxide leave the still as gases where they enter an absorber in which the ammonia is absorbed in a sulfuric acid solution. The carbon dioxide is

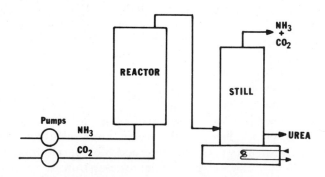

Figure 14.14   Flow diagram for a urea pilot plant.

discharged to the atmosphere from the absorber. Thus, the unit is an example of the simple "once through" process.

The pilot plant reactor consists of a steel shell 14.9 cm i.d. and 2 m long with a 3.2-cm thick wall. The interior is lined with lead 1 cm thick. The carbon steel heads of the reactor are held in place by means of nickel-steel studs and bolts. By using long-stem valves with the seats contained within the autoclave proper, all danger of freezing up the inlets with solid ammonium carbamate is eliminated. The volume of the reactor is 35.8 liters. The dip stem in the exit port provides for a 2-liter vapor space. Therefore, the total volume of the liquid melt in the reactor is 33.8 liters.

The type of data obtained during the pilot plant operation is indicated in Table 14.7. The operating conditions were adjusted to study the effect of temperature, pressure, excess components, and feed rate on the yield of urea under continuous-flow conditions. The urea yields in the pilot plant operation were in general found to be 5 to 10% lower than the yields calculated from Eq. 14.26. For example, the pilot plant yield of urea at a reactor temperature of 185°C, a pressure of 193 atm, and a NH_3/CO_2 ratio of 3.83 is 65.1%; whereas the yield of urea as calculated by Eq. 14.26 should be 72% under these conditions.

Pilot plant studies on the effects of feed rate and excess ammonia on the conversion of carbon dioxide to urea and the urea production rate are summarized in Fig. 14.15. Data were obtained at 160°C and 193 atm. The curves show that increasing the total feed rate decreases the percent conversion of carbon dioxide to urea but increased the urea yield per hour. They also show that excess ammonia increases the percent conversion of carbon dioxide to urea, but that for any specific total feed rate, excess ammonia decreases the urea yield on an hourly basis.

If we were going to operate the urea pilot plant at 160°C and 193 atm of pressure with a NH_3/CO_2 feed ratio of 4:1 so as to maximize conversion of CO_2 to urea, we would feed the reactants at a rate of 8 lb total feed/hr. On the other hand, if we wanted to get the maximum amount of urea production per hour, we would operate the plant with a total feed rate of 26 lb/hr.

The pilot plant data serves to demonstrate the dependence of conversion and yield on total feed rate. The optimum feed rate for the full-scale urea plant will be a function of operating and capital costs. When using the once-through process, the optimum feed rate will also depend on the type and economic value of the by-product produced with the excess ammonia. For example if the by-product is of low economic value, then the urea plant would be operated to maximize the conversion of ammonia to urea. This would call for lower feed rates. On the other hand if the by-product had a reasonably high economic value and it was desirable to maximize urea production, the plant would be operated at a high feed rate.

Table 14.7 Pilot Plant Data for the Production of Urea

Reactor Temp, °C	175	180	180	185	185	185	185
Reactor Pressure, Atm	193	193	193	193	193	193	193
Feeding Rate, kg/Hr							
NH_3	25.1	18.45	18.95	9.1	17.9	22.2	10.66
CO_2	14.5	10.7	11.85	11.6	15.5	15.0	4.45
H_2O	0	0	0	0	0	0	2.82
Total	39.6	29.15	30.8	20.7	33.4	37.2	17.93
Excess NH_3, %	124	124	106	0	49	21.5	210
Mole ratio NH_3/CO_2	4.48:1	4.48:1	4.14:1	2.0:1	2.99:1	3.83:1	6.21:1
Time of Residence* Hrs	1.1	1.5	1.4	2.7	1.5	1.2	2.7
Urea Yield (% of CO_2 converted)	60.2	63.2	57.4	44.1	55.6	65.1	62
Rate of Urea Production, kg/Hr	6.96	9.22	9.26	6.96	11.71	13.34	3.74

*The liquid volume in the reactor is 33.8 liters. If an average density of 1.29 kg/ℓ is used for the liquid melt, then the charge to fill the reactor weighs 43.6 kg. Feeding at a rate of 39.6 kg/hr would result in a residence time of 43.6 kg/39.6 kg/hr or 1.1 hours.

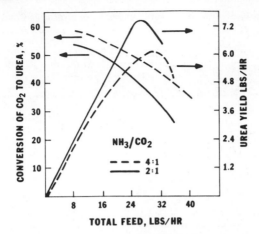

Figure 14.15 Effects of feed rate and excess ammonia on urea production.

Irrespective of the type and amount of pilot plant data that we gather, the final resolution of the optimum operating conditions for the full-scale urea plant can only be decided by actual operation. The pilot plant data can only serve as a guideline for designing the plant.

Both the pilot plant and laboratory batch studies have shown that the *rate* of urea production is increased by:

1. Increased pressure.
2. Increased temperature (so long as the operating pressure is substantially above the dissociation pressure of ammonium carbamate).
3. The use of low excess ammonia.

The question then may be asked, "Why not operate the plant at the maximum pressure and temperature possible and with low excess ammonia?" There are several reasons why this should not be done. One reason is that increased pressure increases the capital and operating costs of compression and reaction equipment. Increased temperature in the absence of excess ammonia accelerates the decomposition of urea to biuret. But most important, all of the above conditions accelerate the corrosion rate in the reaction system. They do this to such a great extent, and so severely under certain conditions, that because of corrosion problems the reaction cannot be conducted under conditions that result in the maximum production of urea.

A considerable amount of data has been gathered during pilot plant operations on the corrosion rates of various materials under urea synthesis conditions. Some representative corrosion data are listed in Table 14.8.

Table 14.8   Corrosion Rates under Urea Synthesis Conditions

Material Tested	Highest Corrosion Rate Obtained Inches/Year	Lowest Corrosion Rate Obtained Inches/Year
410 S.S.	Total loss	
304 S.S.	5.4	.0002
321 S.S.	.9	.003
Nickel	3.0	.....
Pfaudler glass	0.6	
Tantalum	.009	.001
Teflon	Weight gain and blistered	

The highest and lowest corrosion rates obtained are given to indicate the tremendous effect of operating conditions on the rate of corrosion. In general, it can be said that the corrosive conditions of the urea synthesis reaction are very severe. Only a few materials exhibited any semblance of being resistant. Tantalum showed good corrosion resistance for all tests to which it was exposed, but its high cost rules out its use—except perhaps, as a very thin liner. Most urea process equipment is now made of stainless steel where corrosion rates are minimized through passivation of the stainless steel by incorporating oxygen in the range of $0.1 \rightarrow 3\%$ into the process stream. Even with air injection, however, use of stainless steel still limits the reactor temperature to about 190°C and the operating pressure to about 200 atm.

When all factors for a once-through urea process are taken into consideration (the factors to consider involve minimization of corrosion rates, minimization of capital investment, minimization of ammonia usage, and maximumization of urea production), our pilot plant and batch studies suggest that the reactor should be operated at a temperature of approximately 180°C under a pressure of approximately 200 atm. The feed rate and reactor volume should be such that the reaction time (the residence time for the reaction mixture in the reactor) be approximately 1 hr. So as to maximize urea production, but yet minimize ammonia consumption, only a 10% excess of ammonia in the feed stream should be used. Under these operating conditions a 46 to 50% conversion of CO_2 to urea should be possible.

If a total-recycle process is used, the synthesis reactor should be operated under a pressure of 220 atm with the reactor temperature maintained at approximately 190°C. The molar NH_3 to CO_2 ratio in the feed should be about $4:1$. Under these operating conditions, a 64 to 68% conversion of CO_2 to urea should be possible for each cycle through the reactor.

Under these operating conditions a material and heat balance on the synthesis reactor can be obtained. Both of these balances will depend on the particular type of process that is used (i.e., whether the reaction is operated under once-through, partial-recycle or total-recycle conditions). Since we are trying to develop the once-through process, we will restrict our calculations to it.

14.5 MATERIAL AND ENERGY BALANCE FOR THE SYNTHESIS REACTOR IN A ONCE-THROUGH PROCESS

Calculations will be based on the following operating conditions: pressure in the reactor is 190 atm, temperature is 180°C, molar ratio of $NH_3:CO_2$ is $2.2:1$ (a 10% excess), the degree of conversion of ammonium carbamate into urea under these conditions is 50%. The temperature of the reagents on entering the reactor is 25°C, and the temperature of the products leaving the reactor is 180°C. Ammonia is fed to the reactor as a liquid and CO_2 as a gas.

Basis of calculation: The production of 1 metric ton (1000 kg) of urea.

Material Balance. The synthesis of urea from ammonia and carbon dioxide takes place by the following equations:

$$2\,NH_3 + CO_2 \longrightarrow NH_2CO_2NH_4$$

$$NH_2CO_2NH_4 \rightleftharpoons NH_2CONH_2 + H_2O$$

The theoretical consumption of ammonia and carbon dioxide per metric ton of urea is:

$$2\,NH_3 + CO_2 \rightleftharpoons NH_2CONH_2 + H_2O$$
$$2 \times 17 \quad 44 \qquad\qquad 60 \qquad\quad 18$$

Ammonia required is:

$$NH_3 = \frac{(2)(17)(1000)}{60} = 566.7 \text{ kg}$$

and

$$CO_2 = \frac{(44)(1000)}{60} = 733.3 \text{ kg}$$

The theoretical amounts of ammonia and carbon dioxide that have to be introduced into the reactor at a conversion of 50% is:

$$NH_3 \text{ (stoichiometric)} = \frac{566.7 \text{ kg}}{0.5} = 1133 \text{ kg}$$

$$CO_2 \text{ (charged)} = \frac{733.3 \text{ kg}}{0.5} = 1466 \text{ kg}$$

But a $NH_3 : CO_2$ ratio of 2.2 : 1 is used, therefore the amount of ammonia actually charged is:

$$NH_3 \text{ (charged)} = \frac{(2.2)(17)}{(44)(1466)} = 1246 \text{ kg}$$

The amount of ammonium carbamate formed is:

$$NH_2CO_2NH_4 = \frac{(1466)(78)}{44} = 2598 \text{ kg}$$

where 78 denotes the molecular weight of ammonium carbamate. The amount of ammonia consumed to form the carbamate is 1133 kg, therefore, the amount of unchanged ammonia is

$$NH_3 = 1246 - 1133 = 113 \text{ kg}$$

The amount of unchanged ammonia carbamate is

$$NH_2CO_2NH_4 = (0.5)(2598) = 1299 \text{ kg}$$

and the amount of water resulting from the conversion of ammonium carbamate into urea is:

$$H_2O = \frac{(1000 \text{ kg})(18)}{60} = 300 \text{ kg}$$

The material balance for the synthesis reactor for 1 ton of urea produced is shown in Table 14.9. The ammonia shown in Table 14.9 does not remain

Table 14.9   Material Balance for One Ton of Urea Production

Input	Amount (kg)	Output	Amount (kg)	%
Ammonia	1246	Urea	1000	36.8
Carbon dioxide	1466	Ammonium carbamate	1299	47.9
		Water	300	11.1
		Ammonia in excess	113	4.2
Total	2712	Total	2712	

free because the water resulting from the reaction dissolves the excess ammonia by forming ammonium hydroxide:

$$NH_4OH = \frac{113}{17}(35) = 232 \text{ kg}$$

where 35 represents the molecular weight of ammonium hydroxide and 17 the molecular weight of NH_3. The amount of free water left is

$$H_2O \text{ (free)} = 300 - \frac{113}{17}(18) = 180 \text{ kg}$$

The energy balance for the synthesis reactor for the production of 1 metric ton of urea is given in the following.

Energy Balance. Heat input plus heat liberated during reaction is

1. Heat contained in liquid ammonia:

$$Q_1 = (1246 \text{ kg})(1.16 \text{ kcal/kg}°C)(25°C) = 36{,}134 \text{ kcal}$$

2. Heat contained in carbon dioxide:

$$Q_2 = (1466 \text{ kg}/44)(8.22 \text{ kcal/kmole}°C)(25°C) = 6846 \text{ kcal}$$

3. Heat evolved with the formation of liquid ammonium carbamate

$$2NH_3(l) + CO_2(g) \longrightarrow NH_2CO_2NH_4(l) \ \Delta H = -26.5 \text{ kcal}$$

$$Q_3 = \frac{2598 \text{ kg}}{78 \text{ kg/mole}}(26.5 \text{ kcal/kmole})(10 \text{ mole/kmole})$$

$$= 882{,}633 \text{ kcal}$$

4. Heat evolved with the formation of ammonium hydroxide from water and ammonia:

$$NH_3(l) + H_2O(l) \longrightarrow NH_4OH(aq)\Delta H = -3.55 \text{ kcal/mole}$$

$$Q_4 = \frac{232 \text{ kg}}{35 \text{ kg/kmole}}(3500 \text{ kcal/kmole}) = 23{,}531 \text{ kcal}$$

5. Heat associated with the formation of urea—this means we need to calculate the heat for the reaction (at 180°C)

$$NH_2CO_2NH_4(l) \longrightarrow NH_2CONH_2(l) + H_2O(l)\Delta H = ?$$

The following equation may be utilized to obtain the value of ΔH:

$$H_2O(l) + NH_2CONH_2(s) \longrightarrow$$

$$NH_2CO_2NH_4(s)\Delta H_{25°C} = -6.8 \text{ kcal} \quad (a)$$

The change in the heat content of the reactant and the products in this reaction that accompanies the raising of the reaction temperature from 25°C to 180°C can be represented by the following equations.

Using an average specific heat of 0.0364 kcal/mole°C for solid and molten ammonium carbamate, we obtain:

$$NH_2CO_2NH_4(s)_{25°C} \longrightarrow$$
$$NH_2CO_2NH_4(s, 180°C); \quad \Delta H = +5.64 \text{ kcal} \quad (b)$$

$$NH_2CO_2NH_4(s, 180°C) \longrightarrow$$
$$NH_2CO_2NH_4(l, 180°C); \quad \Delta H = +18.5 \text{ kcal} \quad (c)$$

Using an average value of 0.0192 kcal/°mole for solid and molten urea, we obtain

$$NH_2CONH_2(s, 180°C) \longrightarrow$$
$$NH_2CONH_2(s, 25°C); \quad \Delta H = -2.97 \quad (d)$$

$$NH_2CONH_2(l, 180°C) \longrightarrow$$
$$NH_2CONH_2(s, 180°C); \quad \Delta H = -3.6 \text{ kcal} \quad (e)$$

Using an average specific heat of 0.0184 kcal/°mole for liquid water gives

$$H_2O(l, 180°C) \longrightarrow H_2O(l, 25°C); \quad \Delta H = -2.79 \text{ kcal} \quad (f)$$

Adding Eq. a, b, c, d, e, and f gives:

$$NH_2CO_2NH_4(l, 180°C) \longrightarrow$$
$$NH_2CONH_2(l, 180°) + H_2O(l, 180°)\Delta H = -7.98 \text{ kcal}$$

Thus, while the decomposition of ammonium carbamate at 25°C is endothermic, it is exothermic at 180°C. Therefore, the heat evolved in the formation of 1000 kg of urea is:

$$Q_5 = \frac{(1000 \text{ kg})(7980 \text{ kcal/kmole})}{(60 \text{ kg/kmole})} = 134{,}061 \text{ kcal}$$

The total heat input is then:

$$Q_I = Q_1 + Q_2 + Q_3 + Q_4 + Q_5$$

$$= 36{,}138 + 6{,}846 + 882{,}633 + 23{,}531 + 134{,}061 = 1{,}083{,}205 \text{ kcal}$$

Heat Output

1. Heat content of urea

$$Q_1 = \frac{(1000 \text{ kg})(19.2 \text{ kcal/kmole}°C)(180°C)}{(60 \text{ kg/kmole})}$$

$$= 57,600 \text{ kcal}$$

2. Heat content of water

$$Q_2 = \frac{(300 \text{ kg})(18.4 \text{ kcal/kmole}°C)(180°C)}{18 \text{ kg/kmole}}$$

$$= 55,200 \text{ kcal}$$

3. Heat content of unchanged ammonia

$$Q_3 = \frac{(113 \text{ kg})(23.8 \text{ kcal/kmole}°C)(180°C)}{17 \text{ kg/kmole}}$$

$$= 28,476 \text{ kcal}$$

4. Heat content of unchanged ammonium carbamate

$$Q_4 = \frac{(1299 \text{ kg})(36.4 \text{ kcal/kmole}°C)(180°C)}{78 \text{ kg/kmole}}$$

$$= 109,116 \text{ kcal}$$

The total heat output is then:

$$Q_o = Q_1 + Q_2 + Q_3 + Q_4$$
$$= 57,600 + 55,200 + 28,476 + 109,116 = 250,392 \text{ kcal}$$

Since the total heat input was calculated to be $Q_I = 1,083,205$ kcal, the amount of excess heat is:

$$Q = Q_I - Q_o = 1,083,205 - 250,392 = 832,813 \text{ kcal}$$

This is the amount of excess heat produced when 1 ton of urea is produced at the given operating conditions. A small part of this amount of heat is lost through the walls of the reactor, but the remainder must be removed from the system.

In the once-through process, this heat is removed by a cooling jacket and/or a large excess of ammonia. In the partial-recycle process, the excess heat is removed by recycling the excess ammonia. In the total-recycle process, the excess heat is removed by recycling the ammonium carbamate solution.

We now have reached the point in our development program where we can get some estimate of the operating and capital costs associated with a urea plant that utilizes one of the three types of proposed processes.

14.6   CAPITAL AND OPERATING COSTS

Thus far our studies have been aimed at gathering knowledge so that a plant can be designed to make urea. But the urea plant has to make something else besides urea—it has to make money. Therefore, a urea plant will be built only after it has been established that the process is economical and that the plant is capable of making money.

The capital cost for building a urea plant is a function of the size of the plant and the type of process used, that is, whether it is a once-through operation or a total recycle. The capital cost versus capacity can be plotted arithmetically. By plotting the capital cost per daily ton of urea against plant size for a total-recycle process, the reduction in capital costs per ton as the size of the plant is increased becomes visually apparent as Fig. 14.16 shows.

The fact that the capital element per ton of production decreases as the

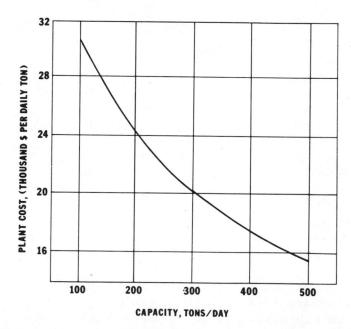

Figure 14.16   Capital cost per ton of urea produced by total-recycle versus plant capacity.

Table 14.10   Comparison of Capital Costs Based upon a Plant
with a Daily Production of 100 Tons of Urea in
Solution Form

Process	Investment Cost per Installed Daily Ton(1976)
Once-through	$15,000
Partial recycle	21,200
Complete recycle	31,000

plant grows bigger is one of the most important driving forces in the chemical industry today. It puts the producer with a larger plant at a competitive advantage because he can sell at a lower price than the producer from a small plant and still meet all his charges including the cost of servicing and repaying the capital.

A comparison of the capital costs for the three major urea processes is given in Table 14.10. The increase in cost of the recycle processes as compared to the once-through process is due to the need for additional pumping equipment for recycling the ammonium carbamate or ammonia solution and larger reactors and decomposers to handle the larger volume of liquid that is fed to the reactor. In addition to the cost of building the plant, another important economic factor is the operating cost of the plant.

The operating requirements for urea plants may be broken down into the normal subdivisions of feedstocks and utilities to show the relative importance of each item. Estimates of these values as obtained from pilot plant data are shown in Table 14.11.

This comparison of operating requirements shows significant differences in the consumption of ammonia and carbon dioxide. Since the ammonia consumption per ton of urea greatly affects the production cost, it is always given careful consideration in designing a plant. In some plants one may think that the CO_2 is available free from an ammonia syntheses gas purification process. However, in most real economic situations, things are free until considered useful, and then accounting will fix a value on them that the process developer must face. The value is often arbitrary or could be as high as market value if purchased outside. Thus while it looks free overall, it is given a cost in calculations.

About 65% of the electric power consumed in a urea plant is due to CO_2 compression. The remaining 35% of the power is consumed by process pumps, conveyors, fans, and so on.

Table 14.11   Comparison of Feedstock and Utility Requirements for the Different Urea Processes (Based on a 100 ton/day plant)

Units/ton of prilled urea

	Once-through	Partial Recycle	Total Recycle
Liquid Ammonia			
Feed, ton	1.25	.87	.64
Consumption, ton	.57	.64	.64
In off-gas, ton	.68	.22	--
Carbon Dioxide			
Feed, ton	1.47	.94	.83
Consumption, ton	.73	.83	.83
In off-gas, ton	.74	.11	--
Electric power, kw-hr	220	154	176
Steam (150 lb/in), lb	--	2,310	2,750
Steam (50 lb/in), lb	880		
Cooling water (11°C, rise)			
gal	7,370	13,200	16,500

The steam and cooling water are used in the decomposition of ammonium carbamate and subsequent condensing of the recovered ammonia gas.

The feedstock and utility costs for the three processes can be calculated from the information provided in Tables 14.11 and 14.12. A comparison of these costs for the three processes is given in Table 14.13. It is quite clear that the most important factor that controls the feedstock and utility costs is the amount of ammonia used in producing a ton of urea. The total cost of producing a ton of urea in a 100 ton day plant is given in Table 14.14.

Table 14.12   Utility Costs (1976)

Item	Unit	Unit Cost, $
Ammonia	ton	64.46
CO_2	ton	0
Steam	1000 lb	2.50
Cooling Water	1000 gal	0.08
Electricity	kwh	0.03

Table 14.13 Comparison of Feedstock and Utility Costs per Ton of Prilled Urea

Process	Ammonia	Steam	Cool. Water	Elec.	Total
Once-through	$80.58	$2.20	$0.59	$6.60	$89.97
Partial recycle	56.08	5.78	1.06	4.62	67.54
Total recycle	41.25	6.88	1.32	5.28	54.73

Urea in 1976 sold for an average price of $135/ton. If only urea is sold, then the average profit before taxes per ton of urea produced is:

Process	Profit [$/ton (based on $135/ton urea)]
Once-through	$35.72
Partial-recycle	55.09
Total-recycle	63.06

One must realize that the actual profit will be considerably lower than these figures indicate, because we have not included in our calculations such costs as labor overhead, sales cost, and so on.

The profit margin for the once-through and partial-recycle processes can be increased by the manufacture of a by-product with the excess ammonia

Table 14.14 Urea Production Cost (Based on a 100 ton/day plant) for the Production of 1 Ton of Prilled Urea

Item	Once-through	Partial Recycle	Total Recycle
1. Feedstock + Utilities	$89.97	$67.54	$54.73
2. Labor, per shift @ $8/hr	1.92	1.92	1.92
3. Capital Charges* @ 10% a year	4.11	5.81	8.49
4. Taxes and Insurance @ 3%/yr of plant	1.24	1.74	2.54
5. Maintenance @ 5%/yr of plant investment	2.04	2.90	4.26
Total Production Cost	$99.28	$79.91	$71.94

*As an example calculation, take the case of the once-through process. Investment is $15,000 per daily ton of urea. 10% of $15,000 is $1,500. Divide this figure by the number of daily tons of urea in a year; $1,500/365 = $4.11.

that is available in these processes. For example, in the once-through process for every ton of urea produced, 0.68 ton of ammonia is available in the off-gas. This amount of ammonia can be turned into 3.2 tons of ammonium nitrate at a cost of $90/ton. The average value of a ton of ammonium nitrate at 1976 prices is $96. Therefore, the profit obtained in converting the excess ammonia into ammonium nitrate is $19.26. When this profit is added to the profit from the sale of urea, we find that the total profit from operating a 100 ton/day once-through urea plant is $54.92/ton of prilled urea.

For the partial-recycle process where 0.22 ton of ammonia is in the off-gas, 1.04 tons of ammonium nitrate can be produced. This results in a profit of about $6.24. When this is added to the profit from the sale of urea, the total profit per ton of urea produced in the partial-recycle plant is $61.33.

At this point, a comparison of the various processes indicates that larger profits per ton of urea produced are achieved with the total-recycle process. Another desirable feature with this process is that there are no additional by-products with which to contend. Also, two integrated processes are more difficult to operate than one. Although the capital investment in a total-recycle plant is greater than for the other urea processes, it is not greater—probably less—than the combined investments in the two integrated processes that are necessary for the once-through urea plant.

In deciding whether to build a urea plant or not, most chemical companies would require that the plant return at *least* 10% on capital before taxes. However, before a company can choose one of the three processes as a basis for plant design, a number of factors as discussed in Chapter 2 must be considered. For example, the question of how much capital is available for investing in the urea plant must be considered. Market conditions for the by-product should be analyzed. An evaluation should be made as to how the by-product fits into the overall product line of the company. Also, the question of how much urea production is required must be answered. Once all of these factors are taken into consideration, then a process can be chosen and a plant designed.

14.7 THE DESIGN OF THE PLANT

Our developmental project has matured to the point where a plant can now be designed for producing urea. For example, our studies have provided information on the following:

1. The general flow scheme of the process and an idea of points where difficulty is likely to be experienced.
2. The likely effects of various impurities in the raw materials on the yield and quality of product.

3. The type of reactor to be employed, and the conversion and yields that can be expected.
4. The quantities of heat to be transferred at various points, and whether these transfers are likely to present problems.
5. The problems of product purification and other separations in the process and how they can be dealt with.
6. The materials of construction.
7. The probable costs of the plant and of operating it.

It is now the responsibility of the engineering group to take our data and proceed to the detailed design of individual plant items, taking into consideration all the factors involved to ensure that the completed plant will operate in a correct and economical way.

14.8 REFERENCES

1. J. P. Egan, J. E. Potts, Jr., and G. D. Potts, *Ind. Eng. Chem.* **38**, 454 (1946).
2. M. Frejacques, *Chim. Ind.* (*Paris*) **60**, 22 (1948).
3. T. Cambon, *Chim. Ind.* (*Paris*) **74**, 917 (1955).
4. K. K. Clark and H. C. Hetherington, *J. Am. Chem. Soc.* **43**, 1909 (1927).
5. G. N. Lewis and G. H. Barrows, *J. Am. Chem. Soc.* **34**, 1515 (1912).
6. E. Terres and H. Behrens, *Z. Phys. Chem.* (*Leipzig*) **A 139**, 695 (1928).
7. K. K. Clark, V. L. Gaddy, and C. E. Rist, *Ind. Eng. Chem.* **25**, 1092 (1933).
8. S. Kawasumi, *Bull. Chem. Soc. Jap.* **24**, 148 (1951).
9. H. J. Krease and V. L. Gaddy, *J. Am. Chem. Soc.* **52**, 3088 (1930).
10. S. Kawasumi, *Bull. Chem. Soc. Jap.* **25**, 227 (1952).
11. S. Kawasumi, *Bull. Chem. Soc. Jap.* **26**, 218, 222 (1953), and **27**, 254 (1954).
12. "Process Survey, Urea," *Eur. Chem. News* (*London*), January 17, 1969.
13. C. E. Redemann, F. C. Riesenfeld, and F. S. LaViola, *Ind. Eng. Chem.* **50**, 633 (1958).
14. M. Toknoka, *J. Agr. Chem. Soc.* **11**, 107 (1935).
15. B. A. Bolotov, A. N. Papova, and Y. K. Sokolova, *J. Chem. Ind.* (*U.S.S.R.*) **14**, 631 (1937).

Appendix A   Units and Conversion Factors

Quantity	SI Unit	Other Units	Conversion Factors
length	meter (m)	inch (in.)	1 in. = 2.54×10^{-2} m = 8.333×10^{-2} ft $= 1.58 \times 10^{-5}$ mile
		foot (ft)	1 ft = 3.048×10^{-1} m = 12 in. $= 1.8939 \times 10^{-4}$ mile
		mile	1 mile = 1.61×10^3 m = 5280 ft
volume	cubic meter (m^3)	liter (l)	1 liter = 61.03 in.^{3} = 3.53×10^{-2} ft^3 $= 2.642$ gal = 1×10^{-3} m^3
		gallon (gal)	1 gal = 2.31×10^2 in.^{3} = 1.337×10^{-1} ft^3 $= 3.785$ liters
		cubic inch (in.^{3})	1 in.^{3} = 5.787×10^{-4} ft^3 = 4.329×10^3 gal $= 1.64 \times 10^{-2}$ liter
		cubic foot (ft^3)	1 ft^3 = 1.728×10^3 in.^{3} = 7.481 gal $= 28.32$ liters
mass	kilogram (kg)	gram (g)	1 g = 3.527×10^{-2} oz = 2.20×10^{-3} lb $= 10^{-3}$ kg
		avoir ounce (oz)	1 oz = 28.35 g = 6.25×10^{-2} lb
		pound (lb)	1 lb = 4.536×10^2 g = 16 oz
pressure	pascal (Pa) $kg \cdot m^2/s^2$	atmosphere (atm)	1 atm = 1.01325×10^5 Pa = 760 torr $= 14.696$ lb/in.^{2}
		torr (mm Hg) pound per square inch (lb/in.^{2})(psi)	1 torr = 133.3 Pa = 1.934×10^{-2} lb/in.^{2} 1 lb/in.^{2} = 6.8947×10^3 Pa = 51.71 torr $= 6.805 \times 10^{-2}$ atm
energy	joule (J) $kg \cdot m^2/s^2$	kilocalorie (kcal)	1 kcal = 4.184×10^3 J = 3.9657 Btu $= 4.267 \times 10^2$ kg·m = 3.086 $\times 10^3$ ft·lb
		British thermal unit (Btu)	1 Btu = 2.52×10^{-1} kcal = 1.055×10^3 J $= 1.0758 \times 10^2$ kg·m $= 7.7816 \times 10^2$ ft·lb
		kilogram·meter (kg·m)	1 kg·m = 9.8067 J = 2.34×10^{-3} kcal $= 9.296 \times 10^{-3}$ Btu = 7.233 ft·lb
		foot·pound (ft·lb)	1 ft·lb = 1.356 J = 3.24×10^{-4} kcal $= 1.285 \times 10^{-3}$ Btu $= 1.383 \times 10^{-1}$ kg·m
power	watt (W)	foot-pound per second (ft·lb/s)(fps)	1 ft·lb/s = 1.356×10^{-3} kW $= 1.818 \times 10^{-3}$ hp
	J/S	kilowatt (kW)	1 kW = 1.341 hp = 737.56 ft·lb/s
		horsepower (hp)	1 hp = 7.457×10^{-1} kW = 550 ft·lb/s

Appendix B   Selected Physical Properties of Various Organic and Inorganic Substances

Compound	Formula	Specific Gravity*	Melting Point (°C)	Boiling Point (°C)
Acetone	C_3H_6O	0.791	-95.35	56.2
Aniline	C_6H_7N	1.022	-6.2	184.4
Benzene	C_6H_6	0.879	5.5	80.1
Butane	$n\text{-}C_4H_{10}$	0.579	-138.2	-0.5
iso-Butane	$i\text{-}C_4H_{10}$	0.557	-159.5	-11.6
Calcium phosphate	$Ca_3(PO_4)_2$	3.14	1670	
Carbon	C	2.26	3600	4200
Chlorobenzene	C_6H_5Cl	1.107	-45	132
Ethyl acetate	$C_4H_8O_2$	0.901	-83.7	77.1
Ethyl benzene	C_8H_{10}	0.867	-94.9	136.3
Ethylene glycol	$C_2H_6O_2$	1.113	-13.1	197.3
Hydrogen fluoride	HF	1.15	-35	120
Methyl cyclohexane	C_7H_{14}	0.769	-126.4	101.1
Napthalene	$C_{10}H_8$	1.145	80.2	210.8
Nitric acid	HNO_3	1.502	-41.5	86
n-Octane	C_8H_{18}	0.703	-56.5	125
n-Pentane	C_5H_{12}	0.630	-129.6	36.1
Sodium chloride	NaCl	2.163	808	1465
Sulfur (rhombic)	S_8	2.07	113	444.7
Sulfur (monoclinic)	S_8	1.96	119	444.7
Toluene	$C_6H_5CH_3$	0.866	-94.9	110.7
m-Xylene	C_8H_{10}	0.864	-47.8	139.2
o-Xylene	C_8H_{10}	0.880	-25.1	144.5
p-Xylene	C_8H_{10}	0.861	13.3	138.4

* The specific gravities are at 20°C relative to water at 4°C.

Appendix C Heat Capacity and Enthalpy Data

The enthalpy change for a reaction ΔH obtained calorimetrically or by calculation corresponds to some one definite temperature. Since ΔH of a reaction is a function of temperature and pressure, a method for calculating the enthalpy change at one temperature from that at another is, therefore, highly desirable. Such calculations can be accomplished by use of heat capacity data. The relationship between enthalpy and heat capacity is normally expressed in the form:

$$\int_{H_1}^{H_2} dH = \Delta H = \int_{T_1}^{T_2} Cp\, dT$$

or as

$$\int_{\Delta H_1}^{\Delta H_3} d(\Delta H) = \Delta H_2 - \Delta H_1 = \int_{T_1}^{T_2} \Delta Cp\, dT$$

The variation of heat capacities with temperature is generally expressed by empirical formulas of the type:

$$Cp = a + bT + cT^2 + dT^3$$

or

$$Cp = a' + b'T + c'T^{-2}$$

Heat capacity data for several compounds are given in Table C1 on page 408.

Table C1 Heat Capacity Data for Various Organic and Inorganic Compounds*

Compound	State	a	b × 10²	c × 10⁵	d × 10⁹	Range (°C)
Benzene	g	17.700	7.875	−6.022	18.54	0–1000
n-Hexane	l	51.702				0–1000
Methyl cyclohexane	g	32.850	9.763	−5.716	13.78	0–1000
	l	45.17				
	l	53.03				
	g	29.00	13.510	−9.016	24.09	0–1000
Nitrogen	g	6.919	0.1365	−0.02271	0	0–3000
Oxygen	g	7.129	0.1407	−0.01791	0	0–3000
n-Pentane	g	27.450	8.148	−4.538	10.10	0–1000
Sulfur dioxide	g	9.299	0.9330	−0.7418	2.057	0–1200
Toluene	l	35.56				0
	l	43.30				100
	g	22.509	9.292	−6.658	19.20	0–1000

* Data obtained from (1) F. D. Rossini et al., *Selected Values of Chemical Thermodynamic Properties*, from "National Bureau of Standards Circular 500," U.S. Government Printing Office, Washington, D.C., 1952, and (2) F. D. Rossini et al., *Selected Values of Physical and Thermodynamic Properties of Hydrocarbons and Related Compounds*, American Petroleum Institute Research Project 44, Carnegie Institute of Technology, Pittsburgh, 1953. Form of expression is $C_p = a + bT + cT^2 + dT^3$. Units are cal/(g-mole)(°C).

INDEX